Materials Science and Technologies

Additional books in this series can be found on Nova's website
under the Series tab.

Additional e-books in this series can be found on Nova's website
under the e-book tab.

Chemical Engineering Methods and Technology

Additional books in this series can be found on Nova's website
under the Series tab.

Additional e-books in this series can be found on Nova's website
under the e-book tab.

MATERIALS SCIENCE AND TECHNOLOGIES

ZEOLITES

SYNTHESIS, CHEMISTRY AND APPLICATIONS

MOISEY K. ANDREYEV
AND
OLYA L. ZUBKOV
EDITORS

Nova Science Publishers, Inc.
New York

For permission to use material from this book please contact us:
Telephone 631-231-7269; Fax 631-231-8175
Web Site: http://www.novapublishers.com

NOTICE TO THE READER

Additional color graphics may be available in the e-book version of this book.

Library of Congress Cataloging-in-Publication Data

Zeolites : synthesis, chemistry, and applications / [edited by] Moisey K. Andreyev and Olya L. Zubkov.
p. cm.
Includes bibliographical references and index.
ISBN 978-1-61942-861-4 (hardcover)
1. Zeolites--Synthesis. 2. Zeolites--Industrial applications. I. Andreyev, Moisey K. II. Zubkov, Olya L.
TP245.S5Z388 2011
549'.68--dc23
2012001022

Published by Nova Science Publishers, Inc. † New York

CONTENTS

PREFACE

Zeolites, mainly consisting of silicon, aluminum, and oxygen atoms that connect in three-dimensional frameworks, are three-dimensional microporous or mesoporous materials. They are widely used in many applications, such as catalysts, catalyst supports, membranes, etc. In this book, the authors present current research in the study of the synthesis, chemistry and applications of zeolites. Topics include the conversion of ethanol to hydrocarbons over zeolite catalysts; air pollution catalytic control by metal promoted zeolites; zeolite from fly ash-iron oxide magnetic nanocomposites; application of zeolite containing rocks in berry crop growing; and dealuminated zeolites in biological systems.

Chapter 1 – Conversion of ethanol to hydrocarbons over zeolite catalysts was investigated. H-ZSM-5 zeolite with low Si/Al_2 ratio showed high activity in ethanol conversion to aromatic compounds, while other zeolites (H-ZSM-5 zeolite with high Si/Al_2 ratio, H-mordenite, H-Beta and Y-type zeolite) exclusively formed ethylene. Effect of metal-addition to H-ZSM-5 (low Si/Al_2 ratio) was investigated. Ga or noble metal/H-ZSM-5 catalysts have high activity for the formation of aromatics. Among them, Au/H-ZSM-5 catalyst can partially inhibit carbon deposition during the reaction, thus maintaining constant catalytic activity for aromatics formation. On the other hand, addition of alkali earth metal (for example, Mg) led to exclusive formation of ethylene. Cr-, Fe- or Ni-modified catalysts showed higher selectivity of C_{3+} olefins than other metal-modified catalysts.

Next, ethanol conversion to C_{3+} olefins over Fe/H-ZSM-5 catalysts was investigated, since Fe is non-toxic and cheap metal. C_{3+} olefins are useful as not only fuels but also chemicals. The initial selectivity of C_{3+} olefins was not affected by the kinds of Fe source and calcination temperature so significantly. In general, selectivity of C_{3+} olefins was slightly decreased with time-on-stream. However, in the cases of $Fe_2(SO_4)_3$- and $FeCl_3$-derived catalysts, calcined at 700 °C, the decrease in C_{3+} olefins selectivity could be suppressed. Selectivity of aromatics was decreased and selectivity of ethylene was increased with time-on-stream. Used catalyst can be regenerated by calcination treatment in air-flow, but the degree of regeneration was dependent on the kind of iron starting materials and the calcination temperature. $FeCl_3$-based catalyst, calcined at 700 °C and catalysts calcined at 900 °C (irrespective of iron source) can be almost completely regenerated, while $Fe(NO_3)_3$- or $Fe_2(SO_4)_3$-based catalyst, calcined at 700 °C, cannot be completely regenerated by this treatment. Effect of Fe-loading and reaction temperature was also investigated. Carbon deposition on catalyst and framework collapse of zeolite support can be considered as cause of the change in selectivity of each product. Lower Fe-loading (1 wt%) could inhibit carbon deposition and framework collapse in some degree, and higher reaction temperature (450 °C) could enhance the formation of C_{3+} olefins. To improve the propylene (C_{3+} olefins) selectivity and catalytic stability, P-addition to Fe/H-ZSM-5 catalyst was tried. The effect of presence of water in ethanol was also investigated. Successive addition of Fe and P (Fe → P and P → Fe) improved initial selectivity of propylene, while sole addition of Fe or P and co-addition of Fe and P showed medium initial selectivity. In general, the catalysts showing higher initial selectivity of propylene shows larger decrease in propylene selectivity with time-on-stream. In the case of neat ethanol used as ethanol, successive addition (P → Fe) to H-ZSM-5 zeolite support improved catalytic performance, which was regenerated by calcination treatment, while the catalytic performance of Fe/H-ZSM-5 catalyst (without P) could not be regenerated completely. In general, presence of water in ethanol led to larger change in selectivity of each product. However, co-impregnation of Fe and P to H-ZSM-5 zeolite support led to high catalytic performance, which could be regenerated by calcination treatment.

Chapter 2 – Air pollution would be achieved as the concentrations of some harmful materials, such as nitrous oxide (N_2O), nitrogen oxides (NO and NO_2), hydrocarbons (HCs), nitrile etc., are high enough to affect the people's health as well as destroy the sustainable ecosystem. Traffic and industry are the most important sources of the air pollutants mentioned above. Catalysis has become indispensable in converting these environmentally harmful molecules into more benign ones such as N_2, H_2O, and CO_2, although the latter now deems as a greenhouse gas. In this chapter, our recent progress on the synthesis and characterization of metal-promoted microporous and mesoporous zeolites as well as their applications as catalysts for the purification of exhaust gases including N_2O from nitric acid or adipic plant, NO and HCs from motor vehicles, and CH_3CN from acrylonitrile plant was systematically introduced.

N_2O direct decomposition over Fe-BEA zeolite prepared by wet ion-exchange (WIE) was investigated. Both isolated iron ions in the charge compensation positions and highly dispersed FeO_x-like crystallite were verified. Two kinds of reaction mechanism, with respective formation of oxygen (O_2) or nitrogen oxide (NO_x) intermediate, were proposed. The corresponding energy calculations for O_2-formation mechanism revealed that O_2 desorption was the rate determining step with an energy barrier of 63.2 kcal/mol. Another mechanism with NO formation was also simulated by density functional theory (DFT) calculations, whose energy calculations indicated that the formation of Fe-Z-$(NO)_2$ was the main pathway for NO generation with an energy barrier of 26.9 kcal/mol.

Aiming at the effective purification of NO and HCs from automobile, a series of mono-functional M/SBA-15 (M = Cu, Fe, Cr, and Al) and bifunctional M'/SBA-15 (M' = Cu-Al, Fe-Al, and Cr-Al) catalysts was thereafter prepared via an incipient wetness impregnation, and further used for $C_3H_6 + NO + O_2$ reaction. Ordered mesoporous structure of SBA-15 was well maintained even after impregnating the various metallic components. The deNOx activities of investigated catalysts follow a trend of Cu-Al/SBA-15 > Cu/SBA-15, Fe/SBA-15 > Cr-Al/SBA-15 > Cr/SBA-15 >> Fe-Al/SBA-15, Al/SBA-15 > SBA-15. Synergistic effect between copper and alumina was achieved showing a satisfactory NO conversion of approximately 80% over Cu-Al/SBA-15 at temperature as low as 350°C under an atmosphere of 3000 ppm NO, 3000 ppm C_3H_6, 1% O_2 and GHSV of 60,000 h^{-1}.

A series of M''/SBA-15 (M''=Cu, Co, Cr and Mn) catalysts was also prepared and used for acetonitrile elimination. The CH_3CN conversion of investigated catalysts follow a trend of Cu/SBA-15 > Co/SBA-15 ≈ Cr/SBA-15 > Mn/SBA-15 > SBA-15. Apart from the target product of N_2, harmful byproducts of NO, NO_2, N_2O, NH_3 and CO were also observed during the combustion process. The corresponding selectivities were verified to be associated with both chemical nature and redox properties of the loaded metallic species. Among the prepared M''/SBA-15 samples, Cu/SBA-15 exhibited a nearly complete conversion of CH_3CN associated with a N_2 selectivity around 80% at T>350°C, which was regarded as a promising material for the catalytic removal of cyanide-containing gas.

Chapter 3 – This chapter presents and discusses current research on the study of the adsorption efficiency of zeolite from fly ash-iron oxide nanocomposites towards removal of hazardous pollutants from aqueous solutions. The magnetic properties of iron oxides have been combined with the adsorption features of zeolites synthesized from fly ash to produce a magnetic nanocomposite adsorbent. Magnetite nanoparticles in suspension were synthesized by precipitating iron ions in a NaOH solution. Zeolite was synthesized from coal fly ash by alkaline hydrothermal treatment. Zeolite-iron oxide composite was prepared by mixing zeolite with magnetite suspension with weight ratios of 3:1, 2:1 and 1:1. The resulting materials were characterized by X-ray diffraction, surface area, magnetization measurements, Fourier transform infrared and X-ray fluorescence analyses. Batch tests were carried out to investigate the adsorption mechanism of dyes and metal ions from aqueous solution onto magnetic nanocomposite. The composite adsorbents after be used to adsorb contaminants in water were removed from the medium by a simple magnetic process. The effect of various experimental parameters such as contact time, pH, adsorbent dose and temperature were investigated. The experimental data were analyzed using the pseudo-first-order, pseudo-second-order and intraparticle diffusion equations adsorption kinetic models. The Langmuir and Freundlich isotherm models were tested for their applicability. The results indicated that zeolite from fly ash-iron oxide magnetic nanocomposite, a low cost adsorbent, exhibited considerable potential for application in removal of contaminants from aqueous solution. Future prospects for research are discussed.

Chapter 4 – HZSM-5/SAPO-34 composite and HBeta/SAPO-11 composite were prepared by one-step hydrothermal crystallization, respectively. The structure, texture, morphology and acidity of these as-prepared samples were comparatively investigated by powder X-ray diffraction (XRD), infrared spectroscopy (IR), solid state nuclear magnetic resonances (NMR), N_2 isothermal adsorption-desorption, scanning electron microscope (SEM), temperature-programmed desorption of adsorbed ammonia (NH_3-TPD) and infrared spectroscopy of adsorbed pyridine (Py-IR), etc.. In addition, the catalytic performance of HZSM-5/SAPO-34 composite in the conversion of ethanol to propylene, and the catalytic performance of HBeta/SAPO-11 composite in the cracking of 2-butylene to propylene were studied. Chemical interaction of aluminosilicate with silicoaluminophosphate zeolites in the composites occurred. Such interaction modified structure, texture, morphology and acidity of these composites, which in turn significantly improved their catalytic reactivity.

Chapter 5 – An efficient and practicable hydrogen storage and delivery system is a key for the success of hydrogen fuel for various ecological energy systems in the future, as the conventional hydrogen storage systems such as gas cylinders and liquid hydrogen have serious limitations in transportation. Consequently, the storages of hydrogen in solid materials have been extensively studied in the last a few decades because of their advantages as potential energy carriers. Among these storage systems, the adsorption in porous materials has attracted increasing interest due to its simplicity, less drastic pressures and temperature requirements. Although attentions have been focused on light carbon microporous materials owing to their potential high gravimetric storage capacities, other microporous materials, such as organic frameworks and zeolites, have also been examined. The consideration of hydrogen storage in zeolite materials is mainly due to their rich variety of porous structures, which compare favorably when adsorption capacities are measured on a volume basis. In addition, the high stability and low cost of zeolites are also important advantages for hydrogen storage media. On the other hand, the selective adsorption properties of zeolites as molecular sieves have also been applied to the separation of the hydrogen isotopes in the potential nuclear fusion reactions for energy generation. In the experimental and theoretical works devoted to the above applications, the physisorption and encapsulation of molecular hydrogen in tailored zeolites are the most available options.

Chapter 6 – Zeolites, crystalline aluminosilicate materials, possess 3-dimensionaly connected framework structures constructed from corner-sharing TO_4 tetrahedra, where T is any tetrahedrally-coordinated cation such as Si and Al. These framework structures are composed of n-rings, where n is the number of T-atoms in the ring (e.g. 4-, 5-, and 6-rings), and large pore openings of 8-, 10-, and 12-rings are framed by these small rings. The sizes of the intracrystalline pores and nanospaces, depending on the type of zeolite providing the framework, are close to the molecular diameters of lighter hydrocarbons. Moreover, strong acid sites exist on the nanopore surfaces within the zeolite crystal and on the external surface of the crystal, enabling the zeolites to be used in reaction processes (shape-selective catalysts such as fluid catalytic cracking of heavy oil, isomerization of xylene and synthesis of ethylbenzene) and separation process (adsorptive- and membrane-separation).

Chapter 7 – Zeolites are widely used as a solid acid catalyst in many industrial processes. The acidity of zeolite could be due to the Bronsted and Lewis acid sites. Exchange of monovalent cations with polyvalent ion creates strong Bronsted centers by the hydrolysis phenomena, which is useful for some of acid catalyst reactions such as alcohol dehydration, acylation, estrification, and Diels-Alder reaction. The number of proton sites can be decreased and increase the number of Lewis sites by dehydration reaction, which is carried out above 875 K°. The time of the cation exchange process can be reduced under microwave irradiation. The Lewis sites can be increased when the metal / zeolite is prepared in the solid state phase ion-exchange under microwave irradiation. Interestingly, in some reactions, including intramolecular acylation, isomerization and conversion of acids to benzimidazoles, the Lewis sites are the active centers. Despite the many reports a good catalytic activities for the solid ion exchanged zeolites, the interaction mechanism of microwave radiation with a mixture of metal ion salt/zeolite has not been yet completely understood. In particular it has been found that, as a result of microwave treatment, the metal ions may occur in at least three different forms over or within zeolites: i) external coating of zeolite crystals by metal clusters , ii) residual metal salts dispersed over the surface of zeolites, and iii) metal ions incorporated in zeolite channels and cages by ion-exchange . FT-IR spectroscopies with various probe molecules, such as ammonia, quinoline, pyridine and carbon monoxide have been used for the characterization of acid sites. Recently, Rietveld refinement method can be used for characterization of these sites. In this chapter, we introduce a selective method for tailoring Bronsted and Lewis sites in zeolite .Also; explain about some methods for characterization these sites. Finally, mechanism of some organic reactions with Bronsted and Lewis sites will be considered.

Chapter 8 – Zeolites, mainly consisting of silicon, aluminum, and oxygen atoms that connect in three-dimensional frameworks, are three-dimensional microporous or mesoporous materials. They are widely used in many applications, such as catalysts, catalyst supports, membranes, etc. Zeolites can be classified in terms of framework type, such as AFI and MFI; or in terms of Si:Al ratio, such as LTA (NaA), ZSM-5; or even in terms of hydrolysis agent, such as LTA (NaA) and KA. The synthetic zeolites can be obtained from sol-gel process followed by a hydrothermal treatment, such as conventional heating, microwave heating or electrophoretic heating technique, to form crystalline-structured materials. The obtained synthetic zeolites are varied, depending on the production technique and raw materials used in synthesis. Many chemical compounds, used as raw materials for zeolite synthesis, produce different properties in the obtained zeolite, leading to different results in applications. In this chapter, home-made atrane materials (silatrane and alumatrane), used as silica and alumina sources, respectively, are shown to be distinguishable raw materials for synthesizing many types of zeolites since they provide good thermal and moisture stability, as compared to commercially available materials. The applications, properties, and performances of those synthetic zeolites are illustrated, as well.

Chapter 9 – The possibilities of application of zeolite-containing rock from Khotynetz deposit (Orel region, Russia) have been investigated in the field and laboratory experiments for the purpose of microelement composition improvement of berry crops: raspberry, gooseberry and black currants. Zeolite (2…30 t/ha) was applied in the grey forest soil, which had gross content of heavy metals 2…6 times higher than background level.

The application of zeolite-containing rock led to the lowering of the contents of mobile forms Pb, Ni, Zn and Cu in the soil by 30…50% depending on the metal and zeolite dose. The reduction of heavy metal compound supply within the reach for plants changed the leaf and fruit microelement composition of berry crops. The character of changes depended on a zeolite dose, plant species, metal nature and its physiological significance. The contents of toxic Pb and Ni in fruit were reduced by 20…70% depending on the crop. The lowing of mobile forms Zn and Cu in the soil by 30…50% did not lead to a deficiency of those biogenic elements for berry crops.

The application of mineral fertilizer effected the action of the zeolite-containing rock as a deactivator of heavy metals in the “soil-plant” system. On $N_{90}P_{90}K_{90}$ background the content of Cu in black currant berries was higher, but the contents of Pb, Ni and Zn were lower than in the variants without fertilizers.

The application of the zeolite-containing rock increases the highest possible hygroscopic soil humidity and accordingly the wilting point of plants. Zeolite application in doses upwards of 16 t/ha positively reduces the supply of productive moisture that may lead to the lowing of berry crop productivity. Possible deterioration of plant water provision is a limiting factor in a choice of doses of zeolite-containing rocks for the purpose of solving other land-reclamation measures.

Chapter 10 – Synthetic 4A and 5A zeolites were studied to explore the feasibility of applying these adsorbents for fluoride removal from drinking water. The zeolites were characterized with SEM/EDS and N_2-adsorption/desorption for their chemical and pore textural properties. The adsorption isotherms were correlated with the Langmuir and Freundlich adsorption equations. The fluoride adsorption isotherms on synthetic zeolite 4A and 5A followed the Freundlich model. The zeolite 5A adsorbent showed the highest fluoride adsorption capacity. It adsorbed up to 0.1 and 71 mg/g when the fluoride equilibrium concentration in water was 4.0 and 514 mg/L, respectively. Zeolite 5A adsorbed seven times more fluoride than zeolite 4A at the same initial concentration.

A pseudo-second order model and an intraparticle kinetic model fitted well the adsorption kinetic data for both synthetic zeolite adsorbents. It was found that both external and intraparticle diffusions contribute to the rate of removal of fluoride by the zeolites. From the experimental data obtained in this work, it was found that synthetic zeolite 5A has potential adsorption properties for fluoride removal from drinking water.

Chapter 11 – Iron ions may occupy various positions in zeolites and their coordination and oxidation states may be changed during different treatments. An apparent distinction among their siting is the framework/extra-framework assignment. In situ Mössbauer spectroscopy is a sensitive and unique tool to analyze the siting of iron in zeolites. A few examples for the identification of iron located in framework or extra-framework sites is presented and their interconversions are correlated with the detected changes in their oxidation and coordination states. Finally, the formation of coupled framework/extra-framework iron pairs is also demonstrated.

Chapter 12 – The synthetic zeolites are characterized by a great variety of structures and compositions, as well as by the different adsorption ability to ions, clusters and the transition metal complexes. Therefore they are interesting as supports for heterogenization of the olefin polymerization catalysts. In addition the natural zeolites due to their considerable content in nature and low cost may be used as fillers for the polymer composite materials. The immobilized metalcomplex catalysts of olefin (ethylene and propylene) polymerization have been developed on the base of zeolites. In order to prepare these catalysts, the anchored aluminoxanes have been synthesized by reaction of aluminum alkyls ($AlMe_3$, $AlEt_3$ and $AliBu_3$) with water contained in intracrystalline cavities and pores of zeolites (NaY, NaZSM-5, HZSM-5, NH_4ZSM-5, mordenite and clinoptilolite-bearing tuff). These aluminoxanes, with zirconocenes, vanadium chlorides and other transition metals, form the heterogenized complexes that are active in the polymerization of ethylene and propylene without addition of other organoaluminum cocatalysts. The nature of zeolites used as a support, determines the content of zeolite water and affects the formation of anchored aluminoxanes and activity of immobilized catalysts. Polyolefins prepared with the immobilized on zeolites zirconocene catalysts demonstrate higher molecular mass and a wider molecular-mass distribution than those of corresponding homogeneous catalysts. The stereoselectivity of supported catalysts depends on the composition, structure of metallocenes and on the zeolite nature. The new composite materials on the base of ultrahigh-molecular-mass polyethylene UHMPE and natural zeolites were developed. The synthesis of composites was carried out by ethylene polymerization on the catalytically activated surface of zeolite fillers. Depending on the amount of zeolite filler in the UHMPE matrix such compositions can be used as wear-resistant, impact-resistant materials for production of units working in conditions of dry friction at high loading or as organomineral ion-exchange sorbents for water purification from radionuclides and heavy metal ions.

Chapter 13 – The determination of pore sizes is very important for the preparation, characterization, and application of porous membranes. Normalized Knudsen-based permeance (NKP) was proposed for pore-size evaluation of microporous membranes, which is based on the gas translation model. In this article, NKP was briefly summarized and verified using the permeance of zeolite membranes such as MFI (0.55×0.56 nm) and DDR (0.36×0.44 nm).

Moreover, we proposed the use of structured alkoxides for the control of amorphous silica networks with loose structure. Silica membranes were prepared via a sol-gel process using two kinds of organoalkoxysilanes with pendant group (methyltriethoxysilane (MTES) and phenyltriethoxysilane (PhTES)) and three kinds of bridged alkoxides (hexaethoxysiloxane (HEDS), tetraethoxydimethyl disiloxane (TEDMDS) and bis(triethoxysilyl) ethane (BTESE)). Judging from the NKP fitting curves, the order of average pore size for silica membranes by organoalkoxysilanes and bridged alkoxides could be estimated as follows: TEOS < MTES < PhTES and TEOS < HEDS < TEDMDS≈BTESE, respectively. It can be concluded that the membrane pore size or structure of silica membrane can be controlled for larger pore sizes with various types of structured alkoxides.

Chapter 14 – The problem of soil pollution and obtaining of safe production in conditions of techno- and agrogenic pollution of the environment is urgent at present. Application of natural nanoporous minerals – zeolites is a prospective direction for solving this problem. Zeolites have high adsorption abilities and ion-selective features; they can irreversibly adsorb cations of heavy metals. The investigation of the peculiarities of zeolite thin structure and its sorbing features is of practical importance – it allows calculating doses necessary for lowing toxic admixtures in soil and other objects. In our opinion, this direction of the nanotechnology in agriculture and horticulture is turned to the development of environment protecting agrotechnologies, conditions and methods which allow conducting of selective adsorption, stabilizing of soil processes, obtaining of ecologically safe production and neutralizing toxicants, allergens and pathogens in the environment.

Our research work is devoted to the investigation of zeolite of Khotynetz deposit in Orel region (Russia). Clinoptilolite, 34% and montmorillonite, 12% are parts of this zeolite. Clinoptilolite is a highly siliceous zeolite. Its crystalline grating consists of aluminosiliconoxygenous framework with nanochanals (pores). Crystals of montmorillonite have a laminated structure.

Montmorillonite and clinoptilolite cause high adsorption features of the rock. The crumbling up significantly increases specific and exchanging volume.

The investigation of the surface structure of different particles of the mineral (0,05-0,02 mm fraction) by scanning electronic microscope made it possible to reveal the peculiarities of the surface microstructure and define pore size. By energy and dispersion spectrometry it has been stated that a mass part of silicon in Khotynetz zeolite is 30,41% on the average. The mineralogical composition stipulates unique features of Khotynetz zeolite as a nanocomposition material with high contents of amorphous silicon.

The process of selective adsorption was studied in the laboratory model experiments by nanoporous natural polymer of lead ions with 6 and 12 mg/l concentration and of nickel ions with 4 and 8 mg/l concentration under constant temperature and different pH and exposition time. The residual quantity of metal ions in the model solutions was determined by HPLC.

It has experimentally been stated that the adsorption of lead and nickel ions in the studied concentration diapasons depends on pH solution and depends little on ion concentration in the solution.

The obtained results were used in the establishing of field experiments with berry crops.

Chapter 15 – Recent direct and indirect evidence have pointed to both biological reactivity and toxicity of micro- and nano-particles. Particle size appears to be a decisive factor and particles can enter cells in two ways; either through endocytosis, were particles enter the cell in an endosome; or by otherwise passing the cell membrane directly into the cytoplasm. Endocytosis covers sampling of small volumes of fluids and soluble molecules from the milieu surrounding the cell, as well as internalization of large particulate matter such as whole cells or cell fragments from the immediate vicinity of the cells. With respect to toxicity, it is important to identify the size and the chemical composition at which biological systems no longer regard the extraneous matter as particles and hence the material enters the cells directly, through other mechanisms than endocytosis.

In: Zeolites: Synthesis, Chemistry and Applications
Editors: Moisey K. Andreyev and Olya L. Zubkov
ISBN: 978-1-61942-861-4

Chapter 1

CONVERSION OF ETHANOL TO HYDROCARBONS OVER ZEOLITE CATALYSTS

Megumu Inaba*

National Institute of Advanced Industrial Science and Technology (AIST)
Energy Technology Research Institute, BTL Catalyst Group
Higashi, Tsukuba, Ibaraki, Japan

ABSTRACT

Conversion of ethanol to hydrocarbons over zeolite catalysts was investigated. H-ZSM-5 zeolite with low Si/Al_2 ratio showed high activity in ethanol conversion to aromatic compounds, while other zeolites (H-ZSM-5 zeolite with high Si/Al_2 ratio, H-mordenite, H-Beta and Y-type zeolite) exclusively formed ethylene. Effect of metal-addition to H-ZSM-5 (low Si/Al_2 ratio) was investigated. Ga or noble metal/H-ZSM-5 catalysts have high activity for the formation of aromatics. Among them, Au/H-ZSM-5 catalyst can partially inhibit carbon deposition during the reaction, thus maintaining constant catalytic activity for aromatics formation. On the other hand, addition of alkali earth metal (for example, Mg) led to exclusive formation of ethylene. Cr-, Fe- or Ni-modified catalysts showed higher selectivity of C_{3+} olefins than other metal-modified catalysts.

Next, ethanol conversion to C_{3+} olefins over Fe/H-ZSM-5 catalysts was investigated, since Fe is non-toxic and cheap metal. C_{3+} olefins are useful as not only fuels but also chemicals. The initial selectivity of C_{3+} olefins was not affected by the kinds of Fe source and calcination temperature so significantly. In general, selectivity of C_{3+} olefins was slightly decreased with time-on-stream. However, in the cases of $Fe_2(SO_4)_3$- and $FeCl_3$-derived catalysts, calcined at 700 °C, the decrease in C_{3+} olefins selectivity could be suppressed. Selectivity of aromatics was decreased and selectivity of ethylene was increased with time-on-stream. Used catalyst can be regenerated by calcination treatment in air-flow, but the degree of regeneration was dependent on the kind of iron starting materials and the calcination temperature. $FeCl_3$-based catalyst, calcined at 700 °C and catalysts calcined at 900 °C (irrespective of iron source) can be almost completely

* TEL and FAX: +81-29-861-4776, PHS: +81-29-861-4751. Email: mg.inaba@aist.go.jp.

regenerated, while $Fe(NO_3)_3$- or $Fe_2(SO_4)_3$-based catalyst, calcined at 700 °C, cannot be completely regenerated by this treatment. Effect of Fe-loading and reaction temperature was also investigated. Carbon deposition on catalyst and framework collapse of zeolite support can be considered as cause of the change in selectivity of each product. Lower Fe-loading (1 wt%) could inhibit carbon deposition and framework collapse in some degree, and higher reaction temperature (450 °C) could enhance the formation of C_{3+} olefins. To improve the propylene (C_{3+} olefins) selectivity and catalytic stability, P-addition to Fe/H-ZSM-5 catalyst was tried. The effect of presence of water in ethanol was also investigated. Successive addition of Fe and P (Fe → P and P → Fe) improved initial selectivity of propylene, while sole addition of Fe or P and co-addition of Fe and P showed medium initial selectivity. In general, the catalysts showing higher initial selectivity of propylene shows larger decrease in propylene selectivity with time-on-stream. In the case of neat ethanol used as ethanol, successive addition (P → Fe) to H-ZSM-5 zeolite support improved catalytic performance, which was regenerated by calcination treatment, while the catalytic performance of Fe/H-ZSM-5 catalyst (without P) could not be regenerated completely. In general, presence of water in ethanol led to larger change in selectivity of each product. However, co-impregnation of Fe and P to H-ZSM-5 zeolite support led to high catalytic performance, which could be regenerated by calcination treatment.

1. INTRODUCTION

Currently, major compounds in the petrochemical industry, for example, ethylene, propylene, benzene and toluene, are synthesized by cracking the naphtha fraction. However, petroleum resources are limited, both quantity-wise and geographically. Moreover, combustion of petroleum produces CO_2, a greenhouse gas. Therefore, much attention has been paid to alternative resources for petroleum. Biomass is one of these candidates, since biomass is renewable resources and can be obtained everywhere on earth. Moreover, the combustion of biomass does not lead to increase of CO_2 in atmosphere, since the biomass can be photo-synthesized from CO_2 in atmosphere, whether directly or indirectly.

On the other hand, it had been known that H-ZSM-5 zeolite is effective for methanol transformation to hydrocarbons [1-3]. Gayubo *et al.* reported that H-ZSM-5 zeolite can be used for catalytic transformation of oxygenate components of biomass pyrolysis oil into hydrocarbons [4, 5].

Ethanol is one of the main biomass products, which can be obtained by fermentation. Berrier *et al.* reported the transformation of ethanol into higher hydrocarbons on Co/Al_2O_3 catalysts [6]. We experimented with several zeolite catalysts and several metal catalysts supported on zeolite for the conversion of ethanol to hydrocarbons, and found that the low Si/Al_2 ratio version of H-ZSM-5 has high activity for the production of higher hydrocarbons, as reported elsewhere [7-15]. Moreover, we found that loading of Ga and noble metals on H-ZSM-5 support enhances the formation of aromatic hydrocarbons, while Cr, Fe or Ni-loading raised the selectivity of C_{3+} olefins (C_3' – C_7') [16], which are useful as not only fuels but also chemicals. C_{3+} olefins, especially propylene, are important compounds, since polymerized olefins can be used as plastics, which can contribute to CO_2 reduction in atmosphere unless these plastics are burned into CO_2. These results are reported in Chapter 2.

Recently, production of olefins from ethanol using modified ZSM-5 has been reported extensively. Murata *et al.* have reported that W- and La-modification of H-ZSM-5 zeolite to be effective for enhancing the formation of lower olefins from ethanol [17]. Inoue *et al.* have reported the conversion of ethanol to propylene using La-modified H-ZSM-5 zeolite with Si/Al_2 ratio of 280, an especially high ratio for production of C_{3+} olefins from ethanol [18]. Sano *et al.* have reported conversion of ethanol to propylene over HZSM-5 type zeolites containing alkaline earth metals [19].

Fujitani *et al.* have reported that P- [20] or Zr- [21] modified H-ZSM-5 catalysts show improved catalytic activity and stability for conversion of ethanol to propylene in comparison with H-ZSM-5 without modification. Moreover, we found that, for aqueous ethanol, Zr-modified H-ZSM-5 catalysts showed especially high yield of propylene and catalytic stability in ethanol conversion [22].

Gayubo *et al.* had investigated ethanol conversion to hydrocarbons over H-ZSM-5 without metals [12-14], and recently reported Ni- [23, 24] or alkali- [25, 26] modified H-ZSM-5 catalysts for ethanol conversion. It had been reported that Ni-modified H-ZSM-5 catalysts are also efficient for hydrogen production by methane decomposition [27] and cellulose gasification [28].

On the other hand, production of olefins from ethanol using catalysts other than H-ZSM-5 has also been reported. Recently, Baba *et al.* reported the highly selective conversion of ethylene to propylene over SAPO-34; they also found SAPO-34 to be effective for ethanol conversion to propylene [29]. Iwamoto *et al.* reported that Ni ion-loaded mesoporous silica MCM-41 was effective for producing lower olefins from ethanol [30].

As mentioned above, we found that Cr, Fe or Ni-addition to H-ZSM-5 raises the selectivity of C_{3+} olefins [16], and investigated the effect of synthesis condition (for example, Fe source and calcination temperature) on the production of C_{3+} olefins over Fe-modified H-ZSM-5 catalysts, since iron is non-toxic and cheap metal [31]. These results are reported in Chapter 3.

Moreover, we examined the effect of Fe-loading and reaction temperature to improve selectivity for C_{3+} olefins and propylene and found that lower loading of Fe (1 wt%) and higher reaction temperature (450 °C) results in higher selectivity of propylene and catalytic stability [32]. These results are reported in Chapter 4.

To improve catalytic performance, P-addition to Fe/H-ZSM-5 was also tried, since it has been reported that phosphorous modification of H-ZSM-5 zeolite can lead to better hydrothermal stability and resistance to coke deposition [20, 33]. We found that P- and Fe-modification to H-ZSM-5 leads to the improved propylene selectivity and stability for water in ethanol in comparison with H-ZSM-5 without modification and Fe-modified H-ZSM-5 catalysts [34].

These results are reported in Chapter 5. Preparation conditions of catalysts and reaction conditions in Chapter 2-5 are summarized in Table 1. Calsavara *et al.* also reported the effect of conditions for preparation of Fe-modified H-ZSM-5 catalysts on catalytic performance in ethanol conversion [35, 36]: in this case, main products are liquid hydrocarbons. As application of Fe-modified ZSM-5 catalysts other than ethanol conversion, selective reduction of NO_x with ammonia [37, 38], decomposition of nitrous oxide [39], oxydehydrogenation of ethane or propane using nitrous oxide as oxidant [40], and selective oxidation of benzene to phenol with nitrous oxide [41, 42] have been reported extensively. Application in wide field is expected for Fe/ZSM-5 catalysts.

Table 1. Preparation conditions of catalysts and reaction conditions

Chapter	Zeolite	Element	Loading / wt%	Calc. temp. / ˚C	EtOH	Reaction temp. / ˚C
2	H-Beta (27) H-ZSM-5 (29) H-ZSM-5 (190) USY (6.3) H-mordenite (18.3)	----	----	500	Neat EtOH	200-600 400
	H-ZSM-5 (29)	Mg, Cr, Fe, Co, Ni, Cu, Ga, Ru, Rh, Pd, Ag, Re, Ir, Pt, Au	2 (Ru, Rh, Pd, Ir, Pt, Au) 10 (Mg, Cr, Fe, Co, Ni, Cu, Ga, Ag, Re)	700	Neat EtOH	200-600 400
3	H-ZSM-5 (29)	Fe	10	500 700 900	Neat EtOH H_2O + Neat EtOH	400
4	H-ZSM-5 (29)	Fe	1 10	700	Neat EtOH	400 450
5	H-ZSM-5 (29)	Fe P	1 (Fe) 1 (P)	(500) 700 (900)	Neat EtOH, Vodka + neat EtOH	450

Number in parenthesis by zeolite stands for Si/Al_2 ratio.

2. Ethanol Conversion to Hydrocarbons by Several Zeolite Catalysts [16]

2.1. Experimental

2.1.1. Catalyst Preparation

As zeolite supports, H-Beta (27), H-ZSM-5 (29, 190), USY (6.3) and H-mordenite (18.3) were used (numbers in parentheses denote the Si/Al_2 ratio). These zeolites were obtained from Tosoh. Before use, the zeolites were calcined at 500 ˚C in a muffle furnace for 6 h (heating rate was 200 °C h^{-1}). Several metals were impregnated in the zeolite supports. Mg, Cr, Fe, Co, Ni, Cu, Ga, Ru, Rh, Pd, Ag, Re, Ir, Pt and Au were used as pure metals. Metal nitrate hydrates were used as sources of Mg, Cr, Fe, Co, Ni, Cu, Ga and Rh. $Ru(NO_3)_3(NO)$, $AgNO_3$, $[Pt(NH_3)_4](NO_3)_2$ and $HAuCl_4{\cdot}4H_2O$ were used as sources of Ru, Ag, Pt and Au, respectively. Aqueous solutions of these metal sources were used for impregnation. $(NH_3)_2Pd(NO_2)_2$, NH_4ReO_4 and $(NH_4)_2IrCl_6$ were used as sources of Pd, Re and Ir, respectively, where aqueous solutions containing dissolved ammonia were used. As a rule, the metal loading was 10 wt%, except in the cases of Ru, Rh, Pd, Ir, Pt and Au, where the loading was 2 wt%. After impregnation, the wet catalysts were dried at 120 ˚C in an oven, followed by calcination at 700 ˚C in air for 3 h.

2.1.2. Catalytic activity

Catalytic activity was measured using a fixed-bed reactor. The weight of catalyst was 0.2 g. Reaction gas was obtained by mixing N_2 and vaporized ethanol, with the flow rate of N_2 set at 60 cm^3 min^{-1}, resulting in the concentrations of N_2 and ethanol being 82.14 and 17.86

vol%, respectively. WHSV was 7.45 h^{-1}. Reaction temperature was changed from 200 °C to 600 °C in 100-°C steps. Reactions at a constant 400 °C were also carried out to investigate the effect of reaction time on change in catalytic performance. The effluent gas was analyzed by gas chromatography (GC) equipped with Molecular Sieve 5A and Parapak Q columns. Two detectors were used: TCD was used to detect light products (hydrogen, carbon monoxide, and carbon dioxide), and FID to detect hydrocarbons, alcohols, aldehydes and ethers. The selectivity for each product was defined as (number of C atoms in the product)/(total number of C atoms in all gaseous products) × 100 (%).

2.1.3. Characterization

NH_3-temperature-programmed-desorption (TPD) analyses (Nihon Bell, TPD-1-AT-SP) of zeolite supports were carried out to measure the solid acidity of the supports. X-ray diffraction (XRD) (Mac Science, M18XHF^{22}-SRA) measurements of metal/zeolite catalysts were carried out to investigate the state of the loaded metals. The size of metal crystallite was calculated by Sherrer's formula. As cause of catalytic deactivation, carbon deposition during reaction can be considered, since in the case of H-ZSM-5 zeolite used as catalyst, carbon deposition caused catalytic deactivation [9, 12-14]. Therefore, in order to estimate carbon deposition on catalysts, thermogravimetry (TG) analyses of catalysts after reaction were carried out (Mac Science, TG-DTA2000) in the flow of air with the temperature varying from room temperature to 800 °C at 10 °C min^{-1}.

2.2. Results and Discussion

2.2.1. Catalytic Performance of Sole Zeolite

First, the catalytic activity of zeolite catalysts with no metal loaded was investigated. The effect of reaction temperature on ethanol conversion and products selectivity is shown in Figure 1: H-ZSM-5 (Si/Al_2 = 29) catalyst was used. In most zeolites, at below 400 °C, the low reaction temperature favored the exclusive formation of diethyl ether by intermolecular dehydration between two ethanol molecules, although the conversion of ethanol was low. The conversion of ethanol and the selectivity of ethylene increased with increasing reaction temperature, while the selectivity of diethyl ether decreased. It seems that ethylene is formed by dehydration of ethanol via diethyl ether. Above 400 °C, higher hydrocarbons were predominantly formed and the selectivity of ethylene was decreased over H-ZSM-5 (29) zeolite, while other zeolites (H-Beta, USY and H-mordenite) produced ethylene at high selectivity, even above 400 °C. Results of reactions using several kinds of zeolites at 400 °C are summarized in Table 2.

For example, H-ZSM-5 (29) yielded xylenes at selectivity of 30.91%, while H-ZSM-5 (190) formed ethylene at high selectivity, suggesting that H-ZSM-5 zeolites with low Si/Al_2 ratio are effective for production of higher hydrocarbons from ethanol, and adequate amount of solid acidity of the catalyst is essential to this reaction. Chaudhuri *et al.* reported that strong brønsted acid sites and more than one proton per unit cell of H-ZSM-5 are important for catalytic activity [10].

Table 2. Catalytic activity over zeolite catalysts without metal loaded and several metal catalysts supported on H-ZSM-5 ($Si/Al_2 = 29$) zeolite

Catalyst	Ethanol conv. / %	C-selectivity / %								TG weight loss / %
		Diethyl ether	Ethylene	C_{3+}olefins	Paraffins	Benzene	Toluene	Xylenes	BTX	
H-Beta (27)	94.77	1.66	86.50	3.18	3.14	0.93	1.02	2.00	3.95	34.34
H-ZSM-5 (29)	92.24	5.79	10.48	15.62	13.22	3.07	18.90	30.91	52.88	3.31
H-ZSM-5 (190)	96.41	0.33	97.58	1.72	0.03	0.00	0.00	0.00	0.00	0.64
USY (6.3)	93.70	0.42	96.21	0.42	1.23	0.00	0.00	0.00	0.00	21.35
H-mordenite (18.3)	79.83	0.00	95.77	1.04	2.57	0.00	0.00	0.00	0.00	14.37
Mg	36.90	6.76	92.49	0.04	0.03	0.00	0.00	0.00	0.00	0.65
Cr	93.85	5.70	23.81	17.52	9.02	1.15	10.52	29.92	41.59	3.01
Fe	96.96	2.31	11.15	29.76	5.63	0.92	12.81	37.42	51.15	2.44
Co	94.82	0.98	76.16	8.90	2.76	0.00	2.06	8.14	10.20	11.97
Ni	95.56	2.59	30.05	21.05	6.21	0.71	10.69	27.35	38.75	82.41
Cu	90.06	0.20	91.89	0.61	0.11	0.06	1.03	3.44	4.53	0.94
Ga	93.39	3.86	4.78	10.79	6.98	2.52	24.49	46.57	73.58	5.46
Ru	91.48	4.82	12.20	17.28	10.09	3.99	20.77	30.59	55.35	3.81
Rh	94.04	4.27	13.04	17.08	9.87	3.20	19.15	32.81	55.16	6.54
Pd	92.89	4.66	10.38	15.09	8.82	3.37	20.88	36.58	60.83	9.69
Ag	97.09	3.76	38.66	14.55	9.31	0.64	8.86	23.18	32.68	2.71
Re	98.58	3.48	56.04	7.09	5.13	0.91	5.43	16.96	23.30	3.76
Ir	91.50	4.09	11.49	17.67	9.34	3.97	22.56	30.83	57.36	4.32
Pt	98.45	1.30	33.60	6.75	6.93	3.36	14.73	32.65	50.74	10.43
Au	92.88	5.02	6.59	19.12	12.43	4.33	21.96	30.55	56.84	2.55

Number in parenthesis stands for Si/Al_2 ratio. Preparation conditions: Loading of Ru, Rh, Pd, Ir, Pt and Au = 2 wt%, Loading of Mg, Cr, Fe, Co, Ni, Cu, Ga, Ag and Re = 10 wt%. Calcination temperature = 500 °C (zeolite catalysts without metal) and 700 °C (metal supported H-ZSM-5 (Si/Al2 = 29) zeolite catalysts). Reaction conditions: Atmospheric pressure, Reaction temperature = 400 °C, Catalyst weight = 0.2 g, $EtOH/N_2 = 0.22$, N_2 flow = 60 ml/min.

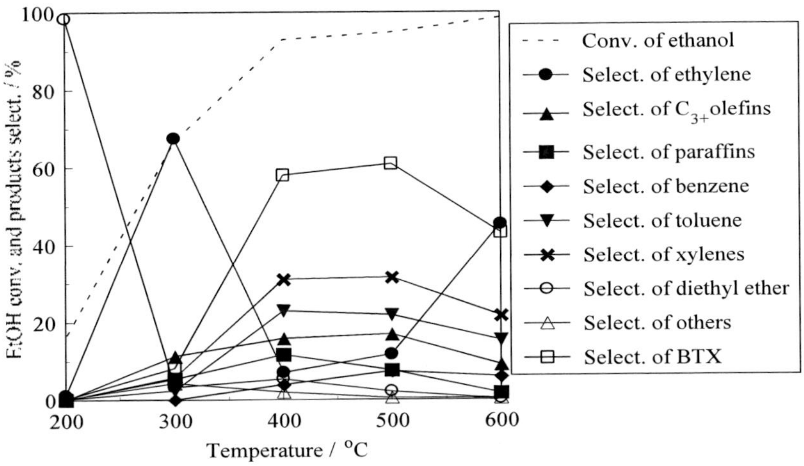

Figure 1. Effect of reaction temperature on conversion of ethanol and selectivity of each product over H-ZSM-5 (Si/Al_2 = 29) zeolite catalyst.

H-ZSM-5 (29) zeolite showed 585 μmol g^{-1} of NH_3 adsorption by NH_3-TPD measurement; this value is smaller than those of H-Beta (27) and H-mordenite (18.3) (1215 and 2453 μmol g^{-1}, respectively), although the Si/Al_2 ratios were not very different among these zeolites. H-ZSM-5 (190) and USY (6.3) showed 269 and 1431 μmol g^{-1} of NH_3 adsorption, respectively. These results suggest that the presence of a large number of acidic sites does not always lead to high BTX selectivity and that adequate strength and quantity of acidity and shape-selectivity are necessary for ethylene polymerization into higher olefins or aromatics. On the other hand, H-mordenite is an efficient catalyst for ethylene formation from ethanol via dehydration even at low temperature (180 °C) [43].

2.2.2. Catalytic Performance of Metal-Modified Zeolite Catalyst

To improve the catalytic activity, we tried several metal/H-ZSM-5 (29) catalysts. Phase and crystallite size of loaded metals obtained from results of XRD measurement is summarized in Table 3. Mg, Cr, Fe, Co, Ni, Cu, Ru, Ag, Ir, Pt and Au exist as MgO, Cr_2O_3, Fe_2O_3, Co_3O_4, NiO, CuO, RuO_2, Ag, IrO_2, Pt and Au, respectively. On the other hand, Ga, Rh, Pd and Re showed no XRD peaks, suggesting that these metals are present at high dispersion.

The results at 400 °C of catalyst are shown in Table 2.

Except for the case of Mg, ethanol conversion was above 90 %, suggesting that the presence of alkali earth metals suppresses the conversion of ethanol. The BTX (Benzene Toluene, Xylenes) selectivity was as follows: Ga >> Pd > Ir > Au > Ru > Rh > blank, where blank means H-ZSM-5 (29) zeolite alone. The selectivity of benzene was as follows: Au > Ru > Ir > Pd > Pt > Rh > blank.

The selectivity of toluene was as follows: Ga > Ir > Au > Pd > Ru > Rh > blank, and that of xylenes was as follows: Ga > Fe > Pd > Rh > Pt > blank. Rh and Pd catalysts showed higher selectivity of all of BTX than H-ZSM-5 zeolite alone. Saha *et al.* reported that Ga-doped H-ZSM-5 has higher activity for the conversion of ethanol to aromatic compounds than H-ZSM-5 alone [11]. On the other hand, other types of metals suppressed the production of aromatic compounds.

Table 3. Phase and crystallite diameter of metal loaded on H-ZSM-5 zeolite

Metal	Phase	Crystallite diameter / nm
Mg	MgO	17.1
Cr	Cr_2O_3	32.1
Fe	Fe_2O_3	24.8
Co	Co_3O_4	27.7
Ni	NiO	23.9
Cu	CuO	20.4
Ga	No peak detected	No peak detected
Ru	RuO_2	22.6
Rh	No peak detected	No peak detected
Pd	No peak detected	No peak detected
Ag	Ag	24.0
Re	No peak detected	No peak detected
Ir	IrO_2	17.4
Pt	Pt	22.6
Au	Au	31.0

Figure 2 shows the plot of BTX selectivity vs. C_{3+} olefins selectivity at 400 °C, and Figure 3 shows the plot of ethylene selectivity vs. C_{3+} olefins selectivity. These results suggest that additions of noble metals or Ga are effective for the formation of aromatic compounds from ethanol. Saha *et al.* reported that Ga-addition enhances the formation of aromatics due to the larger H-transfer abilities in comparison with sole H-ZSM-5. On the other hand, Ali *et al.* [44] and Liebmann *et al.* [45] reported that noble metals have activity for dehydrogenation of hydrocarbons, suggesting that this activity enhances the formation of aromatics.

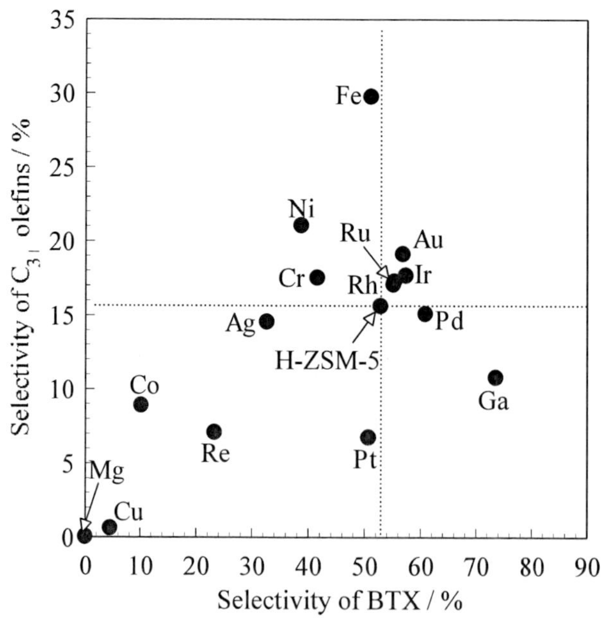

Figure 2. Plots of the selectivity of BTX vs. the selectivity of C_{3+} olefins over H-ZSM-5 catalyst and metal-supported H-ZSM-5 catalysts.

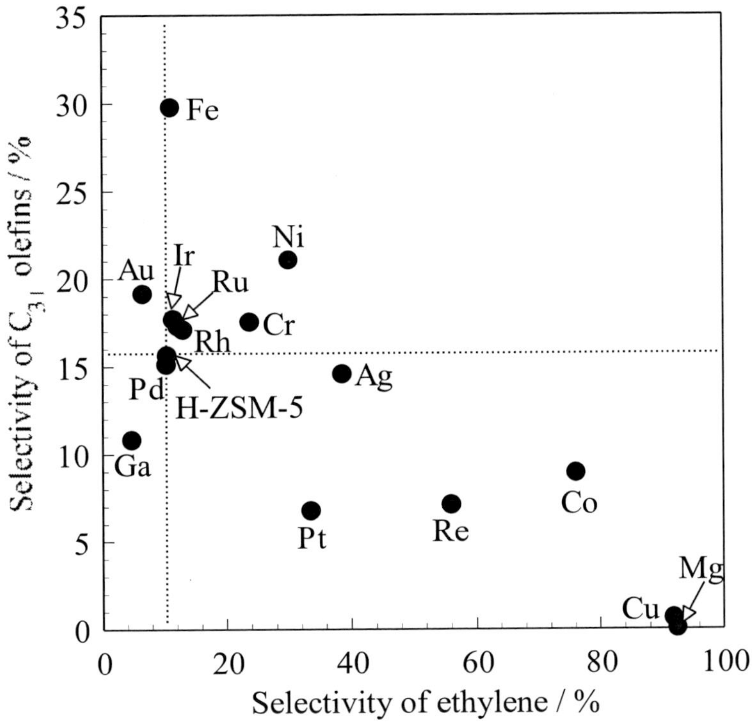

Figure 3. Plots of the selectivity of ethylene vs. the selectivity of C_{3+} olefins over H-ZSM-5 catalyst and metal-supported H-ZSM-5 catalysts.

On the other hand, other metals, such as Cr, Fe and Ni appear to be effective for the formation of C_{3+} olefins and paraffins, which may be precursors of aromatics. Different from the case of ethylene, it is difficult to improve the selectivity of C_{3+} olefins, since they are formed via several steps, for example, dimerization and disproportionation after dehydration of ethanol.

Moreover, in the cases of these metals, moderate selectivity was shown for ethylene and aromatics. Meanwhile, Mg and Cu were effective for ethylene formation. In Sano's report [19], H-ZSM-5 containing Mg catalyst showed high selectivity of propylene. It is suggested that, in this study, loading of Mg was high (10 wt%) and acidic sites of H-ZSM-5, which are necessary for formation of propylene and aromatics, were blocked by Mg species, and as a result, ethylene selectivity was improved.

2.2.3. Carbon Deposition

Carbon deposition on the catalyst, which may lead to catalytic deactivation, occurred during the reaction. The weight loss of catalysts after reaction (200–600 °C) by TG is also shown in Table 2. Moreover, Figure 4 shows the plot of weight loss by TG analysis vs. BTX selectivity at 400 °C. In general, catalysts yielding BTX at high selectivity deposit large amounts of carbon. This result is consistent with a previous report [10]. However, among the zeolites alone, H-Beta (27), USY (6.3) and H-mordenite (18.3) formed no or little BTX, but were prone to heavy deposits of carbon. H-ZSM-5 (190), yielding no BTX, was subject to minor carbon deposits.

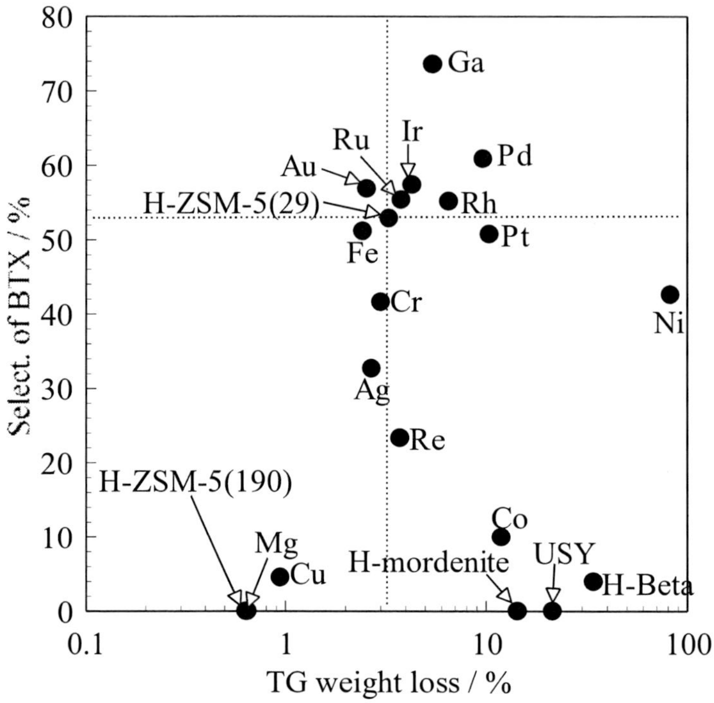

Figure 4. Plots of the weight loss of catalyst after reaction by TG vs. the selectivity of BTX at 400 °C. The numbers in parentheses denote Si/Al_2 ratio.

Meanwhile, H-ZSM-5 (29), which gave high yields of BTX, was subject to a moderate quantity of carbon deposits, suggesting that solid acidity of zeolite accelerates carbon deposition. In Madeira's report [46], in the cases of large pore zeolite (12-membered ring; MR), for example, H-Faujasite and H-Beta, rapid deactivation occurred, since rapid carbon deposition blocks strong brønsted acidic sites, necessary for formation of higher hydrocarbons. However, medium pore zeolite (10MR), for example, H-ZSM-5, could inhibit rapid deactivation even after most of pore was blocked by deposited carbon, suggesting that the reaction proceeds at entrance of micropore of H-ZSM-5. But, in my study, real amount of deposited carbon was larger for H-Beta and USY zeolites possessing 12MR and for H-mordenite with 12MR and 8MR than for H-ZSM-5 possessing 10MR. This result suggests that micropore volume of H-Beta, USY and H-mordenite is larger than that of H-ZSM-5, and this fact leads to large amount of carbon deposition for H-Beta, USY and H-mordenite. In fact, Madeira *et al.* reported that micropore volume of H-Faujasite and H-Beta is larger than that of H-ZSM-5 [46].

On the other hand, in the case of metal/H-ZSM-5 (29) catalysts, there is a correlation between the yield of BTX and the amount of deposited carbon, except for Co and Ni catalysts, which show a large amount of carbon deposition for the selectivity of BTX. Only the Au catalyst showed a slightly higher BTX selectivity and a slightly lower amount of carbon deposition than H-ZSM-5 (29) alone.

2.2.4. Change in Catalytic Activity with Time-on-Stream

The catalytic performance of H-ZSM-5 and metal/H-ZSM-5 catalysts as a function of time was investigated. Figure 5 shows the change in BTX selectivity with time-on-stream.

Reaction temperature was held at 400 °C. A decrease in BTX selectivity was observed over most catalysts, while ethanol conversion was almost constant, and the selectivity of ethylene increased with time.

The difference in BTX selectivity between 50 and 450 min after starting the reaction was calculated and adopted as the decrease in the BTX selectivity. The weight loss of catalyst after reacting at 400 °C for 450 min was obtained by TG analysis.

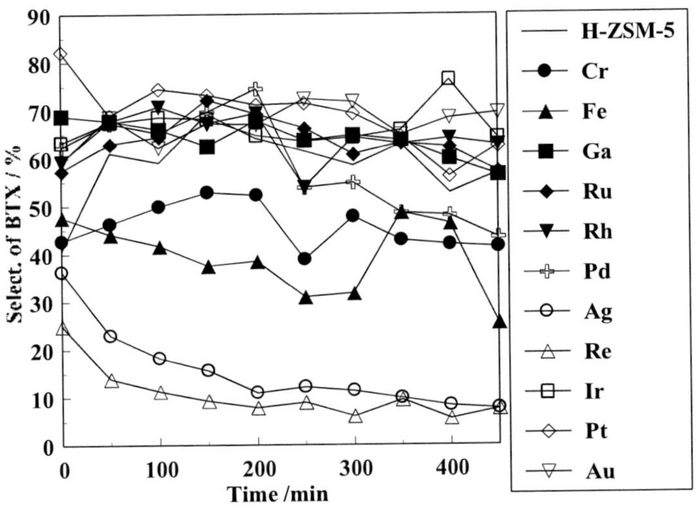

Figure 5. Change in selectivity of BTX (benzene, toluene and xylenes) over H-ZSM-5 and Metal/H-ZSM-5 catalysts at 400 °C with time on stream.

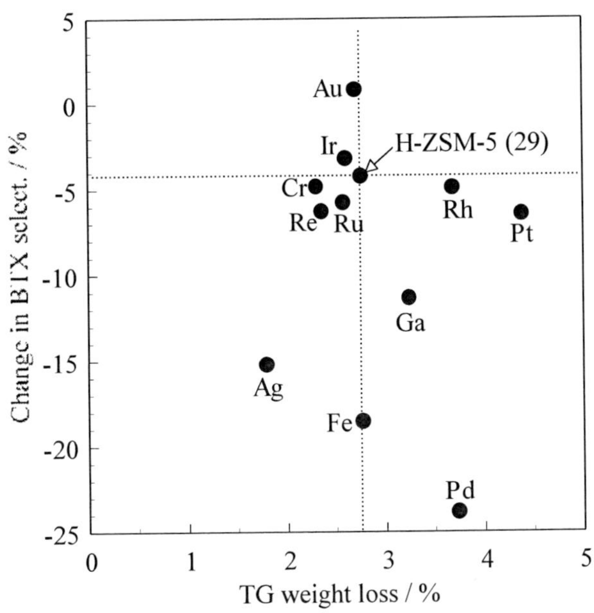

Figure 6. Plots of the weight loss of catalyst after reaction by TG vs. the change in BTX selevtivity at 400 °C. The number in parenthesis denotes Si/Al_2 ratio.

Figure 6 shows a plot of the change in BTX selectivity vs. TG weight loss. Although carbon deposition proceeded less at 400 °C than at above 400 °C, but with no clear correlation found, some metals showed characteristic results: for example, Rh and Pt showed a great deal of deposited carbon, while the decrease in selectivity was small, suggesting that these metals give some immunity to carbon deposition.

Ag showed small amounts of carbon but showed a large decrease in selectivity. Ir and Au showed less deposited carbon and less decrease in selectivity than H-ZSM-5 (29); Au in particular showed no decrease in selectivity.

3. Ethanol Conversion to C_{3+}olefins by Fe-Modified H-ZSM-5 Zeolite Catalysts [31]

3.1. Experimental

3.1.1. Catalyst Preparation

As zeolite support, H-ZSM-5 ($SiAl_2$ = 29, Tosoh) was used. The zeolite was calcined at 500 °C in a muffle furnace for 6 h (heating rate was 200 °C h^{-1}), prior to use (same as 2.1.1.). Iron catalysts were prepared by impregnation method (Fe = 10 wt%). As sources of Fe, iron (III) nitrate enneahydrate ($Fe(NO_3)_3{\cdot}9H_2O$, Wako Pure Chemicals), iron (III) sulfate *n*-hydrate ($Fe_2(SO_4)_3{\cdot}nH_2O$, Wako Pure Chemicals), iron (III) chloride hexahydrate ($FeCl_3{\cdot}6H_2O$, Iwai Chemicals), iron (II) oxalate dihydrate ($FeC_2O_4{\cdot}2H_2O$, Kanto Chemicals) and iron (III) phosphate *n*-hydrate ($FePO_4{\cdot}nH_2O$, Wako Pure Chemicals) were used. Iron (III) sulfate *n*-hydrate contains 60.0-80.0% of $Fe_2(SO_4)_3$. In this study, content of $Fe_2(SO_4)_3$ was assumed to be 70.0%, where value of n was ca. 9.5.

In the case of iron (III) phosphate *n*-hydrate, value of n was assumed to be 2. Iron (II) oxalate dihydrate was dissolved with nitric acid. Iron (III) phosphate *n*-hydrate was dissolved with hydrochloric acid. After impregnation, the wet catalysts were dried at 120 °C, followed by calcinations in flow of air, where calcination temperature was varied from 500 to 900 °C (in any case, heating rate was 200 °C h^{-1}). As mentioned below, in Calsavara's reports, Fe-modified H-ZSM-5 catalysts, prepared by two kind of procedure, were also used for production of hydrocarbons by ethanol conversion [35, 36]. Two kind of preparation procedure were as follows: (1) Fe source was directly added during hydrothermal synthesis using ethanol and seed crystals, and (2) Fe species were introduced by ion-exchange after synthesis of H-ZSM-5. As use other than ethanol conversion, Fe-ZSM-5 catalysts prepared by wetness ion-exchange were used for selective reduction of NO with ammonia as reductant [38] and for oxydehydrogenation of ethane and propane with nitrous oxide as oxidant [40]. Moreover, Fe/H-ZSM-5 catalysts prepared by solid-state ion-exchange were used for decomposition of nitrous oxide [39].

3.1.2. Catalytic Activity

Conditions of reactor and catalyst weight were same as 2.1.2. Reaction gas was obtained by mixing vaporized EtOH and N_2 (flow rate of N_2 was 60 cm^3 min^{-1}), without or with water (N_2 : EtOH = 82.1 : 17.9 vol % and 82.1 : 16.1 : 1.8 vol %, respectively), at a temperature of 400 °C, and the change in catalytic performance with time-on-stream was investigated. The effluent gas was analyzed by gas chromatography (GC): columns and detectors were same as 2.1.2. Catalytic activity may decrease with time-on-stream. Therefore, regeneration treatment of used catalyst after 7h-reaction was carried out by calcinations at 500 °C in flow of air for 1 h on used catalyst, and the secondary catalytic performance was examined.

3.1.3. Characterization

In order to estimate carbon deposition on catalysts, thermogravimetry (TG) analyses of catalysts after 7 h-reaction were carried out: measurement apparatus and conditions were same as 2.1.3. Moreover, as a cause of catalytic deactivation as well as change in product selectivities, sintering and transformation of supported FeO_x species can be considered. Therefore, X-ray diffraction (XRD) measurements of catalysts before and after 7 h-reaction were carried out: measurement apparatus and conditions were same as 2.1.3. Here, also, the size of metal crystallite was calculated by Sherrer's formula.

Framework collapse of zeolite support, due to dealumination of zeolite framework during reaction, may have relation to catalytic deactivation, since solid acidity of zeolite, essential for this reaction, is generated by existence of Al atoms incorporated into zeolite framework. Therefore, in this case, also, XRD measurements were carried out, and intensity of XRD peak of zeolite framework before and after 7 h-reaction was compared to investigate the framework collapse of zeolite during reaction. NH_3-temperature-programmed-desorption (TPD) experiments of H-ZSM-5 support and Fe-modified H-ZSM-5 catalysts were carried out (Nihon Bell, BEL-CAT) to measure the acidity of support and catalyst.

After pretreatment in flowing He at 500 °C for 60 min, adsorption of NH_3 at 100 °C for 90 min, and desorption of weak adsorbed NH_3 in He flow at 100 °C for 30 min, TPD profiles were obtained under conditions of He flow with the temperature varying from 100 to 800 °C at 10 °C min^{-1}. Fourier transform infrared (FT-IR) spectra of zeolite support and Fe catalysts were recorded by a JASCO FT/IR 7000 spectrometer, using KBr pellets under atmospheric conditions, to measure the change of acidic site of zeolite and catalysts.

To eliminate distortions of the fundamental hydroxyl-stretching region due to water adsorption, before analyses, vacuum treatment to sample pellets were carried out with heating at about 200 °C for several days. XPS measurement of Fe catalysts were carried out (ESCA-850, Shimadzu) to investigate the effect of remaining minor element, for example, N, P, S and Cl, contained in Fe sources ($Fe(NO_3)_3 \cdot 9H_2O$, $FePO_4 \cdot nH_2O$, $Fe_2(SO_4)_3 \cdot nH_2O$ and $FeCl_3 \cdot 6H_2O$, respectively).

3.2. Results and Discussion

3.2.1. Catalytic Performance

Conversion of EtOH and selectivity of each product among gaseous products are shown in Table 4: main gaseous products were ethylene, C_{3+} olefins (C_3 – C_7 olefins were detected), paraffins and aromatics (benzene, toluene, xylenes, mesitylene and pseudocumene). In most cases, EtOH conversion was almost 100%.

Table 4. Catalytic activity of H-ZSM-5 ($Si/Al_2 = 29$) zeolite and Fe/H-ZSM-5 ($Si/Al_2 = 29$) catalysts

Fe source	Calc. temp./ °C	Water	Ethanol conv./ %	C-selectivity / %				
				Ethylene	C_{3+}olefins	(Propylene)	Paraffins	Aromatics
No Fe loaded	500	No	100.0	6.9	11.9	(5.0)	20.4	59.6
		$+H_2O$	100.0	6.0	13.1	(5.6)	26.0	53.7
$Fe(NO_3)_3{\cdot}9H_2O$	500	No	100.0	22.4	22.4	(9.5)	20.2	35.0
		$+H_2O$	96.3	21.8	23.6	(11.2)	19.0	35.5
	700	No	97.0	24.9	23.2	(10.1)	20.4	31.5
		$+H_2O$	96.3	24.0	24.6	(11.8)	19.0	31.4
	900	No	91.5	22.9	26.8	(8.6)	37.3	13.0
		$+H_2O$	100.0	45.2	25.0	(11.9)	14.6	14.6
$Fe_2(SO_4)_3{\cdot}nH_2O$	500	No	99.6	22.2	23.3	(10.0)	19.4	35.1
		$+H_2O$	100.0	12.5	22.8	(10.8)	22.5	41.8
	700	No	92.3	18.4	25.8	(11.5)	21.9	33.9
		$+H_2O$	97.0	17.6	24.3	(11.3)	20.7	37.5
	900	No	98.2	52.5	23.7	(11.0)	12.8	10.4
		$+H_2O$	99.9	52.4	24.9	(11.6)	13.5	8.9
$FeCl_3{\cdot}6H_2O$	500	No	99.6	30.9	19.5	(9.0)	18.5	30.8
		$+H_2O$	100.0	24.2	21.7	(9.7)	20.2	34.0
	700	No	100.0	29.8	25.0	(12.0)	15.6	29.6
		$+H_2O$	100.0	30.6	27.1	(12.5)	17.5	24.8
	900	No	99.9	39.0	25.3	(10.8)	14.6	20.9
		$+H_2O$	100.0	45.8	26.5	(11.9)	15.1	12.2
$FeC_2O_4{\cdot}2H_2O$	700	No	95.6	33.9	25.1	(12.0)	15.9	25.2
		$+H_2O$	100.0	32.0	26.5	(12.4)	16.5	24.6
	900	No	100.0	38.2	28.7	(12.8)	16.0	17.2
$FePO_4{\cdot}nH_2O$	700	No	95.60	55.5	17.4	(8.1)	9.6	17.2
		$+H_2O$	100.00	60.9	19.5	(9.1)	11.0	8.5
	900	No	100.00	97.6	1.5	(0.8)	0.7	0.0

Preparation conditions: Fe-loading = 10 wt%. Reaction conditions: Atmospheric pressure, Reaction temperature = 400 °C, Catalyst weight = 0.2 g, $EtOH/N_2 = 0.22$ or $H_2O/EtOH/N_2 = 0.02/0.2/1$, N_2 flow = 60 ml/min.

In general, Fe-loading suppressed the formation of aromatic compounds but enhanced the formations of ethylene and C_{3+} olefins. Paraffins yield not always inhibited, while C_{3+} olefins/paraffins ratio was increased by Fe-loading. In most cases, the selectivity of C_{3+} olefins was within 20 – 28%. Increasing calcination temperature caused increased ethylene selectivity and decreased aromatics selectivity: for example, in the case of $Fe_2(SO_4)_3$ and reaction without water, the selectivity of ethylene was increased (22.2 → 52.5%) with the increase in calcination temperature from 500 to 900 °C. On the other hand, the selectivity of aromatic compounds was decreased (35.1 → 10.4%). The effect of Fe source was not distinguished so much, except for the case of $FePO_4$, which showed especially high selectivity of ethylene (55 – 88%) and low selectivity of other products (the selectivity of C_{3+} olefins was 1-20%).

In general, the higher selectivity of aromatics, the lower selectivity of C_{3+} olefins. On the other hand, the presence of water affected the selectivity of each product, only slightly, probably due to an interaction of surface OH group with water (See Figure 13). The selectivity of propylene seems to be slightly enhanced by adding water:

For example, in the catalyst ($Fe(NO_3)_3$, 500 °C), the selectivity of propylene was increased (9.5 → 11.2%), and in the catalyst ($Fe(NO_3)_3$, 700 °C), the selectivity was increased (10.1 → 11.8%). Moreover, for 500 and 700 °C-calcined catalysts, the selectivity of ethylene was slightly decreased or almost constant by adding water, while the ethylene selectivity was increased for 900 °C-calcined catalysts.

In most cases, the selectivity of C_{3+} olefins and paraffins selectivity was slightly increased by adding water. The selectivity of aromatic compounds was slightly increased by water for 500 °C-calcined catalysts, while the selectivity was decreased by water for most of other catalysts.

3.2.2. Change in Activity with Time-on-Stream and Deactivation

Selectivity of each product changed with time-on-stream, although ethanol conversion remained constant at ca.100%. Time dependence on the product selectivity could also be associated with change in the amount of surface OH group and/or acidity. In general, the selectivity of aromatic compounds was decreased with time-on-stream, while the selectivity of ethylene was increased and in most cases, the selectivity of C_{3+} olefins was slightly decreased. Figure 7 shows the plots of the initial selectivity of C_{3+} olefins vs. the change in the selectivity of C_{3+} olefins between 0 and 7 h after starting reaction. In this reaction, neat ethanol was used. For 500 °C-calcined catalysts, having IR-detectable surface OH groups (Figure 13), there seems to be a correlation between initial selectivity of C_{3+} olefins and decrease in C_{3+} olefins selectivity.

On the other hand, 700 °C- and 900 °C-calcined catalysts showed no correlation: initial selectivity of C_{3+} olefins did not differ so significantly, except for the case of $FePO_4$-derived catalyst showing low initial selectivity. Similarly, two exceptional catalysts ($Fe_2(SO_4)_3$, 700 °C and $FeCl_3$, 700 °C) with weak OH absorptions showed extremely small decrease (between 0 and 7 h after starting reaction) in the C_{3+} olefins selectivity (-0.8 and 0.8%, respectively) even at the high initial selectivity of C_{3+} olefins (25.3 and 25.0%, respectively). Moreover, in most cases, the decrease in C_{3+} olefins selectivity was slightly increased by adding water, while the increase in ethylene selectivity and the decrease in aromatics selectivity were enlarged by adding water.

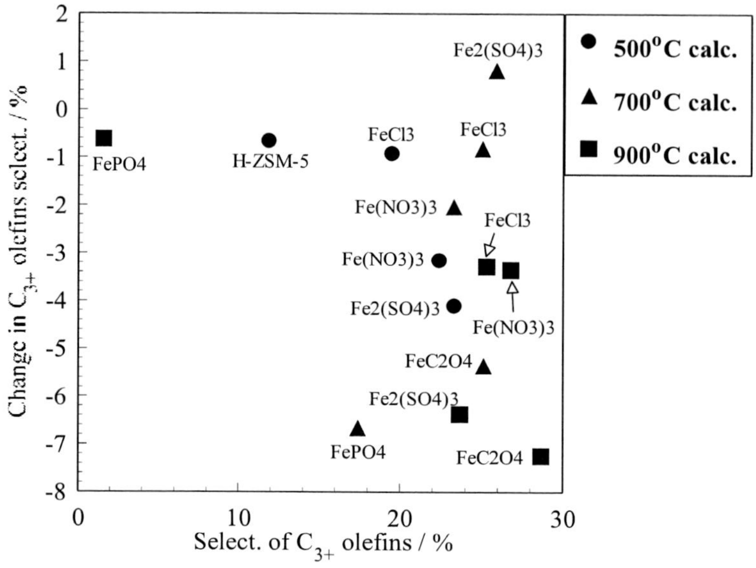

Figure 7. Plots of the initial selectivity of C_{3+} olefins vs. the change in selectivity of C_{3+} olefins over H-ZSM-5 ($Si/Al_2 = 29$) zeolite and Fe/H-ZSM-5 ($Si/Al_2 = 29$) catalysts. Reaction gas: N_2 + EtOH (without H_2O).

3.2.3. Carbon Deposition

In order to estimate carbon deposition on the catalyst surfaces, TG analyses were carried out and mole number of C atoms in deposited carbon was calculated from the weight loss due to combustion of deposited carbon. Figure 8 shows the plots of deposited carbon/total EtOH ratio in mole number of C atoms vs. the change in C_{3+} olefins selectivity between 0 and 7 h after starting reaction. Here, neat ethanol was used. There was a small difference in deposited carbon/total EtOH ratio between each catalyst, ranging from 0.07 to 0.14, except for $Fe_2(SO_4)_3$, 500 °C (0.2%) and $FePO_4$, 900 °C (0.0%).

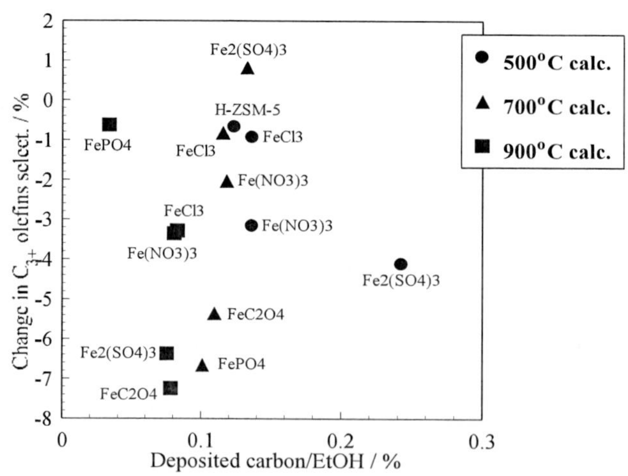

Figure 8. Plots of deposited carbon/total EtOH ratio in mole number of C atoms calculated by result of TG analysis vs. the change in C_{3+} olefins selectivity over H-ZSM-5 ($Si/Al_2 = 29$) zeolite and Fe/H-ZSM-5 ($Si/Al_2 = 29$) catalysts. Reaction gas = N_2 + EtOH (without H_2O).

In general, 500 °C-calcined catalysts showed large amount of deposited carbon: the decrease in C_{3+} olefins selectivity as well as carbon deposition was larger than H-ZSM-5 support. 900 °C-calcined catalysts showed the small amount of carbon deposition (smaller than H-ZSM-5 support) and the large decrease in selectivity except for $FePO_4$-derived catalyst, which showed small decrease in selectivity. On the other hand, 700 °C-calcined catalysts showed the moderate amount of carbon deposition, near H-ZSM-5 support. Moreover, there seems to be weak inversed correlation between amount of carbon deposition and decrease in C_{3+} olefins selectivity. Although deposited carbon/total EtOH ratio was low and negligible for the selectivity of each product and carbon balance, the effect on catalytic activity was not negligible.

Figure 9 shows the plots of deposited carbon/total EtOH ratio in mole number of C atoms vs. the change in the selectivity of aromatic compounds. It is suggested that the lower calcination temperature leads to not only larger amount of carbon deposition but also larger decrease in the selectivity of aromatic compounds: for example, the average decrease in selectivity for 700 °C-calcined catalysts was larger than that for 900 °C-calcined catalysts. As mentioned above, the catalysts yielding aromatic compound in high selectivity tend to deposit large amount of carbon, suggesting that aromatic compounds can be converted to deposited carbon. On the other hand, only weak correlation between the deposited carbon/total EtOH ratio and the initial selectivity of C_{3+} olefins was recognized.

Therefore, carbon deposition can be responsible for catalytic deactivation. However, this is not only cause. In general, acidic sites of zeolite support are essential to formation of aromatic compound [7, 10]. When the acidic sites were covered with deposited carbon, the formation of aromatic compounds was suppressed, but the formation of C_{3+} olefins was not inhibited so significantly. The amount of carbon deposition was slightly decreased by adding water, suggesting that gasification of deposited carbon to H_2 and CO may occur by added water.

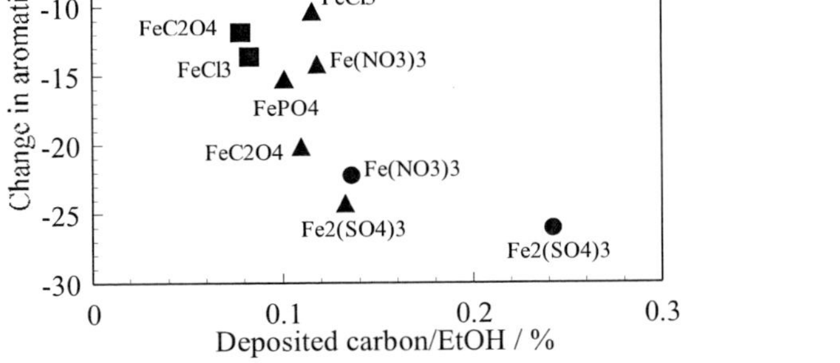

Figure 9. Plots of deposited carbon/total EtOH ratio in mole number of C atoms calculated by result of TG analysis vs. the change in aromatics selectivity over H-ZSM-5 ($Si/Al_2 = 29$) zeolite and Fe/H-ZSM-5 ($Si/Al_2 = 29$) catalysts. Reaction gas = N_2 + EtOH (without H_2O).

3.2.4. Change in Phase and Crystallite Diameter of Loaded Iron

Figure 10 shows XRD patterns of Fe_2O_3, Fe_3O_4 and Fe/H-ZSM-5 catalysts before and after reaction (without water). In the cases of $Fe(NO_3)_3$-derived catalysts, before reaction, loaded Fe exists as Fe_2O_3. However, in the case of 500 and 700 °C-calcined catalysts, Fe_2O_3 species were partially reduced to Fe_3O_4 during reaction, while Fe_2O_3 on 900 °C-calcined catalyst was almost unchanged without reduction even after reaction. On the other hand, catalyst ($Fe_2(SO_4)_3$, 500 °C) does not show XRD pattern of FeO_x before reaction, suggesting that Fe may exist as $FeSO_x$ species, which is effective for the formation of aromatic compounds and carbon deposition, while after reaction, XRD pattern of Fe_3O_4 was observed. The catalyst ($FePO_4$, 700 °C) showed no FeO_x pattern both before and after reaction. Fe may exist as $FePO_x$ species, effective for the ethylene formation, but not effective for the formation of C_{3+} olefins and aromatic compounds.

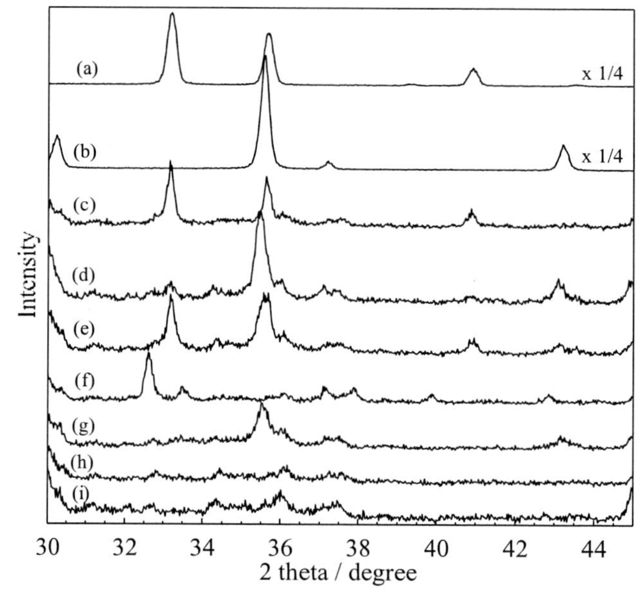

Figure 10. XRD patterns of Fe_2O_3, Fe_3O_4 and Fe/H-ZSM-5 ($Si/Al_2 = 29$) catalysts before and after reaction (without H_2O).(a) Fe_2O_3, (b) Fe_3O_4, (c) $Fe(NO_3)_3$, 900°C, before reaction, (d) $Fe(NO_3)_3$, 700°C, after reaction, (e) $Fe(NO_3)_3$, 900°C, after reaction, (f) $Fe_2(SO_4)_3$, 500°C, before reaction, (g) $Fe_2(SO_4)_3$, 500°C, after reaction, (h) $FePO_4$, 700°C, before reaction, (i) $FePO_4$, 700°C, after reaction.

Phase and crystallite diameter of FeO_x on Fe catalysts before and after 7 h-reaction are summarized in Table 5. Except for the cases of catalyst ($Fe_2(SO_4)_3$, 500 °C) and $FePO_4$-derived catalysts, FeO_x existed as Fe_2O_3 before reaction, while after reaction, Fe_2O_3 was partially reduced to Fe_3O_4 for 500 and 700 °C-calcined catalysts but Fe_2O_3 was unchanged for 900°C-calcined catalysts. In the cases of $Fe(NO_3)_3$-derived and $Fe_2(SO_4)_3$-derived catalysts, higher calcination temperature led to larger crystallite diameter both before and after reaction. In most cases, crystallite diameter of FeO_x on 500 and 700 °C-calcined catalysts became smaller during reaction, perhaps due to removal of O atoms during reaction. On the other hand, in the case of 900 °C-calcined catalysts, crystallite diameter of Fe_2O_3 was not decreased so significantly, suggesting that removal of O atoms scarcely occurred. From these results, it appeared that the sintering of supported Fe scarcely occurred. However, there seems to be a poor correlation between the crystallite diameter before reaction and selectivity of each product. In general, the larger decrease in FeO_x crystallite diameter resulted in the lower decrease in C_{3+} olefins selectivity and the higher decrease in aromatics selectivity.

Table 5. Phase and crystallite size of FeO_x on Fe/H-ZSM-5 catalysts before and after reaction (without H_2O)

Fe source	Calc. temp. / °C	Phase (crystallite size / nm)		
		Before reaction	→	After reaction
$Fe(NO_3)_3{\cdot}9H_2O$	500	Fe_2O_3 (22.2)	→	Fe_3O_4 (18.3)
	700	Fe_2O_3 (29.1)	→	Fe_3O_4 (27.2)
	900	Fe_2O_3 (34.9)	→	Fe_2O_3 (33.2)
$Fe_2(SO_4)_3{\cdot}nH_2O$	500	$FeSO_x$	→	Fe_3O_4 (20.8)
	700	Fe_2O_3 (35.6)	→	Fe_3O_4 (29.4)
	900	Fe_2O_3 (38.4)	→	Fe_2O_3 (40.2)
$FeCl_3{\cdot}6H_2O$	500	Fe_2O_3 (45.1)	→	Fe_3O_4 (42.4)
	700	Fe_2O_3 (45.2)	→	Fe_3O_4 (39.7)
	900	Fe_2O_3 (41.8)	→	Fe_2O_3 (39.7)
$FeC_2O_4{\cdot}2H_2O$	700	Fe_2O_3 (32.3)	→	Fe_3O_4 (25.6)
	900	Fe_2O_3 (29.6)	→	Fe_2O_3 (29.7)
$FePO_4{\cdot}nH_2O$	700	$FePO_x$	→	$FePO_x$
	900	$FePO_x$	→	$FePO_x$

Brandenberger *et al.* reported that in the case of Fe-ZSM-5 catalyst prepared by wetness ion-exchange, Fe migration, formation of FeO_x cluster and increase in size of cluster occurred in hydrothermal conditions [38]. On the other hand, in my work, though reaction condition was similar with hydrothermal condition, since water was formed by ethanol dehydration, the growth of crystallite during reaction did not occur. This seems to be because Fe-loading was larger than Brandenberger's work and crystallite of FeO_x was already large.

3.2.5. Framework Collapse of Zeolite

Figure 11 shows XRD patterns of the catalyst ($Fe(NO_3)_3$, 700 °C) before and after 7 h-reaction (without H_2O), where all XRD peaks are due to framework of H-ZSM-5 zeolite, not including peak of FeO_x: intensity of XRD patterns was weakened during reaction, suggesting that framework collapse, at least in part, proceeded during reaction. After reaction, XRD peak of deposited carbon was not observed. Similar results were obtained by Ouyang's research: bio-ethanol was converted to ethylene over La-modified H-ZSM-5 catalysts with carbon deposition [47]. The arrow marks the peak at 7.9 °. The intensity of this peak was used as index of crystallinity. After reaction/before reaction ratio in intensity of XRD pattern detected at 7.9° is summarized in Table 6: the lower ratio means that framework collapse of zeolite proceeded in more degree.

Table 6 shows that framework collapse scarcely occurred on 500 °C-calcined catalysts and in general, higher calcination temperature led to collapse in more degree. However, catalyst ($Fe(NO_3)_3$, 700 °C) showed especially low ratio (0.42), lower than $Fe_2(SO_4)_3$, 700 °C (0.62) and $FeCl_3$, 700 °C (0.70), suggesting that remaining S or Cl species can inhibit framework collapse in some degree or that remaining N species enhances framework collapse more significantly than S or Cl species. These findings could be consistent with the fact that the decrease in selectivity of C_{3+} olefins in $Fe_2(SO_4)_3$- or $FeCl_3$-derived catalysts was lower than that in $Fe(NO_3)_3$-derived catalysts (Figure 7).

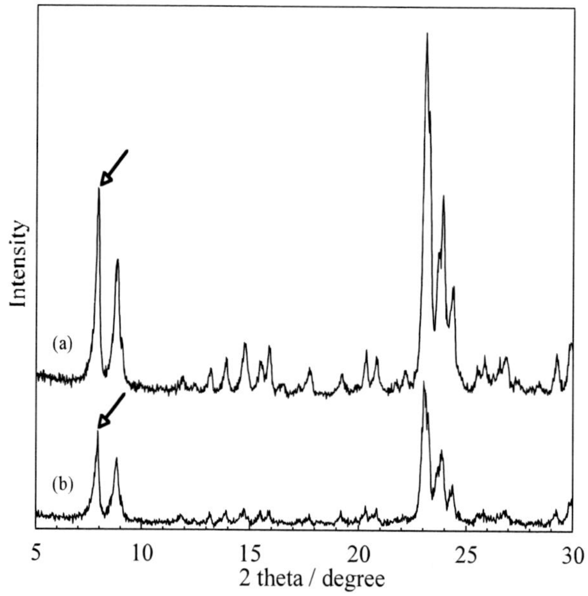

Figure 11. XRD patterns of Fe/H-ZSM-5 (Si/Al_2 = 29)($Fe(NO_3)_3$, 700°C) catalysts before and after reaction (without H_2O). (a) before reaction, (b) after reaction.

Table 6. Ratio of intensity of X-ray diffraction peaks (at 7.9 °) of Fe/H-ZSM-5 (Si/Al_2 = 29) catalysts before and after reaction (without H_2O)

Fe source	Calc. temp. / °C	Ratio of intensity (After reaction/ before reaction)
$Fe(NO_3)_3{\cdot}9H_2O$	500	0.81
	700	0.42
	900	0.61
$Fe_2(SO_4)_3{\cdot}nH_2O$	500	0.98
	700	0.62
	900	0.59
$FeCl_3{\cdot}6H_2O$	500	0.94
	700	0.70
	900	0.47
$FeC_2O_4{\cdot}2H_2O$	900	0.72
$FePO_4{\cdot}nH_2O$	900	0.56

3.2.6. Solid Acidity of Catalysts

NH_3-TPD profiles of zeolite support and Fe catalysts are shown in Figure 12. H-ZSM-5 support (calcined at 500 °C) showed desorption peak at about 400 °C. $Fe(NO_3)_3$-derived catalysts, calcined at 500 and 700 °C, showed NH_3 desorption peak at about 400 °C, while the calcination at 900 °C led to disappearance of NH_3 desorption peak. The catalyst ($Fe_2(SO_4)_3$, 500 °C) showed NH_3 desorption peak at 370 °C: this peak may be due to $FeSO_x$ species. By calcination at higher temperature, NH_3 desorption peak was decreased. In the cases of $FeCl_3$ and FeC_2O_4, NH_3 desorption peaks were not clear. On the other hand, in the case of $FePO_4$,

NH_3 desorption peak disappeared completely, suggesting that $FePO_x$ species block acidic site of zeolite and suppress the formation of aromatic compounds and C_{3+} olefins. From these results, it is suggested that solid acidity showing NH_3 desorption peak at about 400 °C is essential for the formation of aromatic compounds, and calcination at higher temperature may weaken this solid acid by framework collapse or/and ion-exchange with iron cation: the higher calcination temperature resulted in the lower selectivity of aromatic compounds, as shown in Table 4.

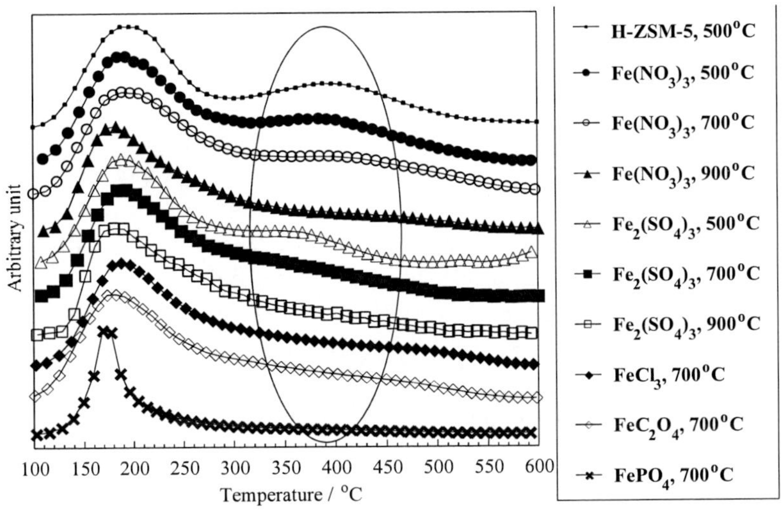

Figure 12. NH_3-TPD profiles of H-ZSM-5 ($Si/Al_2 = 29$) zeolite and Fe/H-ZSM-5 ($Si/Al_2 = 29$) catalysts.

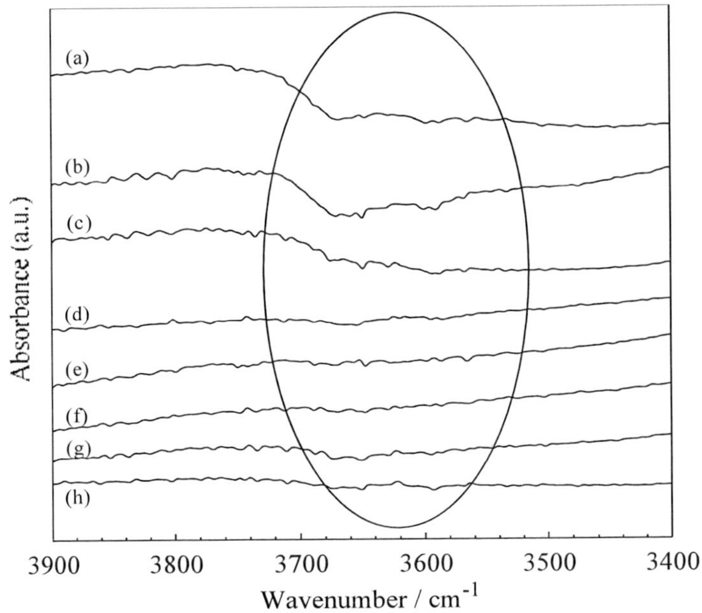

Figure 13. FT-IR spectra of H-ZSM-5 ($Si/Al_2 = 29$) zeolite and Fe/H-ZSM-5 ($Si/Al_2 = 29$) catalysts. (a) H-ZSM-5, 500°C, (b) $Fe(NO_3)_3$, 500°C, (c) $Fe(NO_3)_3$, 900°C, (d) $Fe_2(SO_4)_3$, 500°C, (e) $Fe_2(SO_4)_3$, 900°C, (f) $FeCl_3$, 700°C, (g) FeC_2O_4, 700°C, (h) $FePO_4$, 700°C.

FT-IR spectra of H-ZSM-5 support and Fe catalysts are shown in Figure 13. Absorption peak due to O-H appeared at 3600-3700 cm^{-1}. H-ZSM-5 (calcined at 500 °C) showed clear absorption peaks [48]. On the other hand, catalyst ($Fe(NO_3)_3$, 500 °C) showed same peaks, while catalysts ($Fe(NO_3)_3$, 900 °C) showed weak peaks. Moreover, $Fe_2(SO_4)_3$-, $FeCl_3$-, FeC_2O_4- and $FePO_4$-derived catalysts showed very weak peak. These results suggest that proton in zeolite framework may be exchanged with iron cations and/or framework collapse of zeolite may occur during preparation: calcination at higher temperature led to higher rate of ion-exchange and/or framework collapse, and as a result, solid acidity of H-ZSM-5 zeolite may be reduced in higher degree.

3.2.7. Results of XPS Measurement

XPS measurement of $Fe(NO_3)_3$-derived catalysts showed peak due to N_{1s} at 402 eV of binding energy. Moreover, in the cases of $Fe_2(SO_4)_3$-, $FeCl_3$- and $FePO_4$-based catalysts, S_{2p} (169 eV), Cl_{2p} (199 eV) and P_{2p} (134 eV) peaks were observed, respectively. These results suggest that minor elements due to Fe sources remain even after calcination and these minor elements may affect the natures of Fe catalysts and zeolite frameworks.

3.2.8. Catalyt Performance as a Function of Time and Regeneration Treatment

Selectivity of each product changed with time-on-stream. Therefore, catalytic performance as a function of time and effect of regeneration treatment were investigated in absence of water.

Figure 14 shows the catalytic performance of the catalyst ($Fe(NO_3)_3$, 700 °C) as a function of time. Selectivity of aromatic compounds was decreased (40.7 → 23.7%) and selectivity of ethylene was drastically increased (22.7 → 43.9%). The selectivity of C_{3+} olefins, propylene and paraffins were slightly decreased (20.2 → 19.6%, 9.1 → 8.2% and 14.9 → 12.5%, respectively). To regenerate the used catalyst, combustion treatment was carried out in the air-flow at 500 °C for 1 h. By this treatment, ethylene selectivity was decreased to 32.9% and aromatics selectivity was increased to 34.6%. However, the ethylene selectivity obtained by this treatment was higher than the initial selectivity, and the aromatics selectivity was lower than the initial selectivity. It is suggested that only this air treatment was insufficient to fully regenerate the catalyst. Perhaps, it was due to irreversible deactivation by framework collapse of H-ZSM-5 support.

As shown in Table 6, this catalyst showed lower ratio of intensity of XRD peaks (after reaction/before reaction = 0.42), than other kind of 700 °C-calcined catalysts: $Fe_2(SO_4)_3$- and $FeCl_3$-derived catalysts showed 0.62 and 0.70 of intensity ratio, respectively. By this regeneration treatment, combustion of deposited carbon and re-oxidation of reduced Fe_3O_4 to Fe_2O_3 could be performed almost completely.

Figure 15 shows the catalytic performance of the catalyst ($Fe(NO_3)_3$, 900 °C) as a function of time. The ethylene selectivity was increased (35.5 → 49.3%), and the aromatics selectivity was decreased (15.8 → 8.7%). The selectivity of C_{3+} olefins, propylene and paraffins was slightly decreased (31.1 → 27.4%, 13.7 → 11.7% and 17.2 → 14.3%, respectively). Changes in the selectivity of each product were smaller than the case of 700 °C-calcined catalyst. Moreover, by the combustion treatment, catalytic activity was almost restored: the selectivities of ethylene, C_{3+} olefins, propylene, paraffins and aromatics were recovered to 37.3, 30.4, 13.0, 17.0 and 15.2%, respectively. It is suggested that degree of irreversible deactivation due to framework collapse of H-ZSM-5 support was smaller than the case of the catalyst ($Fe(NO_3)_3$, 700 °C), since moderate framework collapse already occurred during catalyst preparation.

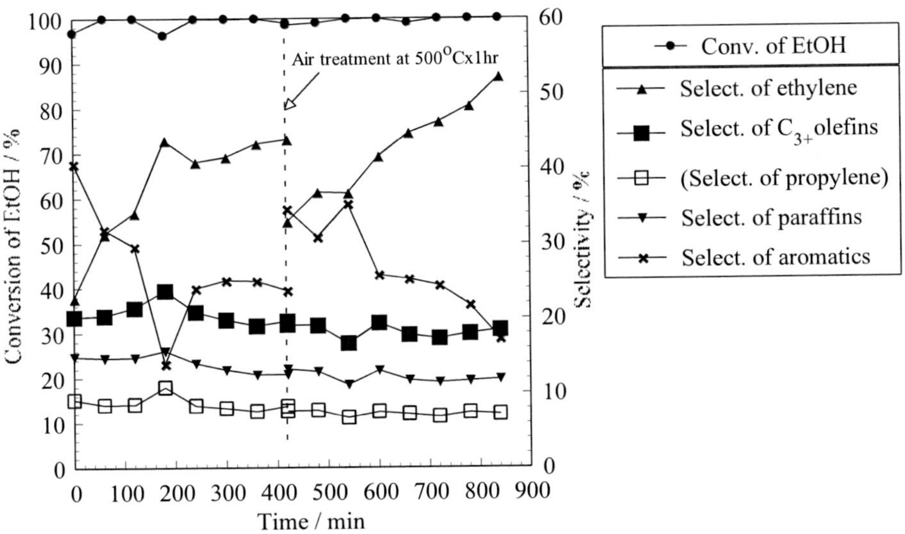

Figure 14. Catalytic performance of Fe/H-ZSM-5 (Si/Al_2 = 29)($Fe(NO_3)_3$-$9H_2O$, 700°Ccalc.) catalyst as a function of time: effect of air treatment at 500 °C on used catalyst. Reaction gas: N_2 + EtOH (without H_2O).

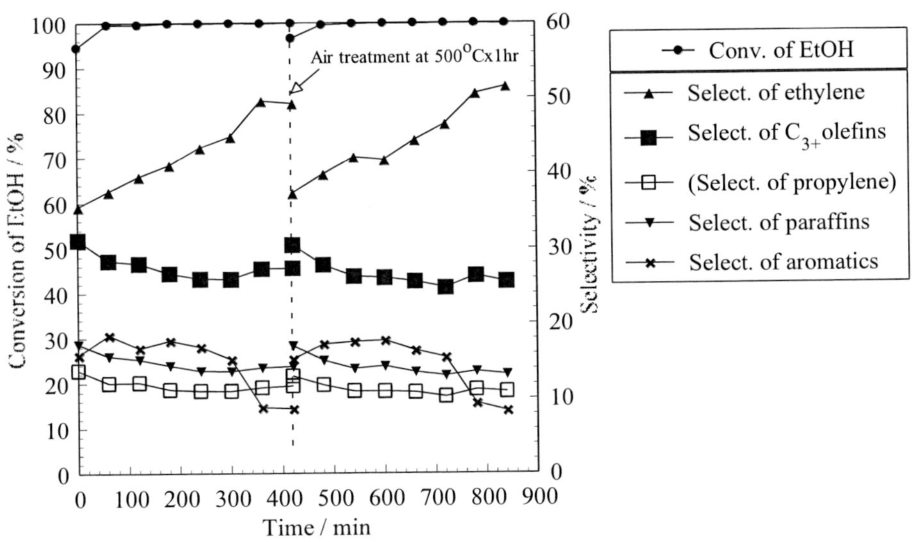

Figure 15. Catalytic performance of Fe/H-ZSM-5 (Si/Al_2 = 29)($Fe(NO_3)_3$-$9H_2O$, 900°Ccalc.) catalyst as a function of time: effect of air treatment at 500 °C on used catalyst. Reaction gas: N_2 + EtOH (without H_2O).

Figure 16 shows the catalytic performance of the catalyst ($Fe_2(SO_4)_3$, 700 °C) as a function of time. The pronounced increase in the ethylene selectivity (12.7 → 44.1%) as well as the marked decrease in aromatics selectivity (47.0 → 15.5%) were observed, and catalytic activity could not be restored by this combustion treatment completely. For example, by the regeneration treatment, the selectivity of ethylene was restored to 28.2%, higher than the initial selectivity in the first reaction (12.7%). However, the selectivity of C_{3+} olefins was increased with time-on-stream in the first reaction (21.8 → 26.2%), and the selectivity of C_{3+} olefins in the second reaction (29.9%) was higher than the first reaction (21.8%).

The selectivity of propylene was also increased in the first reaction (9.6 → 11.8%), and the selectivity of propylene in the second reaction (13.9%) was higher than the first reaction. It is suggested that combustion treatment enhanced the formation on C_{3+} olefins and propylene. In this case also, framework collapse may occur. However, as shown in Table 6, the degree of collapse seems to be smaller than the case of the catalyst ($Fe(NO_3)_3$, 700 °C). Therefore, remaining S may play a role to deposit carbon which cannot be combusted easily.

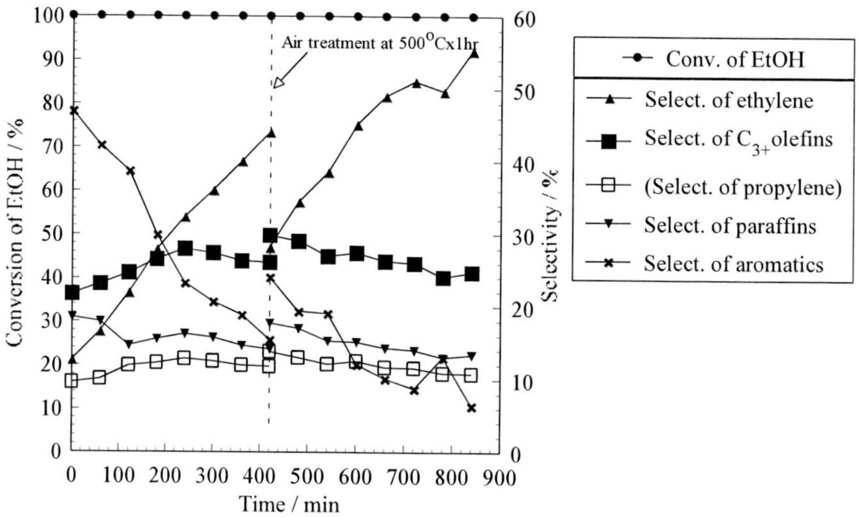

Figure 16. Catalytic performance of Fe/H-ZSM-5 (Si/Al_2 = 29)($Fe_2(SO_4)_3$-nH_2O, 700°Ccalc.) catalyst as a function of time : effect of air treatment at 500 °C on used catalyst. Reaction gas: N_2 + EtOH (without H_2O).

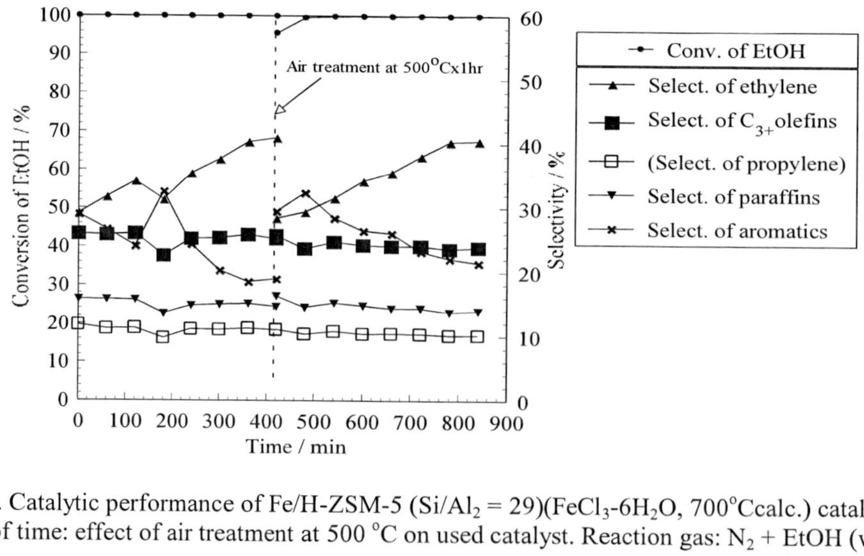

Figure 17. Catalytic performance of Fe/H-ZSM-5 (Si/Al_2 = 29)($FeCl_3$-$6H_2O$, 700°Ccalc.) catalyst as a function of time: effect of air treatment at 500 °C on used catalyst. Reaction gas: N_2 + EtOH (without H_2O).

Figure 17 shows the catalytic performance of the catalyst ($FeCl_3$, 700 °C) as a function of time. The increase in the ethylene selectivity (29.3 → 41.0%) and the decrease in aromatics selectivity (29.0 → 19.0%) were small. The selectivity of C_{3+} olefins, propylene and paraffins were almost constant (26.0 → 25.2%, 11.8 → 11.0% and 15.8 → 14.7%, respectively). Moreover, by the combustion treatment, catalytic activity was restored almost completely: the selectivity of ethylene, C_{3+} olefins, propylene, paraffins and aromatics was recovered to 28.4, 25.7, 11.1, 16.3 and 29.5%, respectively.

As shown in Table 6, the degree of framework collapse was smaller than the cases of $Fe(NO_3)_3$, 700 °C and $Fe_2(SO_4)_3$, 700 °C, suggesting that remaining Cl species played a role to prevent framework collapse or that remaining N or S species enhanced framework collapse more significantly than Cl species.

3.2.9. Reaction Pathway

Scheme 1 shows supposed reaction pathway. First, ethanol is converted to ethylene by dehydration: this reaction can proceed on acidic site. Successive reactions such as oligomerization, cracking, dehydrogenation and H-transformation occur on acidic site of zeolite support [9, 11-13], and C_{3+} olefins and paraffins are formed.

Propylene may be formed by the cracking of larger olefins, especially C_6 olefins, since in this reaction, methane was observed only rarely, suggesting a low likelihood that propylene is formed by addition of methane and ethylene or by division of C_4 olefins into methane and propylene. Ingram reported that propylene is formed from the cracking of larger olefins, possibly 4-methylpentene, and not as an intermediate between C_2 and C_4 production [15].

It is suggested that other olefins are formed via these several reaction. Aromatic compounds are formed by further oligomerization and dehydrogenation of C_{3+} olefins and paraffins. Furthermore, the compounds can undergo carbonization to give deposited carbon. Fe species may inhibit the formation of aromatics in some degree.

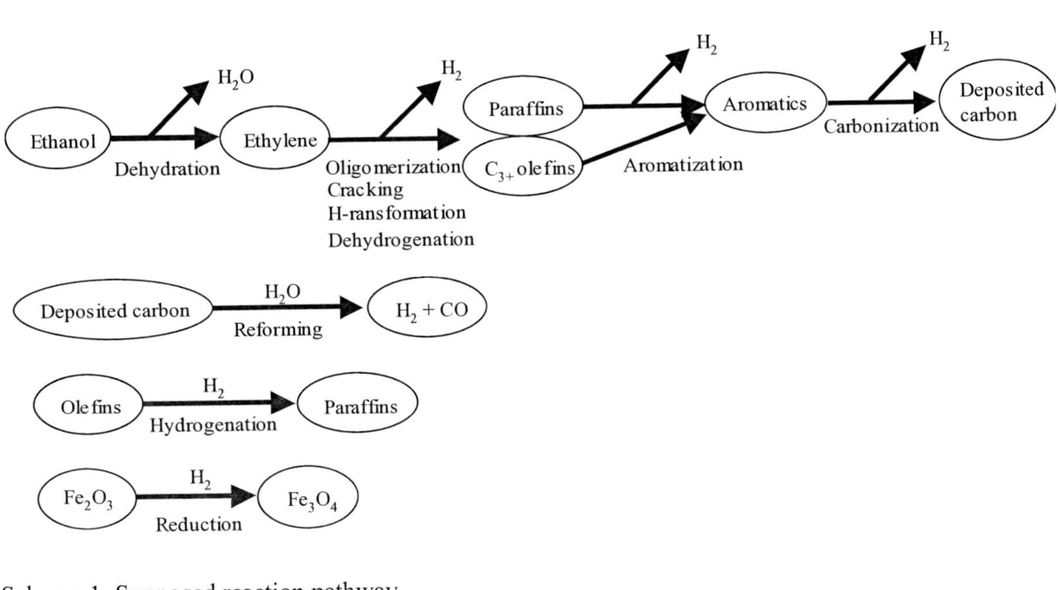

Scheme 1. Supposed reaction pathway.

In the case of catalysts calcined at higher temperature, the amount of acidic site may be decreased by ion-exchange with iron cation or framework collapse of zeolite support, which lead to decrease in both aromatics selectivity and carbon deposition and increase in ethylene selectivity.

Water can be obtained by dehydration of ethanol, easily. This water can react with deposited carbon to CO and H_2. Moreover, H_2 can be formed via dehydrogenation in aromatization and carbonization. H_2 can hydrogenate olefins to paraffins. However, the hydrogenation to paraffins can be suppressed by reduction of Fe_2O_3 to Fe_3O_4. In fact, as shown in Table 4, generally, Fe catalyst showed higher C_{3+} olefins/paraffins ratio in selectivity than H-ZSM-5 support (no metal loaded).

In the case of 900 °C-calcined catalysts, Fe_2O_3 was unchanged without reduction, since the aromatics formation and the carbonization scarcely proceed, resulting in little amount of H_2 formed.

As reaction proceeds, the carbon deposition and the framework collapse lead to significant decrease in aromatics selectivity and slight decrease in selectivity of C_{3+} olefins and propylene. Maybe, the carbon deposition suppresses the aromatics formation predominantly over C_{3+} olefins, since it is suggested that aromatics formation needs strong acidic site, not covered with deposited carbon. In the presence of water, carbon deposition can be inhibited by steam-combustion in some degree, but framework collapse can be enhanced, resulting in enlarged decrease in aromatics selectivity and enlarged increase in ethylene selectivity.

3.2.10. Comparison with Other Reaction System Using Fe-Modified H-ZSM-5 Catalysts

In the case of catalyst prepared by direct incorporation of Fe source during hydrothermal synthesis, reported by Calsavara *et al.* [35], a part of Fe species, incorporated into zeolite framework, removed from framework by calcination and formed cluster.

In production of hydrocarbons by ethanol conversion, main products were liquid hydrocarbons, especially aromatics. Lifetime of catalytic activity was prolonged by Fe-addition. Cause of deactivation was carbon deposition. Fe-addition could weaken solid acidity slightly and could delay the rate of carbon deposition. The best Fe-loading was 0.5 wt% to obtain liquid hydrocarbons in high yield.

Calcination at high temperature (700 °C) led to the decrease in hydrocarbons yield in comparison with low temperature (500 °C). Framework collapse of zeolite was not mentioned here, but it is suggested that framework collapse did not occur, or if collapse occurred, the degree of collapse was very low, since catalytic activity could be recovered by calcination in air.

Moreover, Calsavara *et al.* reported that Fe-modified H-ZSM-5 catalysts where Fe was introduced by ion-exchange was also effective for hydrocarbons production by ethanol conversion [36].

Calsavara's study was different from our study: Fe-loading was low and calcination temperature was low. Moreover, main products were higher hydrocarbons such as aromatics. In our study, Fe-loading was high (10 wt%) and as a rule, calcination temperature was 700 °C. Therefore, it seems that acidic sites were blocked in comparison with Calsavara's catalysts, and as a result, aromatics formation was decreased.

Brandenberger *et al.* prepared Fe-ZSM-5 catalysts for selective reduction of NO with NH_3 by wetness ion-exchange, and reported about the effect of aging in hydrothermal conditions [38]. Aging led to dealumination from zeolite framework, followed by migration of Fe from ion-exchange site, formation of FeO_x cluster and increase in cluster size. In this case, stability of catalytic activity was influenced by stability of Fe species rather than that of zeolite framework. Moreover, existence of proton accelerated dealumination, while aluminum sites possessing Fe ions hardly underwent hydrolysis.

In our study, it was found that in the case of 10 wt% of Fe-loading, loaded Fe species, Fe_2O_3 was reduced into Fe_3O_4 during ethanol reaction and increase in crystallite size was not observed. Therefore, it is suggested that loaded Fe species acted in different way from Brandenberger's report. In the case of 1 wt% of Fe-loading, as mentioned later, formation of FeO_x crystallite was not observed by XRD measurement. In this study also, framework collapse of zeolite was observed, and as mentioned in 5.2.3, framework collapse was inhibited for Fe/H-ZSM-5 catalyst (Fe = 1 wt%) in comparison with sole H-ZSM-5 with no metal. On the other hand, Fe/H-ZSM-5 catalyst (Fe = 10wt%) showed more significant framework collapse than Fe = 1 wt% catalyst, suggesting that excess amount of Fe-loading is not preferable for keeping zeolite structure. Though Fe itself may be harmful for maintenance of zeolite structure, there may be a probability that remaining N, S or Cl species enhance the framework collapse.

Kowalska-Kuœ *et al.* reported oxydehydrogenation of ethane and propane with N_2O as oxidant using Fe-ZSM-5 catalyst prepared by wetness ion-exchange [40]. Here, alkali metal, such as Na,-modified catalysts were also used, and calcination at high temperature (900 °C) reduced brønsted acidic sites more significantly by surface dehydroxylation and partial dealumination than the case of 600 °C-calcination.

Moreover, calcination at high temperature resulted in the formation of Fe-O-Al species. In strong protonic site, oxydehydrogenation and propane cracking occurred simultaneously. However, the removal of brønsted acidic sites by high temperature calcination led to the change in reaction pathway, and after oxydehydrogenation, side reactions such as cracking and total oxidation occurred mostly as consecutive reactions. Carbon deposition also occurred.

Sun *et al.* reported N_2O decomposition over Fe/H-ZSM-5 prepared by solid-state ion-exchange [39]. In the case of calcination at 550 °C in air after ion-exchange, Fe species exist as Fe^{3+} in clustered form, and additional treatment at 900 °C in He changed Fe species into Fe^{2+}, stabilized as extra-framework Fe-O-Al species located in the channels of the zeolite. Both of Fe species have catalytic activity for N_2O decomposition, but the latter species was more active.

Though the framework collapse of zeolite was not observed by treatment in high temperature He, the catalyst after calcination in oxygen reduced the brønsted acidic sites by ion-exchange, and moreover, Al species in framework was extracted to extra-framework position by treatment in high temperature He, and as a result, acidic sites disappeared.

In our study, ethanol conversion over Fe/H-ZSM-5 catalyst calcined at 900 °C after Fe-addition was also investigated. Calcination at 900 °C resulted in the reduction of the brønsted acidic sites, the decrease in aromatics selectivity and the increase in ethylene selectivity: these results were consistent with Kowalska-Kuœ's report. However, framework collapse of zeolite proceeded because of calcination in air, not in He as Sun's report. Formation of Fe-O-Al species could not be confirmed.

4. Effect of Fe-Loading and Reaction Temperature on Ethanol Conversion to C_{3+}olefins by Fe-Modified H-ZSM-5 Zeolite Catalysts [32]

4.1. Exerimental

4.1.1. Catalyst Preparation

Used zeolite support and calcination condition before use were same as 3.1.1. Iron catalysts were prepared by impregnation method. Loading of Fe was 1 and 10 wt%. As sources of Fe, iron (III) nitrate nonahydrate ($Fe(NO_3)_3 \cdot 9H_2O$, Wako Pure Chemicals) was used. After impregnation, the wet catalysts were dried at 120 °C, followed by calcinations in flow of air, where calcination temperature was 700 °C for 3 h.

4.1.2. Catalytic Activity

Conditions of reactor and catalyst weight were same as 2.1.2. Reaction gas was obtained by mixing N_2 and vaporized EtOH, with the flow rate of N_2 set at 60 cm^3 min^{-1}, resulting in the concentration of N_2 and ethanol being 82.1 : 17.9 vol.%, respectively. Reaction temperature was 400 and 450 °C, and the change in catalytic performance with time-on-stream was investigated. The effluent gas was analyzed by gas chromatography (GC): columns and detectors were same as 2.1.2. Catalytic activity may decrease with time-on-stream. Therefore, regeneration treatment of used catalyst after reaction for 7 h was carried out by calcinations at 500 and 600 °C in air-flow for 1 h, and the catalytic performance after this regeneration treatment was also examined.

4.1.3. Characterization

In order to estimate carbon deposition on catalysts, thermogravimetry (TG) analyses of catalysts after 7 h-reaction were carried out: measurement apparatus and conditions were same as 2.1.3. In order to investigate the framework collapse of zeolite during reaction, XRD measurements were carried out, and intensity of XRD peak of zeolite framework before and after 7 h-reaction was compared: measurement apparatus and conditions were same as 2.1.3.

4.2. Results and Discussion

4.2.1. Catalytic Performance

In most cases, EtOH conversion was almost 100%. In general, the higher selectivity of ethylene leads to the lower selectivity of aromatics: Fe-loading suppressed the formation of aromatic compounds but enhanced the formations of ethylene and C_{3+} olefins, as mentioned in 3.2.1. Catalytic performance of Fe/H-ZSM-5 catalysts as a function of time, in conditions that Fe-loading was 10 wt% and reaction temperature was 400 °C, is shown in Fig .18(a). Initial selectivity of C_{3+} olefins was 20.2%, and gradually decreased to 19.6% after reaction for 7 h, and initial selectivity of propylene was 9.1% and gradually decreased to 8.2%. Selectivity of aromatics was decreased from 40.7 to 23.7%, while the selectivity of ethylene was increased from 22.7 to 43.9%, with time-on-stream. In general, selectivity of aromatics was decreased and selectivity of ethylene was increased with time-on-stream.

Figure 18(b) shows the catalytic performance in the case 10 wt% Fe-loading and 450 °C. In this case, the initial selectivity of C_{3+} olefins and propylene was 27.3 and 16.3%, respectively, higher than the case of 400 °C of reaction temperature. However, the selectivity of each product was decreased to 18.3 and 9.9 %, respectively and the decrease in selectivity was larger than the case of 400 °C. On the other hand, the increase in ethylene selectivity (32.9 → 64.2%) and the decrease in aromatics selectivity (24.7 → 6.4%) were also larger than the case of 400 °C.

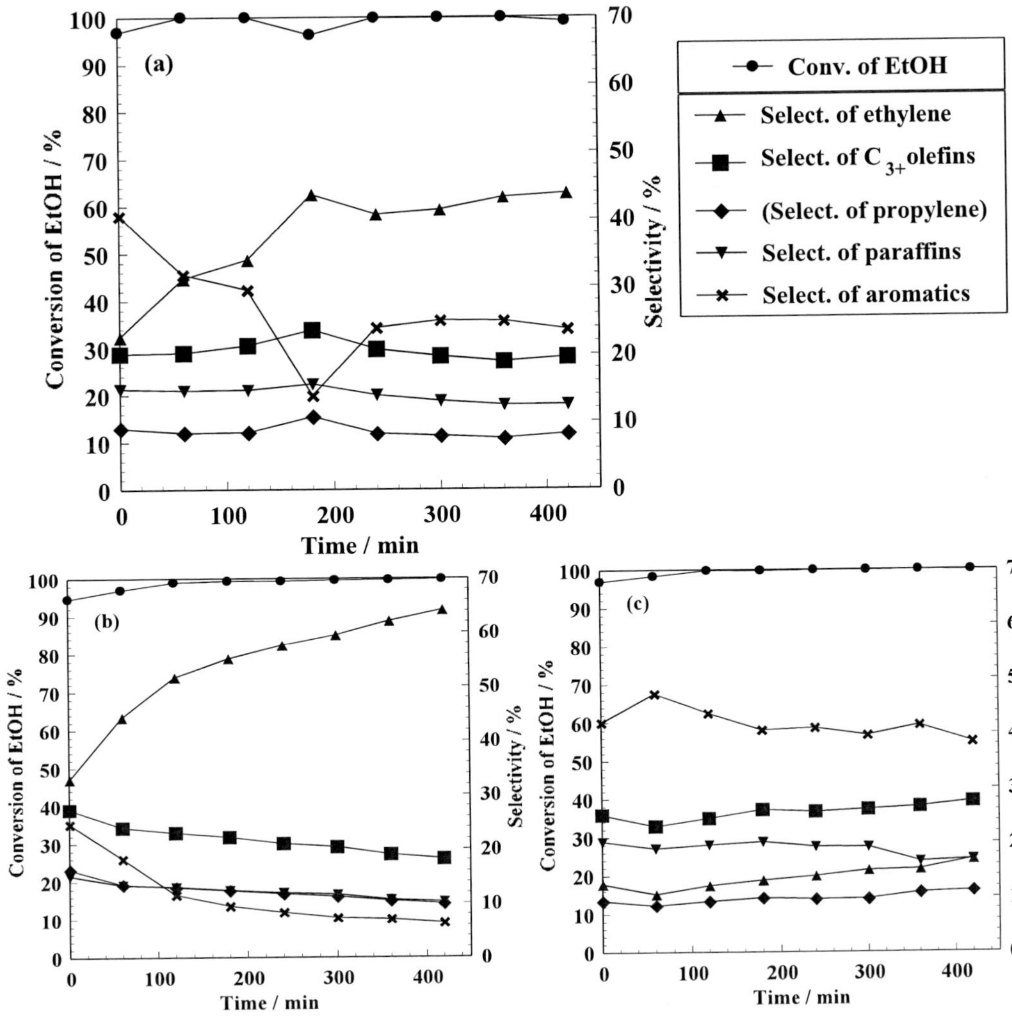

Figure 18. Catalytic performance of Fe/H-ZSM-5 catalysts as a function of time. Conditions: (a) Fe = 10 wt%, Reaction temperature = 400 °C; (b) Fe = 10 wt%, Reaction temperature = 450 °C; (c) Fe = 1 wt%, Reaction temperature = 400 °C.. Preparation conditions: Fe-source = $Fe(NO_3)_3{\cdot}9H_2O$, Calcination temperature = 700 °C. Reaction conditions: Atmospheric pressure, Catalyst weight = 0.2 g, $EtOH/N_2$ = 0.22, N_2-flow = 60 ml/min.

Figure 18(c) shows the catalytic performance in the case 1 wt% Fe-loading and 400 °C. The initial selectivity of C_{3+} olefins and propylene was 25.1 and 9.4%, respectively, higher than the case 10 wt% Fe-loading and 400 °C, but lower than the case 10 wt% Fe and 450 °C.

However, after 7 h-reaction, the selectivity of each product was slightly increased to 27.6 and 11.3%, respectively, which were higher than these former two cases. Moreover, the selectivity of ethylene and aromatics was almost constant: 12.5 → 17.0% and 42.1 → 38.5%, respectively. These results suggest that the lower Fe-loading and the lower reaction temperature can lead to stable catalytic performance (i.e. selectivity of each product). However, in this case, the selectivity of aromatics was relatively high. Since our major targets are lower olefins, so the formation of aromatic compounds is not preferable.

Figure 19 shows the catalytic performance in the case 1 wt% Fe-loading and reaction temperature of 450 °C.

The initial selectivity of C_{3+} olefins and propylene was 25.0 and 13.6%, respectively, and after reaction for 7 h, the selectivity was increased to 30.3 and 16.1 %, respectively. These values were the best among these reaction conditions. Moreover, this reaction condition can reduce the formation of aromatics in some degree. However, in this case also, the ethylene selectivity was increased (14.1 → 37.0%) and the aromatics selectivity was decreased (40.3 → 15.5%). Here, to regenerate the catalyst, regeneration treatment of used catalyst after 7 h-reaction was carried out by calcinations at 500 °C in air-flow for 1 h. After this treatment, the selectivity of C_{3+} olefins and propylene was increased to 32.3 and 17.5%, respectively.

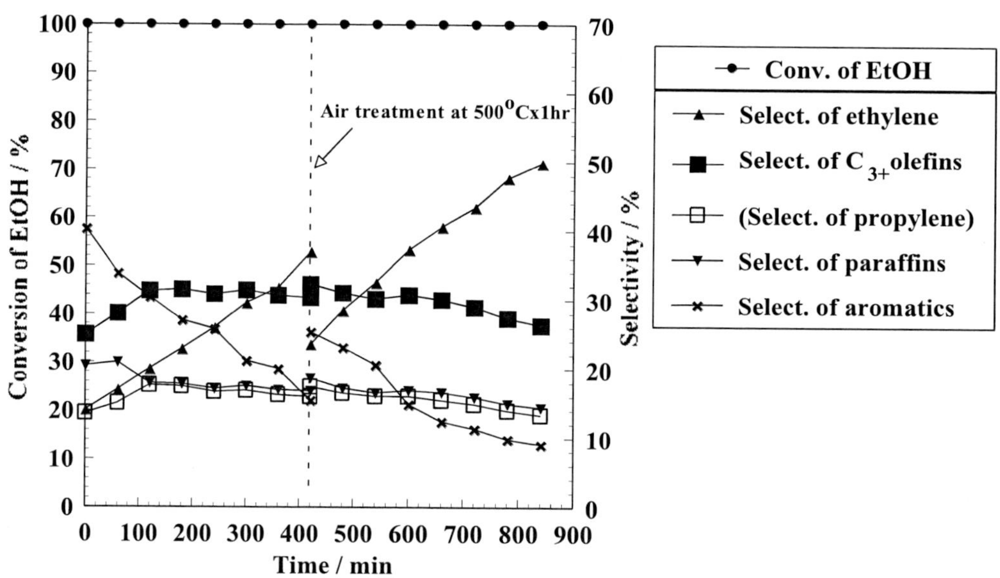

Figure 19. Catalytic performance of Fe/H-ZSM-5 catalysts as a function of time: effect of air treatment at 500 °C on used catalyst. Fe source = $Fe(NO_3)_3$-$9H_2O$, Fe-loading = 1wt%, Reaction temperature = 450 °C.

However, after the second reaction for 7 h, the selectivity of each product was decreased to 26.4 and 13.4 %, respectively, lower than after the first reaction before the regeneration treatment (30.3 and 16.1%, respectively). Moreover, ethylene selectivity was increased (23.6 → 49.9%) and aromatics selectivity was decreased (25.4 → 9.1%) during the reaction after this regeneration treatment: the ethylene selectivity after the second reaction was higher than after the first reaction (37.0%) and the aromatics selectivity after the second was lower than after the first reaction (15.5%). These results show that only this regeneration treatment cannot recover the catalytic performance completely.

4.2.3. Carbon Deposition

The results of TG analyses of used catalysts are summarized in Table 7. The catalyst with 1 wt% Fe-loading showed weight loss below 3%, lower than the catalyst with 10 wt% of Fe, which showed the weight loss above 3%, suggesting that excess amount of loaded Fe-species enhances the carbon deposition. TG analysis of Fe_2O_3, used as catalyst, after 7 h-reaction was also carried out. From the result, it was found that Fe_2O_3 could solely deposit some amount of carbon during reaction, and reaction at 450 °C led to more amount of carbon deposition than reaction at 400 °C. As mentioned previously, deposited carbon may block the acidic site of zeolite and inhibit the aromatics formation, and as a result, ethylene formation was enhanced.

Table 7. Weight loss by TG analyses of used catalysts and ratio of intensity of X-ray diffraction peaks of catalysts before and after reaction

Fe-loading	Reaction temp. / °C	Weight loss by TG / %	Ratio of XRD intensity (After reaction/ before reaction)
1	400	2.78	0.80
1	450	2.40	0.88
10	400	3.10	0.42
10	450	3.46	0.51

4.2.4. Framework Collapse of Zeolite

After reaction/before reaction ratio in intensity of XRD pattern detected at 7.9° is also summarized in Table 7.

In the case of the catalyst with 1 wt% of Fe, the ratio was above 0.8, higher than the case of the catalyst with 10 wt% of Fe (about 0.5). There seems to be no clear difference affected by reaction temperature (400 and 450 °C).

Further work is in progress to clarify a correlation between reaction temperature and framework collapse. XRD peak of loaded Fe species appeared at 35.4 ° for the catalysts with 10wt% of Fe, while this peak did not appear for the catalysts with 1 wt% of Fe. These results suggest that in the case of 1 wt% Fe-loading, Fe species dispersed almost completely; on the other hand, for 10 wt% of Fe, the excess amount of Fe species could not disperse completely, and formed the crystallite of FeO_x, which may enhance the carbon deposition as mentioned above.

On the other hand, as mentioned in 3.2.10, Fe species or remaining N species may enhance the framework collapse. In any case, the catalyst with 10 wt% of Fe contains more amount of the species which may be harmful for keeping zeolite framework than the catalyst with 1 wt% of Fe, and as a result, the former catalyst showed the more pronounced framework collapse of zeolite than the latter one. Gayubo *et al.* reported that in ethanol conversion using Ni- [23, 24] or alkali metal- [25, 26] modified H-ZSM-5 catalysts, the higher reaction temperature or the more amount of water in ethanol leads to the framework collapse in more degree.

In our study also, the reaction at 450 °C may enhance the framework collapse more significantly than 400 °C-reaction, but it seems that the effect of Fe-loading on the collapse is larger than the effect of reaction temperature.

4.2.5. Regeneration Treatment at High Temperature

Next, in order to recover selectivity of products completely, the regeneration treatment by calcinations at 600 °C in air-flow for 1 h was also carried out.

After this treatment, the selectivity of C_{3+} olefins and propylene changed during 7 h-reaction as follows: 28.6 → 31.4% and 13.6 → 15.4%, respectively: these values were almost same as the reaction before this regeneration treatment (25.0 → 30.3 and 13.6 → 16.1%, respectively), and the values after the second reaction for 7 h were higher than after the second reaction in the case of regeneration treatment at 500 °C (26.4 and 13.4%, respectively).

On the other hand, after this treatment, the selectivity of ethylene and aromatics changed as follows: 18.0 → 35.8% and 30.2 → 14.2%, respectively, similar with the first reaction (14.1 → 37.0% and 40.3 → 15.5%, respectively).

The ethylene selectivity after the second reaction was lower than after the second reaction in the case of regeneration treatment at 500 °C (49.9%), while the aromatics selectivity was higher than in the case of 500 °C (9.1%). These results suggest that the regeneration treatment at 600 °C leads to the improvement of recovery of catalytic performance in more degrees than that at 500 °C.

5. Ethanol Conversion to Propylene over Fe and/or P-Modified H-ZSM-5 Zeolite Catalysts [34]

5.1. Experimental

5.1.1. Catalyst Preparation

Used zeolite support and calcination condition before use were same as 3.1.1. H-ZSM-5 (Si/Al_2 = 29) was used as both support and catalyst: when used as catalyst, additional calcination at 700 °C in a muffle furnace was carried out after calcination at 500 °C. Two kinds of H-ZSM-5 were used as catalysts: (1) H-ZSM-5 was calcined at 500 °C (denoted as H-ZSM-5 (500)), and (2) H-ZSM-5 was calcined at 700 °C after 500 °C-calcination (H-ZSM-5 (700)). Fe or P-modified H-ZSM-5 zeolites, used as both support and catalyst, were prepared by impregnation. $FeCl_3{\cdot}6H_2O$ (Iwai Chemical) was used as the Fe source, and $(NH_4)_3PO_3{\cdot}3H_2O$ (Wako Pure Chemicals) was used as the P source. Loading of Fe or P was 1 wt%. After impregnation, the wet catalyst/support was dried at 120 °C, followed by calcination at 700 or 900 °C in flowing air for 3 h (heating rate 200 °C h^{-1}). Fe and P-modified H-ZSM-5 catalysts were also prepared. The source and loading of Fe and P and calcination process were the same as for Fe or P/H-ZSM-5. Four kinds of Fe and P-modified H-ZSM-5 catalysts were prepared: (1) P-modified H-ZSM-5 support calcined at 700 °C, followed by Fe-modification to P/H-ZSM-5 support and calcination at 700 °C (shown as Fe/P(700)/H-ZSM-5), (2) P-modified H-ZSM-5 support calcined at 900 °C, followed by Fe-modification to P/H-ZSM-5 support and calcination at 700 °C (Fe/P(900)/H-ZSM-5), (3) Fe-loaded H-ZSM-5 support calcined at 700 °C, followed by P-modification to Fe/H-ZSM-5 support and calcination at 700 °C (P/Fe/H-ZSM-5); and (4) Fe- and P-modified H-ZSM-5 catalysts calcined at 700 °C were prepared by co-impregnation (Fe-P/H-ZSM-5). To investigate the effect of calcination temperature, two kinds of P-modified H-ZSM-5 support, calcined at 700 and 900 °C, were prepared.

5.1.2. Catalytic Activity

Conditions of reactor and catalyst weight were same as 2.1.2. Reaction gas was obtained by mixing N_2 and vaporized EtOH, with the flow rate of N_2 set at 60 cm^3 min^{-1}, resulting in the concentration of N_2 and ethanol being 82.1 : 17.9 vol%, respectively. As ethanol, pure ethanol and bioethanol were used: bioethanol was obtained by mixing pure ethanol with commercial vodka (alcohol = 50 %) at a ratio of 3:1, giving a concentration of ethanol of approx. 87.5%. Reaction temperature was 450 °C, and changes in catalytic performance with on-stream time were observed. Analysis of vodka by gas chromatography (GC) confirmed the chief organic compound in vodka to be ethanol, with only traces of other organic compounds present. The effluent gas was analyzed by gas chromatography (GC): columns and detectors were same as 2.1.2.

Catalytic activity may decrease with time on-stream, so regeneration treatment of used catalyst after 7 h-reaction was carried out by calcination at 500 °C in flowing air for 1 h. The secondary catalytic performance was then examined.

5.1.3. Characterization

To estimate carbon deposition on the catalysts, thermogravimetric (TG) analyses of catalysts after 7 h-reaction were carried out: measurement apparatus and conditions were same as 2.1.3.

In order to investigate the framework collapse of zeolite during reaction, XRD measurements were carried out, and intensity of XRD peak of zeolite framework before and after 7 h-reaction was compared: measurement apparatus and conditions were same as 2.1.3.

NH_3-temperature-programmed-desorption (TPD) experiments of H-ZSM-5 and (P and/or Fe)/H-ZSM-5 catalysts were carried out to measure the acidity of supports and catalysts before and after 7 h-reaction using neat ethanol. NH_3-TPD experiments on H-ZSM-5 (500) catalyst after 7 h-reaction and calcination treatment at 500 °C in air for 1 h were also carried out to investigate whether the acidity of H-ZSM-5 (500) could be recovered by calcination. Measurement apparatus and conditions were same as 3.1.3.

SEM measurements were carried out (Hitachi, S-4300) to investigate the change in catalyst morphology/structure before and after reaction. H-ZSM-5 (500) and Fe/H-ZSM-5 before and after reaction using pure ethanol were analysed. H-ZSM-5 samples were coated with Pt and Pd before observations, while Fe/H-ZSM-5 samples were observed without coating. X-ray fluorescent (XRF) analyses of H-ZSM-5 and Fe/H-ZSM-5 before and after reaction using neat ethanol were carried out using an XRF instrument (Rigaku, EDXL300) under vacuum conditions, to measure mass % of Si and Al. For reference, H-ZSM-5 samples with Si/Al_2 ratio of 23 and 80, calcined at 500 °C in air for 6 h, were also analysed.

5.2. Results and Discussion

5.2.1. Catalytic Performance

Initial conversion of ethanol and initial selectivity for each product among gaseous products are shown in Table 8. Selectivity for each product changed with time on-stream, although ethanol conversion remained constant at approx. 100%. There appears to be a correlation between ethylene selectivity and propylene selectivity; a weak inverse correlation between paraffins selectivity and propylene selectivity, especially in the case of aqueous ethanol; and an inverse correlation between selectivity for aromatics and propylene selectivity. H-ZSM-5 (500) showed higher selectivity for aromatics and lower selectivity for ethylene, C_{3+} olefins and propylene, than H-ZSM-5 (700).

Calcination of H-ZSM-5 zeolite at higher temperatures led to reduced formation of aromatics and enhanced formation of ethylene and C_{3+} olefins. In general, addition of both Fe and P suppressed the formation of paraffins and aromatics but enhanced the formation of ethylene and C_{3+} olefins (propylene); while loading of Fe or P alone did not significantly affect overall selectivity. Fe/P(900)/H-ZSM-5 showed higher selectivity for ethylene and lower selectivity for C_{3+} olefins, propylene, paraffins and aromatics. Calcination of P/H-ZSM-5 support at higher temperatures led to reduced formation of C_{3+} olefins, propylene, paraffins and aromatics, and enhanced formation of ethylene.

Table 8. Catalytic activity of H-ZSM-5 and (Fe and/or P)/H-ZSM-5 zeolite catalysts

Loaded element	EtOH	EtOH conv. / %	C-selectivity / %				
			Ethylene	C_{3+}olefins	(Propylene)	Paraffins	Aromatics
No (500)	Neat EtOH	100.0	9.5	17.7	(9.6)	20.8	52.0
	Vodka+EtOH	92.7	9.0	15.5	(8.1)	27.7	47.8
No (700)	Neat EtOH	100.0	18.2	27.2	(14.4)	22.4	32.1
	Vodka+EtOH	98.1	24.2	34.7	(17.9)	15.6	25.2
P	Neat EtOH	99.8	17.0	25.3	(13.2)	24.1	33.3
	Vodka+EtOH	97.5	11.2	24.2	(12.3)	24.5	39.7
Fe	Neat EtOH	94.9	22.0	26.5	(13.2)	21.8	29.4
	Vodka+EtOH	96.0	15.7	27.5	(13.8)	22.6	33.9
Fe/P (700)	Neat EtOH	97.7	26.5	33.2	(18.3)	21.2	18.3
	Vodka+EtOH	96.2	34.0	32.7	(18.3)	14.9	17.7
Fe/P (900)	Neat EtOH	99.7	40.3	31.9	(15.3)	17.6	9.8
	Vodka+EtOH	100.0	53.1	27.8	(13.8)	11.8	6.4
P/Fe	Neat EtOH	97.8	30.9	32.6	(17.2)	18.9	17.4
	Vodka+EtOH	99.7	34.4	29.4	(15.8)	17.3	18.0
Fe-P	Neat EtOH	98.6	33.8	23.5	(11.5)	19.3	22.9
	Vodka+EtOH	100.0	26.4	31.4	(15.1)	24.3	17.9

Numbers in parentheses stand for calcination temperature (°C).
Preparation conditions: P-source = $(NH_4)_3PO_3 \cdot 3H_2O$, P-loading = 1 wt%, Calcination temperature = 700 °C, Fe-source = $FeCl_3 \cdot 6H_2O$, Fe-loading = 1 wt%, Calcination temperature = 700 °C.

Fe/P: addition (P → Fe) to H-ZSM-5 zeolite, P/Fe: addition (Fe → P) to H-ZSM-5 zeolite, Fe-P: co-addition of Fe and P to H-ZSM-5 zeolite. Reaction conditions: Atmospheric pressure, Reaction temperature = 450 °C, Catalyst weight = 0.2 g, EtOH/N_2 = 0.22, Vodka/EtOH = 0.33, N_2 flow = 60 ml/min.

Selectivity for C_{3+} olefins was within 15–35%. Fe/P(700)/H-ZSM-5, Fe/P(900)/H-ZSM-5 and P/Fe/H-ZSM-5 catalysts showed especially high selectivity for C_{3+} olefins (> 30%) for pure ethanol. H-ZSM-5(700), Fe/P(700)/H-ZSM-5 and Fe-P/H-ZSM-5 catalysts showed high selectivity (> 30%) for aqueous ethanol. On the other hand, selectivity for propylene was within 8–19%. Fe/P(700)/H-ZSM-5 and P/Fe/H-ZSM-5 catalysts showed especially high selectivity for propylene (> 17%) with pure ethanol. H-ZSM-5(700) and Fe/P(700)/H-ZSM-5 catalysts showed high selectivity for propylene (> 17%) with aqueous ethanol.

The presence of water in ethanol enhanced the formation of C_{3+} olefins and propylene with H-ZSM-5(700), Fe/H-ZSM-5 and Fe-P/H-ZSM-5, and reduced the formation with H-ZSM-5(500), P/H-ZSM-5, Fe/P(700)/H-ZSM-5, Fe/P(900)/H-ZSM-5 and P/Fe/H-ZSM-5. Choudhary *et al.* have reported that the presence of water in ethanol causes a small decrease in the extent of aromatization but no significant change in product distribution [9]. In our study, however, there seems to be no clear correlation between presence of water and initial selectivity for each product.

Figure 20 shows the plots of the ratio of initial selectivity (propylene/paraffins) vs. the ratio of initial selectivity (propylene/aromatics). In general, there proved to be a correlation between these two ratios: catalysts showing a higher propylene/paraffins ratio showed a higher propylene/aromatics ratio. Calcination of to H-ZSM-5 (H-ZSM-5 (700)) at higher temperatures improved the ratios of propylene/paraffins and propylene/aromatics in comparison to H-ZSM-5 calcined at lower temperatures (H-ZSM-5 (500)). In the case of H-ZSM-5 (700), the values of these two ratios for aqueous ethanol were higher than those for neat ethanol. Sole addition of Fe or P to H-ZSM-5 support did not improve these two ratios to any extent, while co-addition of Fe and P (Fe-P) improved only the propylene/aromatics ratio for aqueous ethanol, and did not improve the two ratios for neat ethanol. On the other hand, successive addition of (Fe/P(700), Fe/(900) and P/Fe) improved the ratios for both neat and aqueous ethanol. Fe/P(900)/H-ZSM-5 catalysts showed especially high ratios for both.

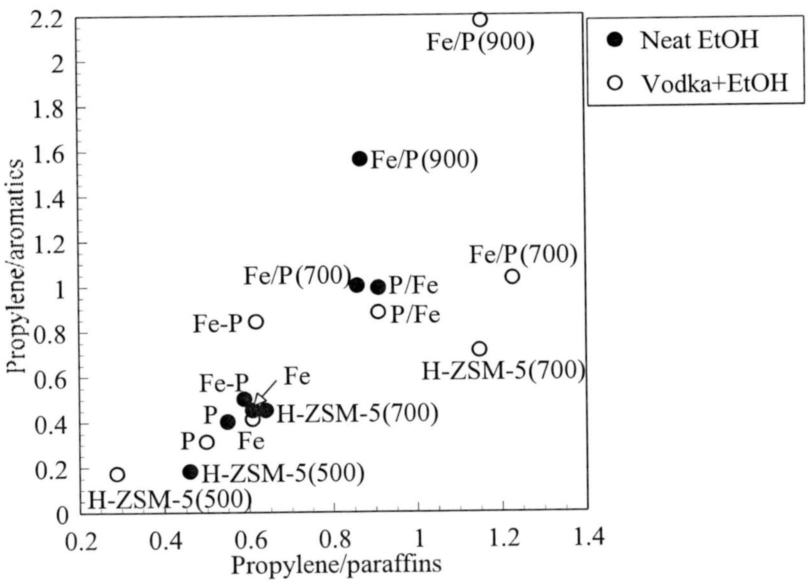

Figure 20. Plots of the ratio of propylene/paraffins vs. the ratio of propylene/aromatics over H-ZSM-5 and (Fe and/or P)/H-ZSM-5 catalysts. Numbers in parentheses stand for calcination temperature (°C).

It appears likely that, during the formation of aromatics from ethanol, hydrogen is released which contributes to the hydrogenation of olefins to paraffins. Calcination of the H-ZSM-5 and P/H-ZSM-5 at higher temperatures, and Fe- and P-modification to H-ZSM-5 may reduce the number of acidic sites that are essential for the formation of aromatics, and as a result, the ratios of propylene/paraffins and propylene/aromatics may be improved. On the other hand, it appears unlikely that ethylene is hydrogenated to ethane, since the formation of ethane was rarely observed.

The selectivity for each product changed with time-on-stream. Since the selectivity for C_{3+} olefins and propylene may decrease with time, we observed changes in the selectivity for propylene between 0 and 7 h after starting the reaction.

Figure 21 shows plots of the initial selectivity for propylene vs. the change in propylene selectivity over H-ZSM-5 and (Fe and/or P)/H-ZSM-5 catalysts. In general, catalysts showing higher initial selectivity for propylene show a greater decrease in propylene selectivity after 7 h-reaction. H-ZSM-5 (500) showed low initial selectivity for propylene, but the selectivity increased during the 7 h period. On the other hand, H-ZSM-5 (700) showed higher selectivity for propylene than H-ZSM-5 (500), and selectivity was higher for aqueous ethanol than for neat ethanol.

Moreover, the selectivity increased only slightly for neat ethanol, but decreased significantly for aqueous ethanol. Successive addition of Fe and P and calcination at 700 °C (Fe/P(700) and P/Fe) improved initial selectivity but led to a steeper decrease in selectivity with time, while calcination at 900 °C (Fe/P(900)) did not much improve initial selectivity but led to a greater decrease in selectivity. On the other hand, sole addition of Fe or P and co-addition of Fe and P (Fe-P) showed medium initial selectivity and increase in selectivity. In most cases, the presence of water in ethanol enhanced the change in propylene selectivity. With H-ZSM-5(500), P/H-ZSM-5 and Fe/H-ZSM-5, the increase in selectivity was greater for aqueous ethanol than for neat ethanol, while for Fe/P(700)/H-ZSM-5, the decrease in selectivity was greater for aqueous ethanol. A similar pattern was observed for the case of C_{3+} olefins.

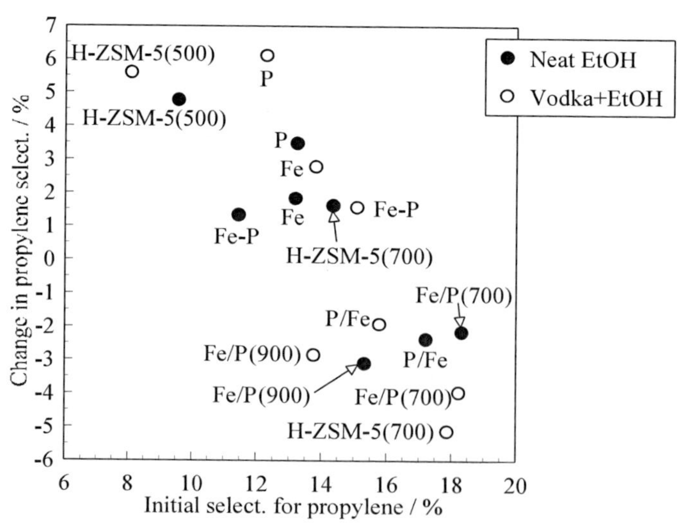

Figure 21. Plots of the initial selectivity for propylene vs. the change in propylene selectivity over H-ZSM-5 and (Fe and/or P)/H-ZSM-5 catalysts.

On the other hand, the selectivity for aromatic compounds and paraffins decreased with time-on-stream, while the selectivity for ethylene increased. In general, catalysts yielding aromatics with higher initial selectivity showed a larger decrease in selectivity for aromatics: addition of both of Fe and P (Fe/P(700), Fe/P(900), P/Fe and Fe-P) led to lower initial selectivity and a smaller decrease in selectivity, than for sole addition of Fe or P and no addition. Moreover, for H-ZSM-5 (500), H-ZSM-5 (700) and Fe/H-ZSM-5, water contained in ethanol led to the larger decrease in selectivity, while for (Fe and P)/H-ZSM-5, the effect of water in ethanol was far less significant. With respect to the formation of ethylene, except in the cases of H-ZSM-5 (700) for both neat ethanol and aqueous ethanol and Fe/H-ZSM-5 for aqueous ethanol, which showed an especially large increase in ethylene selectivity, there seems to be no significant difference in ethylene selectivity behavior. Moreover, there was a weak correlation between initial selectivity for paraffins and the decrease in paraffins selectivity: catalysts showing higher selectivity for paraffins showed a greater decrease in paraffins selectivity. Figure 22 shows plots of the change in selectivity for aromatics vs. the change in propylene selectivity during the 7 h-reaction. There appears to be an inverse correlation between the two: catalysts showing a large decrease in selectivity for aromatics, for example H-ZSM-5 (500) and Fe/H-ZSM-5, showed an increase in propylene selectivity. Catalysts showing little decrease in selectivity for aromatics such as Fe/P(700)/H-ZSM-5, Fe/P(900)/H-ZSM-5 and P/Fe/H-ZSM-5, showed a decrease in propylene selectivity. In the cases of P/H-ZSM-5 and Fe-P/H-ZSM-5, propylene selectivity increased while the decrease in selectivity for aromatics was not large. For H-ZSM-5 (700), propylene selectivity increased very slightly for neat ethanol, but decreased significantly for aqueous ethanol. Figure 23 shows plots of the change in paraffins selectivity vs. the change in propylene selectivity with time-on-stream. Except for H-ZSM-5 (700), which shows a large decrease in selectivity, and Fe/H-ZSM-5, which shows an increase in selectivity for neat ethanol, there was little difference in paraffins selectivity behavior: there appears to be a weak inverse correlation between change in paraffins selectivity and change in propylene selectivity. For example, H-ZSM-5 (500) and P/H-ZSM-5 showed a large decrease in paraffins selectivity and increase in propylene selectivity, while Fe/P(700)/H-ZSM-5 and P/Fe/H-ZSM-5 showed little decrease in paraffins selectivity and a large decrease in propylene selectivity.

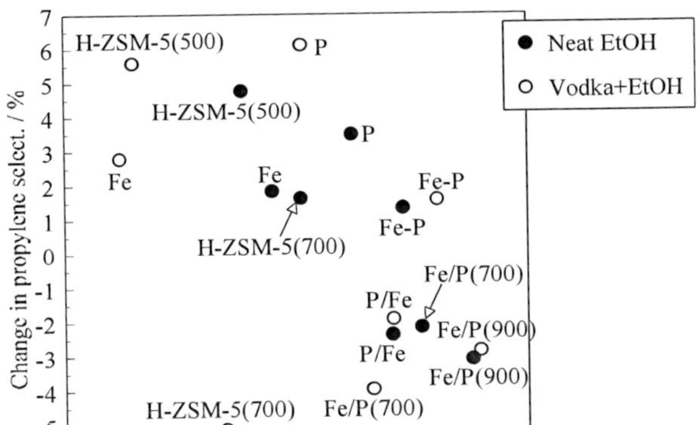

Figure 22. Plots of the change in aromatics selectivity vs. the change in propylene selectivity over H-ZSM-5 and (Fe and/or P)/H-ZSM-5 catalysts.

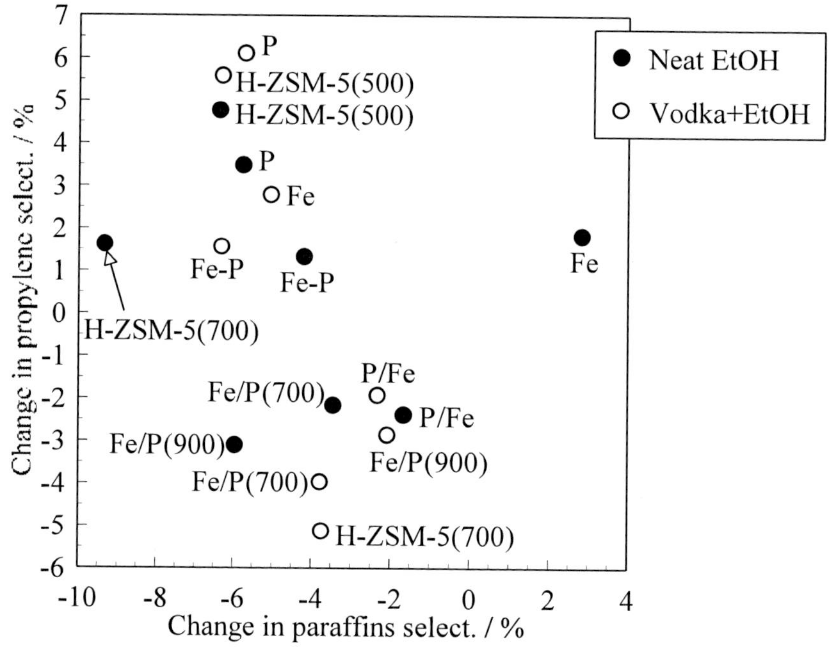

Figure 23. Plots of the change in paraffins selectivity vs. the change in propylene selectivity over H-ZSM-5 and (Fe and/or P)/H-ZSM-5 catalysts.

On the other hand, there seems to be a weak inverse correlation between change in ethylene selectivity and in propylene selectivity, except for H-ZSM-5 (500) and H-ZSM-5 (700), which showed a large increase in ethylene selectivity per unit change in propylene selectivity. Thus, in general, decreased selectivities to aromatics as well as to paraffins after 7 h appear to result in increased in propylene selectivity.

5.2.2. Carbon Deposition

Figure 24 shows plots of weight loss by TG analysis of catalysts after 7 h-reaction vs. change in propylene selectivity between 0 and 7 h. The moles of C deposited were calculated from the weight loss due to combustion of the deposited carbon. Even for H-ZSM-5 (500), when used for neat ethanol conversion, the mole number of deposited carbon atoms was only 0.07% of total carbon atoms in the ethanol used. The amount of deposited carbon had a negligible effect on selectivity for each product and carbon balance.

Calcination at higher temperatures led to less carbon deposition than calcination at lower temperatures: H-ZSM-5(500) > H-ZSM-5(700), Fe/P(700)/H-ZSM-5 > Fe/P(900)/H-ZSM-5. Addition of Fe or P alone to H-ZSM-5 did not contribute to the reduction of carbon deposition, while the addition of both of Fe and P inhibited carbon deposition to some degree. In the case of neat ethanol, an apparently close correlation between TG weight loss and change in propylene selectivity appears to reflect the greater sensitivity of a decrease in aromatic selectivity to carbon deposition, as shown below in Figure 25.

For aqueous ethanol, the amount of deposited carbon was lower than for the neat ethanol used, especially for H-ZSM-5 (700), P/H-ZSM-5 and Fe/H-ZSM-5 catalysts: there seems to be no clear correlation between TG weight loss and change in propylene selectivity. Figure 25 shows the plots of weight loss by TG analysis vs. the change in selectivity for aromatics between 0 and 7 h after starting reaction: there does, in fact, appear to be a correlation between TG weight loss and a decrease in selectivity for aromatics. Notably for H-ZSM-5 (500), H-ZSM-5 (700) and Fe/H-ZSM-5 catalysts, the presence of water in ethanol enhanced the decrease in selectivity for aromatics.

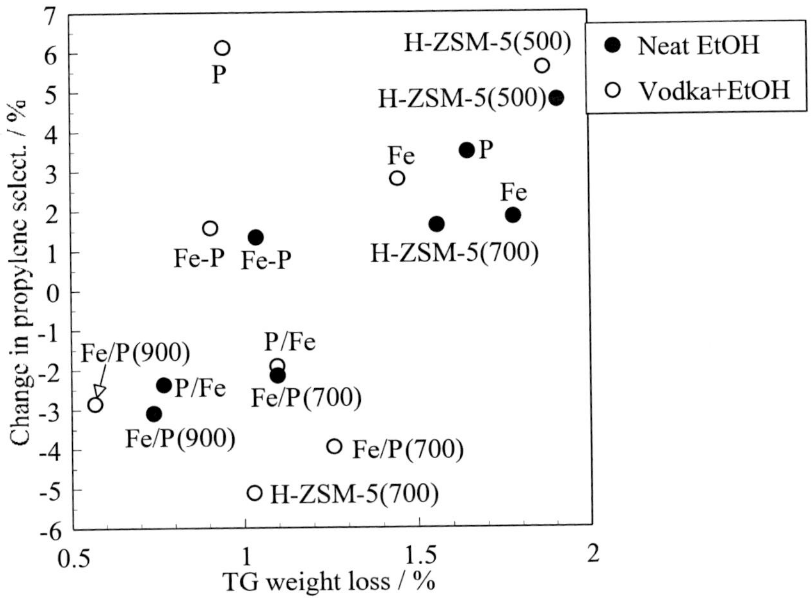

Figure 24. Plots of the weight loss by TG analysis vs. the change in propylene selectivity over H-ZSM-5 and (Fe and/or P)/H-ZSM-5 catalysts.

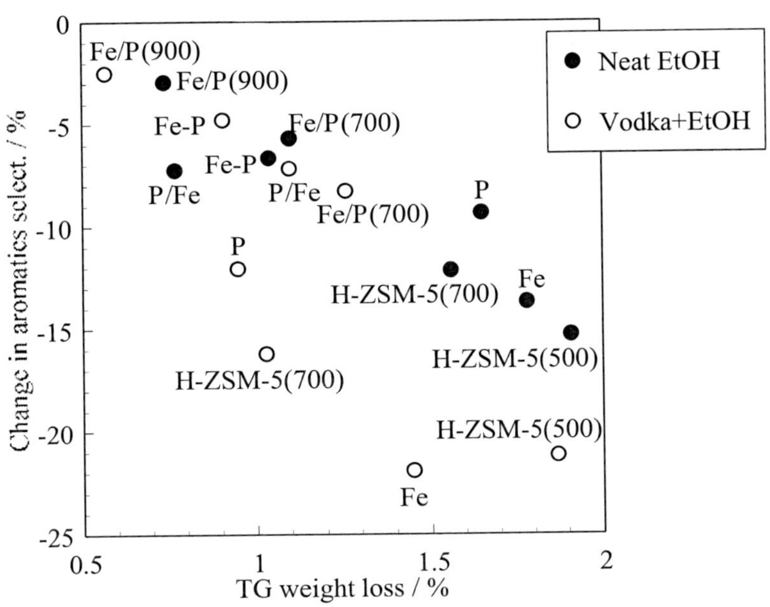

Figure 25. Plots of the weight loss by TG analysis vs. the change in aromatics selectivity over H-ZSM-5 and (Fe and/or P)/H-ZSM-5 catalysts.

Carbon deposition can be responsible for changes in selectivity. In general, acidic sites of the zeolite support are essential to the formation of aromatic compounds.

When the acidic sites were covered with deposited carbon, the formation of aromatic compounds was suppressed: the inhibition of formation of aromatic compounds predominated over the inhibition of formation of propylene.

The amount of carbon deposition slightly decreased on addition of water, suggesting that gasification of deposited carbon to H_2 and CO may occur if water is added. Here, similar results with 3.2.3 were obtained. Carbon deposition does not seem to affect the change in ethylene selectivity.

5.2.3. Framework Collapse of Zeolite

Figure 26 shows XRD patterns of H-ZSM-5 (500) catalyst (a) before and (b) after reaction using neat ethanol, (c) H-ZSM-5 (700) catalyst before reaction and (d) Fe/H-ZSM-5 catalyst before reaction, where all XRD peaks are due to the framework of H-ZSM-5 zeolite, not including peaks of FeO_x species.

The intensity of XRD patterns of H-ZSM-5 (500) was slightly weakened during reaction, suggesting that framework collapse, at least in part, proceeded during reaction. H-ZSM-5 (700) before reaction showed a similar intensity of XRD patterns with H-ZSM-5 (500) before reaction. However, even before reaction, the intensity of XRD patterns of Fe/H-ZSM-5 was weaker than H-ZSM-5 (500) after reaction, suggesting that framework collapse of the zeolite support proceeded more significantly during preparation of Fe/H-ZSM-5 catalyst than during reaction over H-ZSM-5 (500). The arrow marks the peak at 23.1°.

The intensity of this peak was used as an index of crystallinity. Changes in intensity of XRD peaks of the zeolite support were observed. Figure 27 shows the plot of the ratio of XRD peak intensity (after reaction/before reaction) vs. the change in propylene selectivity over H-ZSM-5 and (Fe and/or P)/H-ZSM-5 catalysts: catalysts after 7 h on-stream were analyzed. Framework collapse of H-ZSM-5 (500) and (700) proceeded more significantly than of (Fe and/or P)/H-ZSM-5. Before reaction, the order of intensity of XRD peaks was as follows: H-ZSM-5 (500) > H-ZSM-5 (700) > P/H-ZSM-5 > Fe/P(700)/H-ZSM-5 > Fe-P/H-ZSM-5 > Fe/H-ZSM-5 > Fe/P(900)/H-ZSM-5 > P/Fe/H-ZSM-5.

This result suggests that calcination at higher temperatures and P- or Fe-modification (especially Fe-modification) leads to framework collapse during the calcination and preparation process.

However, P-modification with calcination at adequate temperatures confers thermal stability during Fe-addition, since the intensity of XRD peaks of Fe/P(700)/H-ZSM-5 and Fe-P/H-ZSM-5 was stronger than that of Fe/H-ZSM-5, while the intensity of XRD peaks of Fe/P(900)/H-ZSM-5 was weaker than that of Fe/H-ZSM-5. As mentioned above, in H-ZSM-5 (500) and (700), framework collapse during reaction proceeded more significantly than (Fe and/or P)/H-ZSM-5 catalysts, suggesting that Fe- [38] or P- [20, 33] modification could inhibit the framework collapse in reaction condition.

In most cases, the presence of water in ethanol led to a lower ratio of XRD peak intensity (after reaction/before reaction), suggesting that the presence of water in ethanol worsens framework collapse. However, there seems to be no clear correlation between the ratio of peak intensity and the change in propylene selectivity.

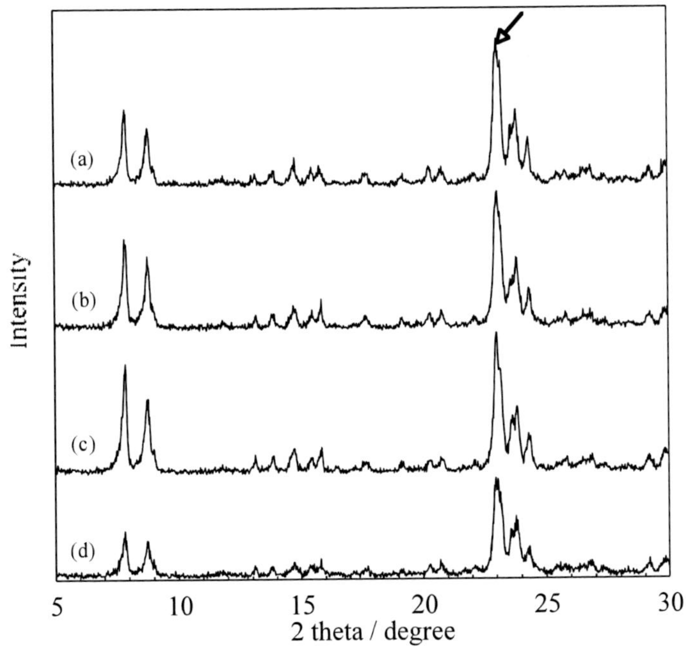

Figure 26. Images of XRD patterns of catalysts before and after reaction using neat ethanol. (a) H-ZSM-5 (500) before reaction, (b) H-ZSM-5 (500) after reaction, (c) H-ZSM-5 (700) before reaction, (d) Fe/H-ZSM-5 before reaction.

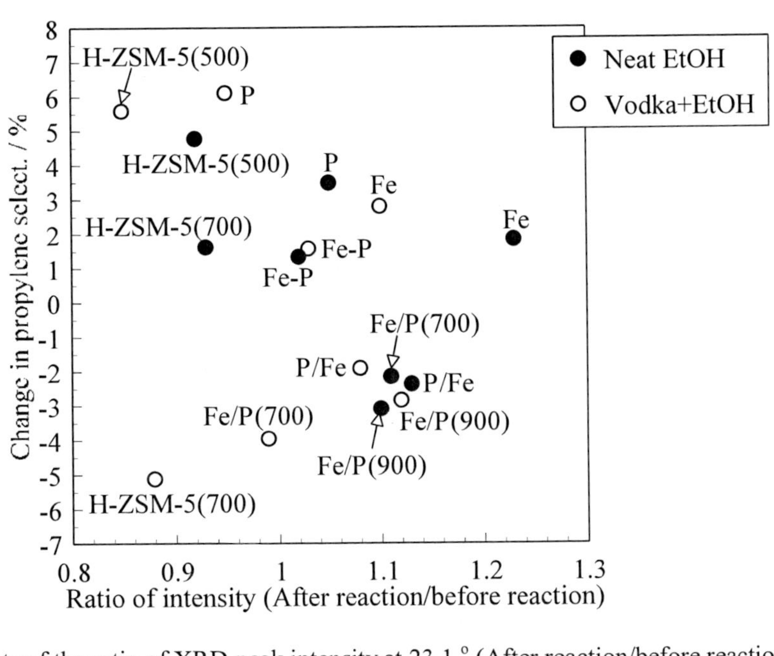

Figure 27. Plots of the ratio of XRD peak intensity at 23.1 ° (After reaction/before reaction) vs. the change in propylene selectivity over H-ZSM-5 and (Fe and/or P)/H-ZSM-5 catalysts.

Figure 28 shows the plot of the ratio of XRD peak intensity (after reaction/before reaction) vs. the change in selectivity for aromatics during 7 h-reaction over H-ZSM-5 and (Fe and/or P)/H-ZSM-5.

There appears to be a clear correlation between the ratio of peak intensity and the change in selectivity for aromatics, except for Fe/H-ZSM-5, showing an especially large decrease in selectivity of aromatics for the ratio of peak intensity for both the neat ethanol and aqueous ethanol used. These results suggest that more framework collapse leads to decreased selectivity for aromatics. Solid acidity, due to aluminum atoms incorporated in the zeolite framework, is essential for the formation of aromatics, and the decrease in solid acidity due to framework collapse may lead to a decrease in formation of aromatics. It is therefore suggested that framework collapse is one of the causes of change in selectivity for each product.

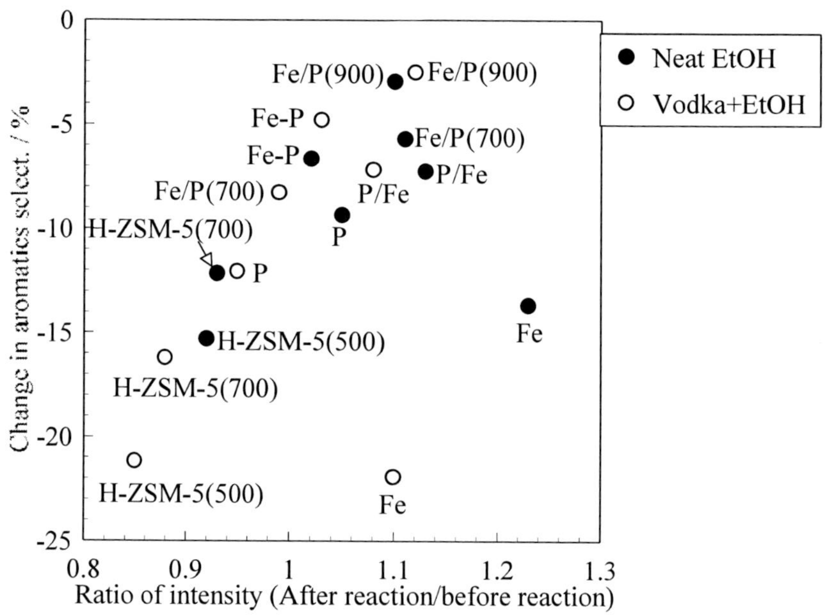

Figure 28. Plots of the ratio of XRD paek intensity at 23.1 ° (After reaction/before reaction) vs. the change in aromatics selectivity over H-ZSM-5 and (Fe and/or P)/H-ZSM-5 catalysts.

5.2.4. Results of NH_3-TPD Measurements

Figure 29(a) shows the NH_3-TPD profiles of H-ZSM-5 and (Fe and/or P)/H-ZSM-5 catalysts before reaction, and Figure 29(b) shows the NH_3-TPD profiles after 7 h-reaction using neat ethanol.

In Figure 29(a), H-ZSM-5 (500) shows a sharp NH_3-desorption peak at about 400 °C: the acidic site showing this desorption peak is essential for the formation of aromatic compounds. On the other hand, for H-ZSM-5 (700), the desorption peak disappeared, while for P/H-ZSM-5, the desorption peak remained to some degree, suggesting that P-addition can inhibit the decrease in acidic sites due to framework collapse during the two calcinations at 500 °C and 700 °C. Although added P species may block the acidic sites of the H-ZSM-5 support, in this case, protection of the zeolite framework by P-addition appears to be more significant, since the amount of added P was not very large. In 3.2.6, by using $FePO_4{\cdot}nH_2O$ as Fe source, NH_3 desorption peak of Fe/H-ZSM-5 catalyst disappeared completely, suggesting that in this case, Fe-loading was high (10 wt%) and amount of P atoms equal to Fe atoms was sufficient to block acidic sites completely. Fe/H-ZSM-5, P/Fe/H-ZSM-5 and Fe-P/H-ZSM-5 catalysts showed a decreased NH_3-desorption peak. Fe/P(700)/H-ZSM-5 catalyst showed a scarcely remaining peak, suggesting that the first addition of P to the H-ZSM-5 support protected the acidic sites of the support during Fe-addition and calcination.

It appears that solid acidity showing an NH_3 desorption peak at about 400 °C is essential for the formation of aromatic compounds, but that calcination at higher temperatures may weaken this solid acidity due to framework collapse. P-addition to the H-ZSM-5 zeolite support can inhibit the loss of the zeolite's acidic sites.

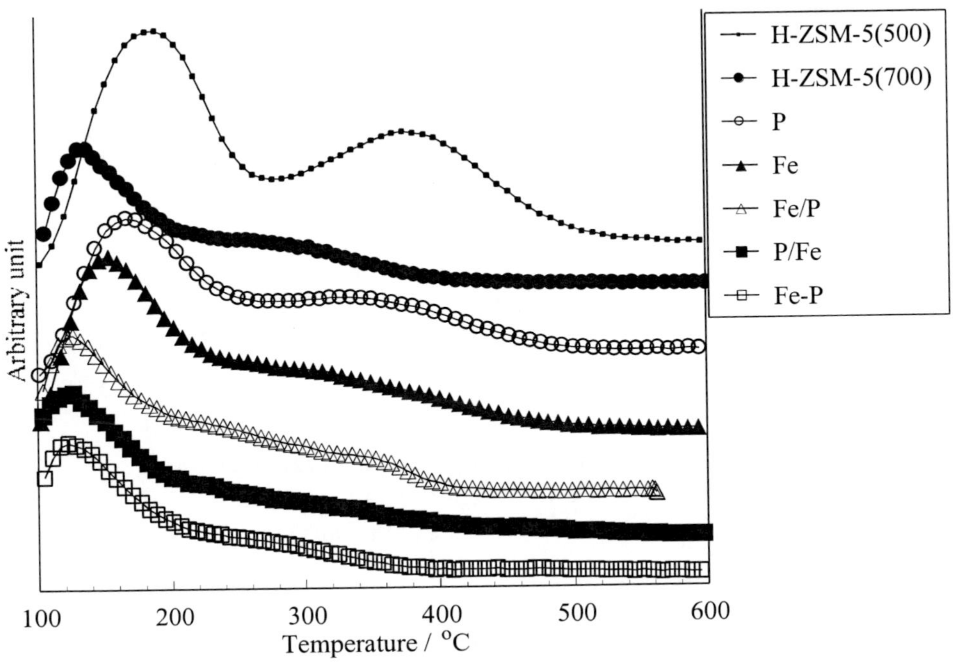

Figure 29. (a) NH_3-TPD profiles of H-ZSM-5 zeolites and (Fe and/or P) /H-ZSM-5 catalysts before reaction.

From Figure 29(b), after a reaction using neat ethanol, the NH_3 desorption peak at about 400 °C for H-ZSM-5 (500) has weakened (a1), suggesting that carbon deposition on the catalyst or framework collapse of the zeolite support may cause the observed decrease in solid acidity of H-ZSM-5 (500).

To remove deposited carbon, calcination treatment at 500 °C in air for 1 h to H-ZSM-5 (500) after reaction was carried out. TPD profile of H-ZSM-5 (500) after calcination was also shown as (a2) in Figure 29(b): intensity of NH_3 desorption peak at about 400 °C was similar to that of H-ZSM-5 (500) before calcination (a1), suggesting that the decrease in solid acidity of H-ZSM-5 (500) was caused by framework collapse of zeolite rather than carbon deposition, since framework collapse is irreversible.

On the other hand, for other catalysts after reaction, no clear decrease in solid acidity of catalysts was observed. H-ZSM-5 (500) and P/H-ZSM-5 catalysts, showing a sharp NH_3 desorption peak, which possess more acidic sites than H-ZSM-5 (700), Fe/H-ZSM-5 and (Fe and P)/H-ZSM-5 catalysts, showed an increase in propylene selectivity.

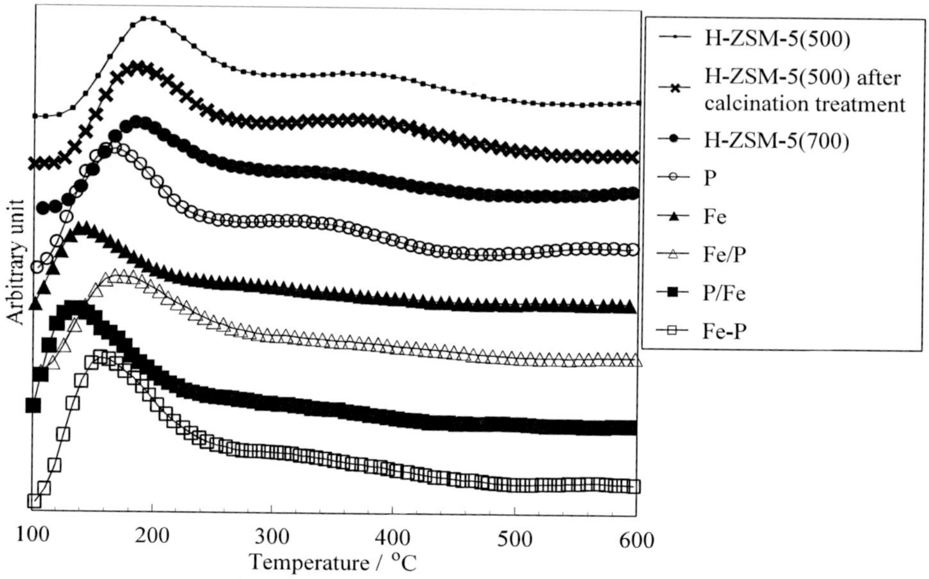

Figure 29. (b) NH_3-TPD profiles of H-ZSM-5 zeolites and (Fe and/or P) /H-ZSM-5 catalysts after 7 h-reaction using neat ethanol.

5.2.5. Results of SEM Observations

Figure 30 shows SEM images of catalysts before and after reaction: (a) H-ZSM-5 (500) before reaction, (b) H-ZSM-5 (500) after reaction with neat ethanol, (c) Fe/H-ZSM-5 before reaction, (d) Fe/H-ZSM-5 after reaction using neat ethanol; the scale bar represents 2μm in each case.

Figure 30. SEM images of catalysts before and after reaction using neat ethanol. (a) H-ZSM-5 (500) before reaction, (b) H-ZSM-5 (500) after reaction, (c) Fe/H-ZSM-5 before reaction, (d) Fe/H-ZSM-5 after reaction.

From Figure 30(a), one can observe rectangular-shaped crystals 1-2 μm in size with several steps and irregularly-shaped crystals smaller than 1 μm. However, the shape and size of these crystals in H-ZSM-5 (500) did not change significantly during reaction, as shown in Figure 30(b), suggesting that the decrease in crystallinity is not clearly evident in the morphology and size of the crystals. On the other hand, Figure 30(c), showing rectangular-shaped crystals 1-2 μm in size with several steps and cuboid crystals smaller than 0.5 μm, suggests that Fe-addition and calcination enhance the collapse of H-ZSM-5 crystals into smaller cubic crystals. This collapse of the zeolite structure may explain the decrease in intensity of XRD peaks. Moreover, Figure 30(d), showing Fe/H-ZSM-5 after reaction, suggests that the shape and size of Fe/H-ZSM-5 does not change during reaction.

5.2.6. Results of XRF Analyses

Table 9 shows the results of XRF analyses. There is no significant difference in mass % of Si and Al for H-ZSM-5 (29) zeolites and Fe/H-ZSM-5 (29) catalyst (numbers in parentheses denote the Si/Al_2 ratio).

Table 9. Results of XRF analyses

Si/Al_2 ratio	Loaded element	Calc. temp. / °C		mass %	
				Si	Al
29	No	500	Before reaction	95.5	4.38
			After reaction	95.8	4.13
29	No	700	Before reaction	95.6	4.29
			After reaction	94.8	4.97
29	Fe	700	Before reaction	93.5	4.37
			After reaction	92.6	4.26
23	No	500	Before reaction	94.3	5.58
80	No	500	Before reaction	98.6	1.38

Neat EtOH was used for reaction.

On the other hand, H-ZSM-5 (23) zeolite showed higher value of Al than H-ZSM-5 (29), while H-ZSM-5 (80) showed lower value of Al. These results suggest that calcination temperature, Fe-loading and ethanol reaction did not affect the Si/Al_2 ratio. It is proposed that any loss of Al species from the zeolite framework by dealumination can still remain associated with the surface of the zeolite, and as a result, Si/Al_2 ratio did not change so significantly.

5.2.7. Catalytic Performance as a Function of Time and Regeneration Treatment

Figure 31 shows the catalytic performance of Fe/H-ZSM-5 catalyst as a function of time, where neat ethanol was used. Selectivity of aromatic compounds decreased (29.4 → 15.7%) and selectivity of ethylene increased (22.0 → 28.4). After the regeneration treatment, aromatics selectivity decreased (26.9 → 16.3%) and ethylene selectivity drastically increased (27.1 → 44.4%). The ethylene selectivity obtained by this treatment was higher than the first reaction before regeneration treatment, and the selectivity for aromatics after this treatment was generally lower than in the first reaction. The initial selectivity, the selectivity after 7 h-reaction, the initial selectivity after regeneration treatment and the selectivity after a further 7 h-reaction of C_{3+} olefins were 26.5, 31.2, 26.1 and 25.1%, respectively, and the selectivity for propylene changed as follows: 13.2 → 15.0 → 13.3 → 12.7%. The selectivity of C_{3+} olefins and propylene did not show any significant change. These results suggest that this regeneration treatment alone cannot completely recover the selectivity of each product, since carbon deposited on the catalysts cannot be completely removed by this regeneration treatment. Figure 32 shows the catalytic performance of Fe/P(700)/H-ZSM-5 catalyst as a function of time, where neat ethanol was used. The initial selectivity, the selectivity after 7 h-reaction, the initial selectivity after regeneration treatment and the selectivity after a further 7 h-reaction of C_{3+} olefins were 33.2, 31.5, 34.1 and 31.1%, respectively. The selectivity for propylene at each stage was as follows: 18.3 → 16.2 → 17.8 → 15.6%. The selectivity of ethylene changed as follows: 26.5 → 37.5 → 26.7 → 37.2%, and that of aromatics changed as follows: 18.3 → 12.6 → 17.7 → 11.9%. In this case, selectivity of C_{3+} olefins and propylene was slightly improved in comparison with Fe/H-ZSM-5 catalyst, and the change in selectivity of each product was smaller than for Fe/H-ZSM-5.

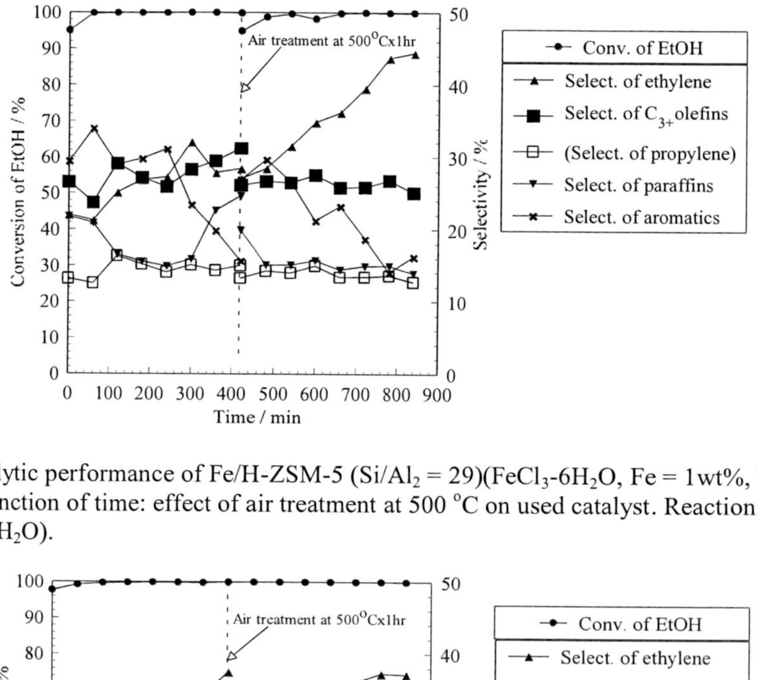

Figure 31. Catalytic performance of Fe/H-ZSM-5 ($Si/Al_2 = 29$)($FeCl_3$-$6H_2O$, Fe = 1wt%, 700°Ccalc.) catalysts as a function of time: effect of air treatment at 500 °C on used catalyst. Reaction gas: N_2 + EtOH (without H_2O).

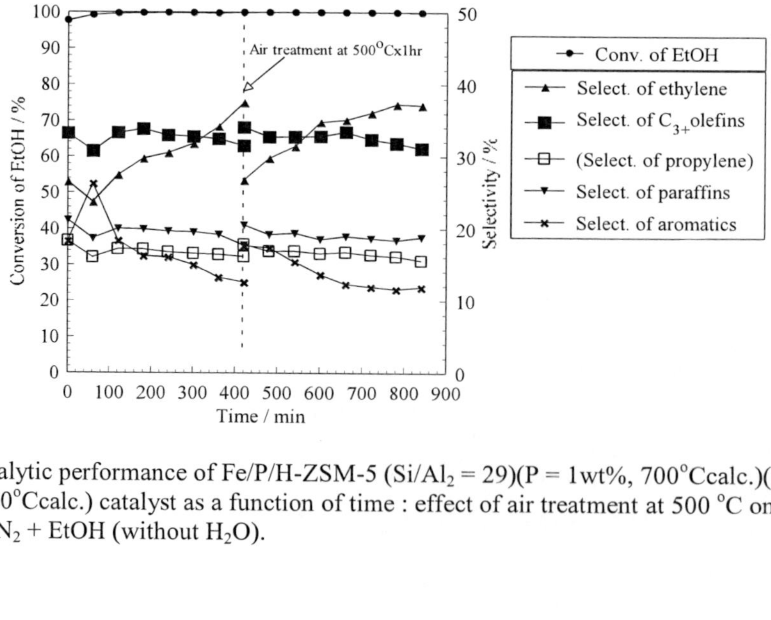

Figure 32. Catalytic performance of Fe/P/H-ZSM-5 ($Si/Al_2 = 29$)(P = 1wt%, 700°Ccalc.)($FeCl_3$-$6H_2O$, Fe = 1wt%, 700°Ccalc.) catalyst as a function of time : effect of air treatment at 500 °C on used catalyst. Reaction gas: N_2 + EtOH (without H_2O).

Moreover, this regeneration treatment completely restored the catalytic activity. It is suggested that deposited carbon, which causes a change of selectivity for each product, can be almost completely removed by this treatment, and that P-addition improves catalytic activity and stability.

In the case of aqueous ethanol, most catalysts showed a greater change in selectivity of each product than with neat ethanol, and by calcination treatment, catalytic activity could not be recovered easily. However, as shown in Figure 33, Fe-P/H-ZSM-5 catalyst showed high selectivity of C_{3+} olefins and propylene, and small changes in selectivity even for aqueous ethanol. Moreover, catalytic activity could be recovered almost completely by the regeneration treatment.

The initial selectivity, the selectivity after 7 h-reaction, the initial selectivity after regeneration treatment and the selectivity after a further 7 h-reaction of C_{3+} olefins were 31.4, 32.2, 30.8 and 31.4%, respectively. The selectivity of propylene at each stage was as follows: 15.1 → 16.6 → 16.1 → 16.1%.

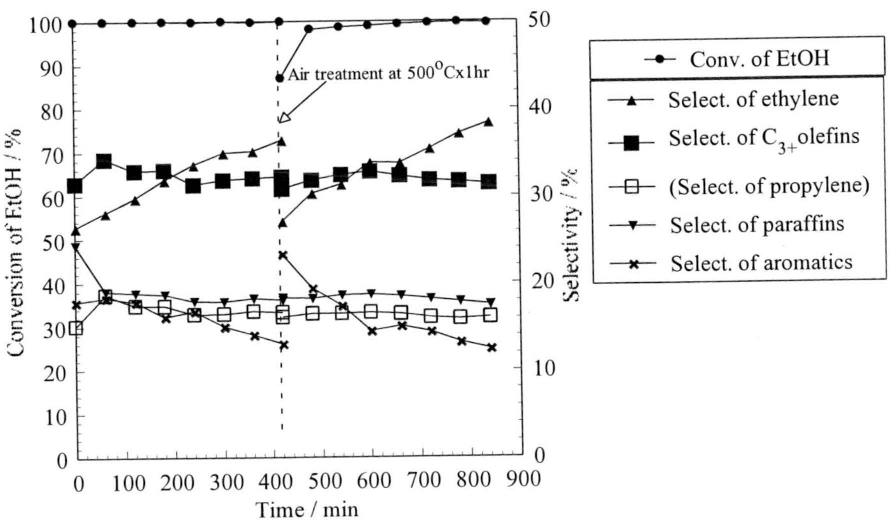

Figure 33. Catalytic performance of Fe-P/H-ZSM-5 (Si/Al_2 = 29)($FeCl_3$-$6H_2O$, Fe = P = 1wt%, 700°Ccalc.) catalyst as a function of time : effect of air treatment at 500 °C on used catalyst. Reaction gas: N_2 + EtOH (with H_2O).

The selectivity of ethylene changed as follows: 26.4 → 36.4 → 27.1 → 38.3%, and that of aromatics changed as follows: 17.9 → 13.0 → 23.3 → 12.5%. These results suggest that co-addition of Fe and P to H-ZSM-5 zeolite can lead to higher catalytic activity and stability, even for aqueous ethanol.

5.2.8. Reaction Pathway

The reaction pathway is assumed to be similar with 3.2.9. These reactions can proceed on acidic sites of the zeolite support, and Fe- and/or P-modification may to some degree inhibit the formation of aromatics and carbon deposition. As the reaction proceeds, deposited carbon may cover the acidic sites in the zeolite support, and framework collapse of zeolite may reduce the number of acidic sites, which are responsible for the formation of aromatics; as a result, the yield of aromatics decreased with time-on-stream. Suppression of formation of aromatics leads to enhanced ethylene formation. Deposited carbon and framework collapse both suppress the formation of aromatics, and this suppression may lead to two responses: (1) increased selectivity for C_{3+} olefins and propylene, and (2) decreased selectivity for C_{3+} olefins and propylene. In the former case, recovery of catalytic performance by regeneration treatment may be difficult, since removal of deposited carbon is problematic and framework collapse of the zeolite support is irreversible. Therefore, for catalysts showing significant framework collapse, for example H-ZSM-5 (500), it may be impossible to recover catalytic performance completely. In the latter case, however, recovery may be easy, since removal of deposited carbon is easy and framework collapse during reaction did not proceed so rapidly.

CONCLUSION

In ethanol conversion to hydrocarbons by sole zeolite catalyst, H-ZSM-5 ($Si/Al_2 = 29$) zeolite showed especially high selectivity of BTX (benzene, toluene, xylenes), while other zeolites (H-ZSM-5 (190), H-mordenite, H-Beta and Y-type zeolite) exclusively formed ethylene. Effect of metal-addition to H-ZSM-5 (29) was investigated. Ga- or noble metal-modified H-ZSM-5 catalysts have high activity for the formation of BTX. Among them, Au-modified H-ZSM-5 catalyst can partially inhibit carbon deposition during the reaction, thus maintaining constant catalytic activity for BTX formation. On the other hand, addition of alkali earth metal (for example, Mg) led to exclusive formation of ethylene. Cr-, Fe- or Ni-modified catalysts showed higher selectivity of C_{3+} olefins than other metal-modified catalysts.

Scheme 2 shows the image of ethanol conversion by (Fe, Ga or noble metal)/ H-ZSM-5 catalysts. Fe/H-ZSM-5 catalysts were efficient for EtOH conversion to C_{3+} olefins, which are useful as not only fuels but also chemicals.

Moreover, Fe is cheap and non-toxic metal. Initial selectivity of C_{3+} olefins over Fe catalysts was little affected by adding water, suggesting that these Fe catalysts are fully effective even in the presence of water in EtOH. In most cases, the initial selectivity of C_{3+} olefins was not affected by the kind of iron source and calcination temperature so significantly.

However, in general, selectivity of C_{3+} olefins was slightly decreased with time-on-stream, and degree of the decrease was affected by these conditions. Among Fe catalysts, $Fe_2(SO_4)_3$- or $FeCl_3$-based catalyst calcined at 700 °C showed low decrease in C_{3+} olefins selectivity. As causes of catalytic deactivation, carbon deposition on catalyst and framework collapse of zeolite support can be considered. Acidic site of support was essential for the formation of olefins, paraffins and aromatics, and Fe species can inhibits the aromatics formation in some degree. H_2 was formed by steam-combustion of deposited carbon and dehydrogenation via aromatization and carbonization, and can hydrogenate olefins to paraffins.

However, Fe_2O_3 can suppress the olefins hydrogenation by reduction to Fe_3O_4, resulting in higher olefins/paraffins ratio than H-ZSM-5 support (without Fe). Air treatment at 500 °C to used catalysts can restore catalytic activity and the degree of regeneration was affected by the kind of catalysts: in the $FeCl_3$-based catalysts calcined at 700 °C and 900 °C-calcined catalysts (irrespective of iron source), catalytic activity can restore almost completely. On the other hand, the $Fe(NO_3)_3$- or $Fe_2(SO_4)_3$-derived catalysts calcined at 700 °C cannot restore the catalytic activity by this treatment completely. At any case, Fe/H-ZSM-5 catalysts are promising candidate for the production of C_{3+} olefins from ethanol.

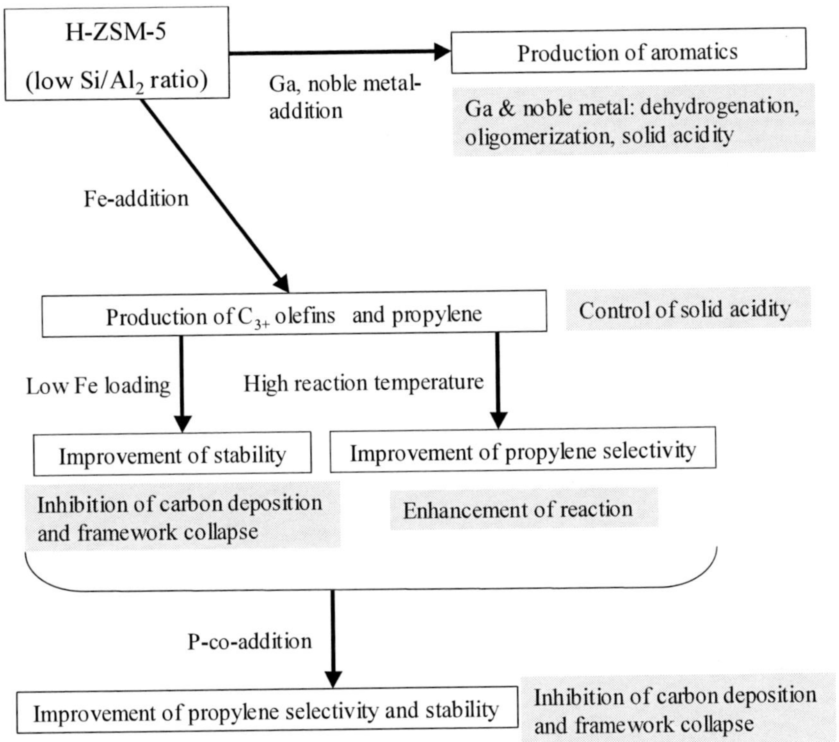

Scheme 2. Image of ethanol conversion by (Fe, Ga or noble metal)/H-ZSM-5 catalysts.

Moreover, lower loading of Fe (1 wt%) and higher reaction temperature (450 °C) led to higher selectivity of C_{3+} olefins and propylene and catalytic stability. Lower Fe-loading could inhibit carbon deposition and framework collapse in some degree, and higher reaction temperature could enhance the formation of olefins. Among the causes of change in selectivity of products, the effect of carbon deposition can be removed by calcinations at higher temperature in air-flow. On the other hand, the effect of framework collapse of zeolite supports cannot be removed, since this is irreversible reaction, however, the smaller amount of Fe loaded on zeolite supports can be favorable in the framework collapse.

Fe- and/or P-modified H-ZSM-5 zeolite catalysts were found to be efficient for the production of olefins (especially propylene) from ethanol. There appears to be a correlation between ethylene selectivity and propylene selectivity, a weak reverse correlation between paraffins selectivity and propylene selectivity, and a reverse correlation between selectivity for aromatics and propylene selectivity. The selectivity for each product changed with time on-stream. In general, catalysts showing higher initial selectivity for propylene show a steeper decrease in propylene selectivity during reaction. H-ZSM-5 calcined at low temperature showed low initial selectivity for propylene but selectivity increased during the reaction period. H-ZSM-5 calcined at high temperature showed higher selectivity for propylene than H-ZSM-5 calcined at low temperature, and the selectivity increased only slightly for neat ethanol, but decreased significantly for aqueous ethanol. Successive addition of Fe and P (Fe → P and P → Fe) improved initial selectivity but led to a steeper decrease in selectivity with on-stream time, while single addition of Fe or P and co-addition of Fe and P showed medium initial selectivity followed by increased selectivity. Our results suggest that H-ZSM-5 zeolite is the active component, while added P and Fe act as zeolite modifiers. In the case of the neat ethanol used, Fe/P(700)/H-ZSM-5 catalyst, which was prepared by successive addition (P → Fe) to the H-ZSM-5 zeolite support, showed high selectivity for C_{3+} olefins and propylene. Catalytic performance was regenerated by calcination treatment in flowing air, while the catalytic performance of Fe/H-ZSM-5 catalyst without P could not be fully regenerated. In general, the presence of water in ethanol enhanced the change in selectivity for each product. However, Fe-P/H-ZSM-5 catalyst, which was prepared by co-impregnation of Fe and P to the H-ZSM-5 zeolite support, showed high catalytic performance that could be regenerated by calcination. Our study suggests that the cause of changed selectivity for each product may be carbon deposition on catalysts and framework collapse of the zeolite support during reaction. Deposited carbon and framework collapse suppress the formation of aromatics, and this suppression may result in: (1) increase in selectivity for C_{3+} olefins and propylene; (2) decrease in the selectivity for C_{3+} olefins and propylene.

Acknowledgments

The authors would like to express sincere thanks to Dr. Okabe of AIST for his kind help in XPS analyses of catalysts.

This work was partly supported by a New Energy and Industrial Technology Development Organization (NEDO) grant.

References

[1] Chang, C. D., and Silvestri, A. J. (1977). The conversion of methanol and other O-compounds to hydrocarbons over zeolite catalysts. *J. Catal.,* 47(2), 249-259.

[2] Kaeding, W. W. and Butter, S. A. (1980). Production of chemicals from methanol: I. Low molecular weight olefins. *J. Catal.:* 61(1), 155-164.

[3] Dejaifve, P., Védrine, J. C., Bolis, V., and Derouane, E. G. (1980). Reaction pathways for the conversion of methanol and olefins on H-ZSM-5 zeolite. *J. Catal.,* 63(2), 331-345.

[4] Gayubo, A. G., Aguayo, A. T., Atutxa, A., Prieto, R., and Bilbao, J. (2004). Deactivation of a HZSM-5 Zeolite Catalyst in the Transformation of the Aqueous Fraction of Biomass Pyrolysis Oil into Hydrocarbons. *Energy Fuels,* 18(6), 1640-1647.

[5] Gayubo, A. G., Aguayo, A. T., Atutxa, A., Aguado, R., and Bilbao, J. (2004). Transformation of Oxygenate Components of Biomass Pyrolysis Oil on a HZSM-5 Zeolite. I. Alcohols and Phenols. *Ind. Eng. Chem. Res.,* 43(11), 2610-2618.

[6] Ph. de Werbier d'Antigneul, Chami, J., Berrier, C., Blanchard, M., and Canesson, P. (1988). Selective transformation of ethanol into propene on cobalt catalysts. *Catal. Lett.,* 1(5), 169-176.

[7] Derouane, E. G., Nagy, J. B., Dejaifve, P., Van Hooff, J. H. C., Spekman, B. P., Védrine J. C., and Naccache, C. (1978). Elucidation of the Mechanism of Conversion of Methanol and Ethanol to Hydrocarbons on a New Type of Synthetic Zeolite. *J. Catal.*, 53(1), 40-55.

[8] Oudejans, J. C., Van den Oosterkamp, P. F., and Van Bekkum, H. (1982). Conversion of ethanol over zeolite h-zsm-5 in the presence of water. *Appl. Catal.*, 3(2), 109-115.

[9] Choudhary, V. R., and Sansare, S. D. (1984). Conversion of ethanol to aromatics on HZSM-5 in a pulse microreactor. *Appl. Catal.*, 10(2), 147-153.

[10] Chaudhuri, S. N., Halik, C., and Lercher, J. A. (1990). Reactions of ethanol over HZSM-5. *J. Mol. Catal.*, 62(3), 289-295.

[11] Saha, S. K., and Sivasanker, S. (1992). Influence of Zn-and Ga-doping on the conversion of ethanol to hydrocarbons over ZSM-5. *Catal. Lett.*, 15(4), 413-418.

[12] Gayubo, A. G., Tarrío, A. M., Aguayo, A. T., Olazar, M., and Bilbao, J. (2001). Kinetic modelling on the transformation of aqueous ethanol into hydrocarbons on a HZSM-5 zeolite. *Ind. Eng. Chem. Res.*, 40(16), 3467-3474.

[13] Aguayo, A. T., Gayubo, A. G., Atutxa, A., Olazar, M., and Bilbao, J. (2002). Catalyst deactivation by coke in the transformation of aqueous ethanol into hydrocarbons, kinetic modeling and acidity deterioration of the catalyst. *Ind. Eng. Chem. Res.*, 41(17), 4216-4224.

[14] Aguayo, A. T., Gayubo, A. G., Tarrío, A. M., Atutxa, A., and Bilbao, J. (2002). Study of operating variables in the transformation of aqueous ethanol into hydrocarbons on an HZSM-5 zeolite. *J. Chem. Technol. Biotechnol.*, 77(2), 211-216.

[15] Ingram, C. W., and Lancashire, R. J. (1995). On the formation of C_3 hydrocarbons during the conversion of ethanol using H-ZSM-5 catalyst. *Catal. Lett.*, 31(4), 395-403.

[16] Inaba, M., Murata, K., Saito, M., and Takahara, I. (2006). Ethanol conversion to aromatic hydrocarbons over several zeolite catalysts. *React. Kinet. Catal. Lett.*, 88(1), 135-142.

[17] Murata, K., Inaba, M., and Takahara, I. (2008). Effects of surface modification of H-ZSM-5 catalysts on direct transformation of ethanol into lower olefins. *J. Jpn. Petro. Inst.*, 51(4), 234-239.

[18] Inoue, K., Inaba, M., Takahara, I., and Murata, K. (2010). Conversion of ethanol to propylene by H-ZSM-5 with Si/Al_2 ratio of 280. *Catal. Lett.*, 136(1-2), 14-19.

[19] Goto, D., Harada, Y., Furumoto, Y., Takahashi, A., Fujitani, T., Oumi, Y., Sadakane, M., and Sano, T. (2010). Conversion of ethanol to propylene over HZSM-5 type zeolites containing alkaline earth metals. *Appl. Catal. A: Gen.*, 383(1-2), 89-95.

[20] Song, Z., Takahashi, A., Nakamura, I. and Fujitani, T. (2010). Phosphorus-modified ZSM-5 for conversion of ethanol to propylene. *Appl. Catal. A: Gen.*, 384(1-2), 201-205.

[21] Song, Z., Takahashi, A., Mimura, N., and Fujitani, T. (2009). Production of Propylene from Ethanol Over ZSM-5 Zeolites. *Catal. Lett.*, 131(3-4), 364-369.

[22] Inaba, M., Murata, K., Takahara, I., and Inoue, K. (2012). Production of C_{3+} olefins and propylene from ethanol by Zr-modified H-ZSM-5 zeolite catalysts. *Advances in Materials Science and Engineering:* in press.

[23] Gayubo, A. G., Alonso, A., Valle, B., Aguayo, A. T., Olazar, M., and Bilbao, J. (2010). Hydrothermal stability of HZSM-5 catalysts modified with Ni for the transformation of bioethanol into hydrocarbons. *Fuel*, 89(11), 3365-3372.

[24] Gayubo, A. G., Alonso, A., Valle, B., Aguayo, A. T., Olazar, M., and Bilbao, J. (2011). Kinetic modelling for the transformation of bioethanol into olefins on a hydrothermally stable Ni–HZSM-5 catalyst considering the deactivation by coke. *Chem. Eng. J.,* 167(1), 262-277.

[25] Gayubo, A. G., Alonso, A., Valle, B., Aguayo, A. T., and Bilbao, J. (2010). Kinetic Model for the Transformation of Bioethanol into Olefins over a HZSM-5 Zeolite Treated with Alkali. *Ind. Eng. Chem. Res.,* 49(21), 10836-10844.

[26] Gayubo, A. G., Alonso, A., Valle, B., Aguayo, A. T. and Bilbao, J. (2010). Selective production of olefins from bioethanol on HZSM-5 zeolite catalysts treated with NaOH. *Appl. Catal. B: Environ.,* 97(2-1), 299-306.

[27] Levinbuk, M. I., and Usachev, N. Y. (1998). The production of hydrogen through methane conversion over reagent catalysts. An evaluation of the feasibility of catalytic cracking unit utilization for methane conversion. *Stud. Surf. Sci. Catal.,* 119, 391-396.

[28] Inaba, M., Murata, K., Saito, M., and Takahara, I. (2006). Hydrogen production by gasification of cellulose over Ni catalysts supported on zeolites. *Energy Fuels,* 20(2), 432-438.

[29] Oikawa, H., Shibata, Y., Inazu, K., Iwase, Y., Murai, K., Hyodo, S., Kobayashi, G., and Baba, T. (2006). Highly selective conversion of ethene to propene over SAPO-34 as a solid acid catalyst. *Appl. Catal. A: Gen.,* 312,181-185.

[30] Iwamoto, M., Kasai, K., and Haishi, T. (2011). Conversion of Ethanol into Polyolefin Building Blocks: Reaction Pathways on Nickel Ion-loaded Mesoporous Silica. *Chem Sus Chem,* 4(8), 1055-1058.

[31] Inaba, M., Murata, K., Saito, M., and Takahara, I. (2007). Production of olefins from ethanol by Fe-supported zeolite catalysts. *Green Chem.,* 9(6), 638-646.

[32] Inaba, M., Murata, K., and Takahara, I. (2009). Effect of Fe-loading and reaction temperature on the production of olefins from ethanol by Fe/H-ZSM-5 zeolite catalysts. *React. Kinet. Catal. Lett.,* 97(1),19-26.

[33] Ramesh, K., Hui, L. M., Han, Y.-F., and Borgna, A. (2009). Structure and reactivity of phosphorous modified H-ZSM-5 catalysts for ethanol dehydration. *Catal. Commun.,* 10(5), 567-571.

[34] Inaba, M., Murata, K., Takahara, I., and Inoue, K. (2011). Production of Olefins from Ethanol by Fe and/or P-Modified H-ZSM-5 Zeolite Catalysts. *J. Chem. Technol. Biotechnol.,* 86(1), 95-104.

[35] Fernandes-Machado, N. R. C., Calsavara, V., Astrath, N. G. C., Neto, A. M., and Baesso, M. L. (2006). Hydrocarbons from ethanol using [Fe,Al]ZSM-5 zeolites obtained by direct synthesis. *Appl. Catal. A: Gen.,* 311, 193-198.

[36] Calsavara, V., Baesso, M. L., and Fernandes-Machado, N. R. C. (2008). Transformation of ethanol into hydrocarbons on ZSM-5 zeolites modified with iron in different ways. *Fuel,* 87(8-9), 1628-1636.

[37] Toops, T. J., Nguyen, K., Foster, A. L., Bunting, B. G., Ottinger, N. A., Pihl, J. A., Hagaman, E. W., and Jiao, J. (2010). Deactivation of accelerated engine-aged and field-aged Fe–zeolite SCR catalysts. *Catal. Today,* 151(3-4), 257-265.

[38] Brandenberger, S., Kröcher, O., Casapu, M., Tissler, A., and Althoff, R. (2011). Hydrothermal deactivation of Fe-ZSM-5 catalysts for the selective catalytic reduction of NO with NH_3. *Appl. Catal. B: Environ.,* 101(3-4), 649-659.

[39] Sun, K., Xia, H., Hensen, E., Van Santen, R., and Li, C. (2006). Chemistry of N_2O decomposition on active sites with different nature: Effect of high-temperature treatment of Fe/ZSM-5. *J. Catal.,* 238(1), 186-195.

[40] Kowalska-Kuœ, J., Held, A., and Nowiñska, K. (2010). Oxydehydrogenation of Ethane and Propane over Alkaline Metal Modified Fe-ZSM-5 Zeolites. *Catal. Lett.,* 136(3-4), 199-208.

[41] Hensen, E. J. M., Zhu, Q., and Van Santen, R. A. (2003). Extraframework Fe-Al-O species occluded in MFI zeolite as the active species in the oxidation of benzene to phenol with nitrous oxide. *J. Catal.,* 220(2), 260-264.

[42] Hensen, E. J. M., Zhu, Q., and Van Santen, R. A. (2005). Selective oxidation of benzene to phenol with nitrous oxide over MFI zeolites: 2. On the effect of the iron and aluminum content and the preparation route. *J. Catal.,* 233(1), 136-146.

[43] Takahara, I., Saito, M., Inaba, M., and Murata, K. (2005). Dehydration of Ethanol into Ethylene over Solid Acid Catalysts. *Catal. Lett.,* 105(3-4), 249-252.

[44] Ali, L. I., Ali, A.-G. A., Aboul-Fotouh, S. M., and Aboul-Gheit, A. K. (1999). Dehydrogenation of cyclohexane on catalysts containing noble metals and their combinations with platinum on alumina support. *Appl. Catal. A: Gen.,* 177(1), 99-110.

[45] Liebmann, L. S., and Schmidt, L. D. (1999). Oxidative dehydrogenation of isobutane at short contact times. *Appl. Catal. A: Gen.,* 179(1-2), 93-106.

[46] Madeira, F. F., Gnep, N.S., Magnoux, P., Maury, S., and Cadran, N. (2009). Ethanol transformation over HFAU, HBEA and HMFI zeolites presenting similar Brønsted acidity. *Appl. Catal. A: Gen.,* 367(1-2), 39-46.

[47] Ouyang, J., Kong, F., Su, G., Hu, Y., and Song, Q. (2009). Catalytic Conversion of Bio-ethanol to Ethylene over La-Modified HZSM-5 Catalysts in a Bioreactor. *Catal. Lett.,* 132 (1-2), 64-74.

[48] Moser, W. R., Chiang, C.-C., and Thompson, R. W. (1989). Infrared diffuse reflectance study of the silicon-rich H-ZSM-5 catalysis of ethanol conversion. *J. Catal.,* 115(2), 532-541.

In: Zeolites: Synthesis, Chemistry and Applications
Editors: Moisey K. Andreyev and Olya L. Zubkov
ISBN: 978-1-61942-861-4

Chapter 2

AIR POLLUTION CATALYTIC CONTROL BY METAL PROMOTED ZEOLITES

Runduo Zhang and Biaohua Chen
State Key Laboratory of Chemical Resource Engineering,
Beijing University of Chemical Technology, Beijing, China

ABSTRACT

Air pollution would be achieved as the concentrations of some harmful materials, such as nitrous oxide (N_2O), nitrogen oxides (NO and NO_2), hydrocarbons (HCs), nitrile etc., are high enough to affect the people's health as well as destroy the sustainable ecosystem. Traffic and industry are the most important sources of the air pollutants mentioned above. Catalysis has become indispensable in converting these environmentally harmful molecules into more benign ones such as N_2, H_2O, and CO_2, although the latter now deems as a greenhouse gas. In this chapter, our recent progress on the synthesis and characterization of metal-promoted microporous and mesoporous zeolites as well as their applications as catalysts for the purification of exhaust gases including N_2O from nitric acid or adipic plant, NO and HCs from motor vehicles, and CH_3CN from acrylonitrile plant was systematically introduced.

N_2O direct decomposition over Fe-BEA zeolite prepared by wet ion-exchange (WIE) was investigated. Both isolated iron ions in the charge compensation positions and highly dispersed FeO_x-like crystallite were verified. Two kinds of reaction mechanism, with respective formation of oxygen (O_2) or nitrogen oxide (NO_x) intermediate, were proposed. The corresponding energy calculations for O_2-formation mechanism revealed that O_2 desorption was the rate determining step with an energy barrier of 63.2 kcal/mol. Another mechanism with NO formation was also simulated by density functional theory (DFT) calculations, whose energy calculations indicated that the formation of Fe-Z-$(NO)_2$ was the main pathway for NO generation with an energy barrier of 26.9 kcal/mol.

Aiming at the effective purification of NO and HCs from automobile, a series of mono-functional M/SBA-15 (M = Cu, Fe, Cr, and Al) and bifunctional M'/SBA-15 (M' = Cu-Al, Fe-Al, and Cr-Al) catalysts was thereafter prepared via an incipient wetness impregnation, and further used for C_3H_6 + NO + O_2 reaction. Ordered mesoporous structure of SBA-15 was well maintained even after impregnating the various metallic components. The deNOx activities of investigated catalysts follow a trend of Cu-Al/SBA-

15 > Cu/SBA-15, Fe/SBA-15 > Cr-Al/SBA-15 > Cr/SBA-15 >> Fe-Al/SBA-15, Al/SBA-15 > SBA-15. Synergistic effect between copper and alumina was achieved showing a satisfactory NO conversion of approximately 80% over Cu-Al/SBA-15 at temperature as low as 350°C under an atmosphere of 3000 ppm NO, 3000 ppm C_3H_6, 1% O_2 and GHSV of 60,000 h^{-1}.

A series of M''/SBA-15 (M''=Cu, Co, Cr and Mn) catalysts was also prepared and used for acetonitrile elimination. The CH_3CN conversion of investigated catalysts follow a trend of Cu/SBA-15 > Co/SBA-15 ≈ Cr/SBA-15 > Mn/SBA-15 > SBA-15. Apart from the target product of N_2, harmful byproducts of NO, NO_2, N_2O, NH_3 and CO were also observed during the combustion process. The corresponding selectivities were verified to be associated with both chemical nature and redox properties of the loaded metallic species. Among the prepared M''/SBA-15 samples, Cu/SBA-15 exhibited a nearly complete conversion of CH_3CN associated with a N_2 selectivity around 80% at T>350°C, which was regarded as a promising material for the catalytic removal of cyanide-containing gas.

1. The Direct Decomposition of Nitrous Oxide over Fe-Beta Zeolites

Nitrous oxide is classified as one kind of greenhouse gases with global warming potential (GWP) as much as 310, making a major contribution (about 6%) to global climate change. A large and increasing amount of N_2O is generated from the chemical industries for the production of adipic acid and nitric acid. N_2O emission control has therefore become one of the great challenges for environmental protection.

The directly catalytic decomposition is regarded as the simplest and most effective way for N_2O abatement with noble metals as the common catalysts due to their high intrinsic activities for N_2O decomposition. However, precious metal catalysts were readily deactivated by unfavorable molecules (e.g. H_2O, NO) present in the tail gases of adipic acid and nitric acid plants [1]. Recently Fe-beta zeolite has attracted great attention because of its excellent catalytic performance for N_2O decomposition [2-4]. It is generally accepted that surface iron species form the core of catalytic sites for N_2O decomposition. Actually, a wide variety of Fe species has been identified in Fe-zeolites, such as isolated Fe^{3+} and FeO^+ cations, di-iron oxo-cations with possible μ-hydroxo or de-η-hydroxo groups and peroxide bridges $(FeO_x)_n$ nanoclusters and small Fe_2O_3 aggregates [5, 6].

The mechanism of N_2O decomposition likely involves an adsorbed oxygen intermediate called "α-oxygen" [7]. Additionally, the desorption of O_2 is considered as the rate-determining step [8]. Although all types of iron species contribute to O_2 desorption, a fast pathway of O_2 desorption was dominated upon isolated iron sites, whereas a slow pathway of O_2 desorption was accordingly associated with the high iron loadings. Two points for N_2O direct decomposition based on Fe-ZSM-5 zeolite were commonly accepted: (1) oxygen formation was through the migration of α-O from two active sties; (2) N_2O decomposition and oxygen evolution occurred at the same active site.

N_2O decomposition behaviors over Fe-beta prepared by wet ion-exchanged (WIE) method were herein investigated [2]. The corresponding decomposition mechanisms were studied by not only experimental methods such as N_2O-temperature programmed desorption (TPD), NO_2-TPD, N_2-TPD, N_2O-temperature programmed surface reaction (TPSR), diffuse reflectance infrared Fourier transform spectra (DRIFTS), and activity test, but also by theoretical study via computational chemistry based on a density functional theory (DFT) in which energies for adsorption/desorption, activation energy barrier and reaction enthalpy were calculated and the energy profiles were depicted.

1.1. Characterization of Fe Species Loaded on the Fe-Beta Zeolites

The Fe content, BET surface area, crystallite size, and pore volume and average diameter of the prepared samples were listed in Table 1. Fe contents of the prepared samples established by AAS were close to the nominal values. A typical structure of beta-typed zeolite for all samples prepared by WIE was confirmed by XRD patterns (not shown). This crystal structure was well maintained even after Fe ion-exchange. Furthermore, no characteristic peak belonging to iron oxides was observed in all the diffraction patterns with the Fe loading varying from 0.25% to 3%, indicating that the Fe component was well dispersed over the molecular sieve. H_2-TPR experiments were conducted over both H-beta and Fe-beta zeolites with Fe contents of 0.25%, 0.5%, 1%, 1.5%, 2%, and 3%, as depicted in Figure 1. The absence of TPR peak in the case of H-beta sample suggested good stability of the zeolite framework under a reducing atmosphere at temperatures as high as 1273 K. TPR spectra for the Fe-containing zeolites (apart from Fe-beta-2% sample) show similar profiles representing two superimposed H_2 reducing peaks between 423 and 773 K. These two peaks were likely corresponding to the reduction of Fe^{3+} to Fe^{2+} in ion-exchange positions of zeolites according to the literature [9]. The additional peaks between 773 and 1073 K were probably due to the reduction of iron-oxides clusters [10]. As reported by Munteanu et al. [11], the peaks between 773 and 900 K in the TPR curves of Fe-beta-2% and Fe-beta-3% samples were assigned to the reduction of Fe_3O_4 to FeO and the peak at 940 K in the curve of Fe-beta-3% sample was assigned to the further reduction of FeO to Fe^0.

H_2-TPR revealed that two kinds of Fe species existed on the zeolite: isolated Fe^{3+} ions, formed via ion-exchange with proton hydrogen of Brönsted acid by valence bond compensation through the WIE, and fine FeO_x species (e.g. Fe_3O_4, oxygen-bridged binuclear Fe species or mononuclear Fe species).

Table 1. Chemical composition, BET surface areas, pore volume, pore average diameter, and crystallize size of the investigated samples [2]

Sample	Fe (wt%)[a]	BET surface area (m^2/g)	V_{micro} (cm^3/g)	Pore average diameter (nm)	Crystallite size (nm)[b]
H-beta	0	406	0.2997	1.437	7.587
Fe-beta-0.25%	0.23	419	0.3032	1.258	8.488
Fe-beta-0.5%	0.42	436	0.3494	1.344	9.129
Fe-beta-1%	0.92	438	0.3182	1.274	9.493
Fe-beta-1.5%	1.35	440	0.3428	1.642	8.279
Fe-beta-2%	1.82	463	0.3694	1.258	8.349
Fe-beta-3%	2.75	501	0.4026	1.292	8.932

a: Measured by AAS.

b: Calculate by Scherrer equation.

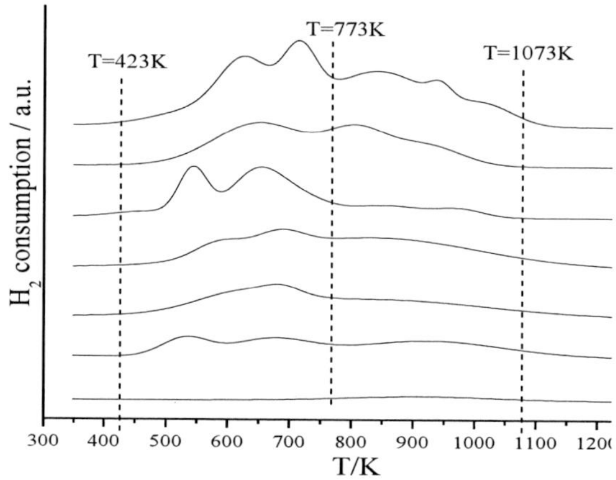

Figure 1. H_2-TPR profiles for (a) H-beta and Fe-beta zeolites with the iron content of (b) 0.25%, (c) 0.5%, (d) 1%, (e) 1.5%, (f) 2%, (g) 3% [2].

As the Fe content was low, it mainly existed in the form of Fe^{3+} ions on the zeolite as shown in the TPR plot of Fe-beta-0.25% and Fe-beta-0.5%. Further increasing the Fe loading, the reducing peak of $FeO_x \rightarrow FeO$ became significant as discerned from the TPR curves of Fe-beta-2% (the peak at 800 K) and Fe-beta-3% (the peak at 840 K). In the spectrum of Fe-beta-3% a special peak with a maximum at 940 K is probably due to a further reduction of FeO to metallic Fe on the catalyst surface.

1.2. Catalytic Activity Evaluation

The conversion of N_2O as a function of temperature for H-beta and Fe-beta zeolites was shown in Figure 2. It was found that the catalytic activity of beta-type zeolites could be significantly promoted by means of Fe loading showing an obvious enhancement in N_2O conversion with the full-conversion temperature of Fe-beta-1% zeolite being 130 K lower than that of H-beta zeolite.

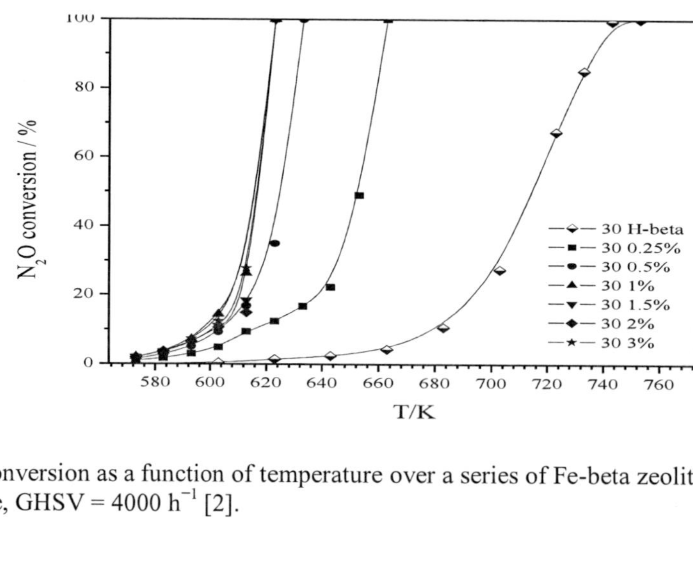

Figure 2. N_2O conversion as a function of temperature over a series of Fe-beta zeolites, conditions: 35 vol% N_2O/He, GHSV = 4000 h^{-1} [2].

In addition, further increase in deN_2O activity upon the increasing loading was marginal for Fe-beta samples with iron contents higher than 1%. Accordingly, we correlated the catalytic behaviors with the status of Fe species. Fe species might be well dispersed as an ionic form at its content below 1%, as verified by H_2-TPR, leading to a monotonous increase of N_2O conversion along with more Fe loading. The increase of loading more than 1% could lead to the appearance of FeO_x phase, which was believed to be less active for N_2O decomposition, as the amount of iron exceeds that needed for monolayer coverage.

1.3. Mechanism Study

1.3.1. In-situ Diffuse Reflectance Infrared Fourier Transform Spectra (DRIFTS)

The DRIFTS of N_2O (4vol% in He) adsorbed on the Fe-beta-1% zeolite at 573 K as a function of exposure time were plotted in Figure 3A. Once, Berlier et al. [12], respectively, assigned the bands at 2218 and 2235 cm^{-1} (with a shoulder at 2000 cm^{-1}) to the physisorbed N_2O and the chemisorbed N_2O on the Fe^{2+} centers. In the literature, the bands at 2208 and 2238 cm^{-1} were ascribed to the ν(N–N) stretching modes [13] and the band at 1300 cm^{-1} was associated with ν(N–O) stretching modes [14]. It was clearly seen from Figure 3A that the intensities of bands at 2208–2218 and 2235 cm^{-1} were increased as the exposure time was prolonged, implying some chemisorbed species were formed and further accumulated over the zeolite, although some reports suggested that these bands might correspond to the infrared vibrations of the N_2O molecule in the gas phase [15]. Accordingly, the bands at 2208–2218 and 2235 cm^{-1} appearing in Figure 3A were attributed to the weakly chemisorbed (possibly overlapping with some physisorbed) N_2O. The slight bands at 1275 and 1300 cm^{-1} were likely due to a contribution of ν(N–O) stretching modes from either adsorbed N_2O or some newly generated NO species after N_2O chemisorption. The DRIFTS for Fe-beta-1% zeolite with an exposure to 4 vol% N_2O/He at different temperatures were plotted in Figure 3B. Subsequently, the area of IR bands at 2208–2218 and 2235 cm^{-1} as a function of temperature was also shown as an inserted figure.

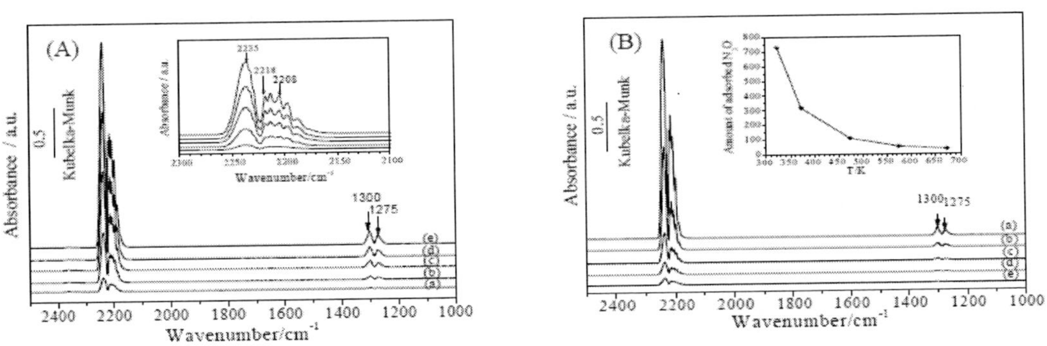

Figure 3. (A) DRIFTS of Fe-beta-1% zeolite with an exposure to N_2O (4 vol% in He, 20 ml/min) at 573 K for (a) 1 min, (b) 2 min, (c) 3 min, (d) 4 min, (e) 5 min. (B) DRIFTS of Fe-beta-1% zeolite with an exposure to N_2O (4 vol% in He, 20 ml/min) at (a) 323 K, (b) 373 K, (c) 473 K, (d) 573 K, (e) 673 K for 1 h [2].

The intensities of infrared bands of N_2O (2208–2218, 2235 cm^{-1}) diminished gradually with the temperature rising from 323 to 673 K (especially at <473 K) and these bands almost vanished at 673 K. This suggests that the physisorbed and weakly chemisorbed N_2O can be easily desorbed even at low temperature (<473 K) with only stronger absorbed species left. At higher temperatures (>573 K), the disappearance of N_2O infrared bands was mainly caused by the decomposition of remaining chemisorbed N_2O species. According to the in situ N_2O-DRIFTS studies, the first step for the direct decomposition of N_2O on the Fe-beta zeolite was thought to involve a reversible equation:

$$N_2O(g) + \sigma \leftrightarrow \sigma N_2O \quad (1)$$

where σ represented active center of Fe-beta (noted as Fe-Z) after He pretreatment at high temperature. σN_2O delegated the cluster of Fe-beta with adsorbed N_2O (noted as Fe-Z-N_2O).

1.3.2. N_2O-TPD Studies

Figure 4 shows the N_2O-TPD profile for Fe-beta-1% zeolite pretreated by N_2O (4 vol% in He, 20 ml/min) at 573 K. The response signal of N_2O dropped rapidly at about 523 K, simultaneously with the appearance of N_2 peak (at 575 K) and O_2 peak (at 588 K). N_2 could be desorbed more easily (exhibiting as a desorption peak at slightly lower temperature) than O_2. The mass signals of both NO and NO_2 were observed with a desorption peak at 623 K, although the uptake of NO_2 was just minor.

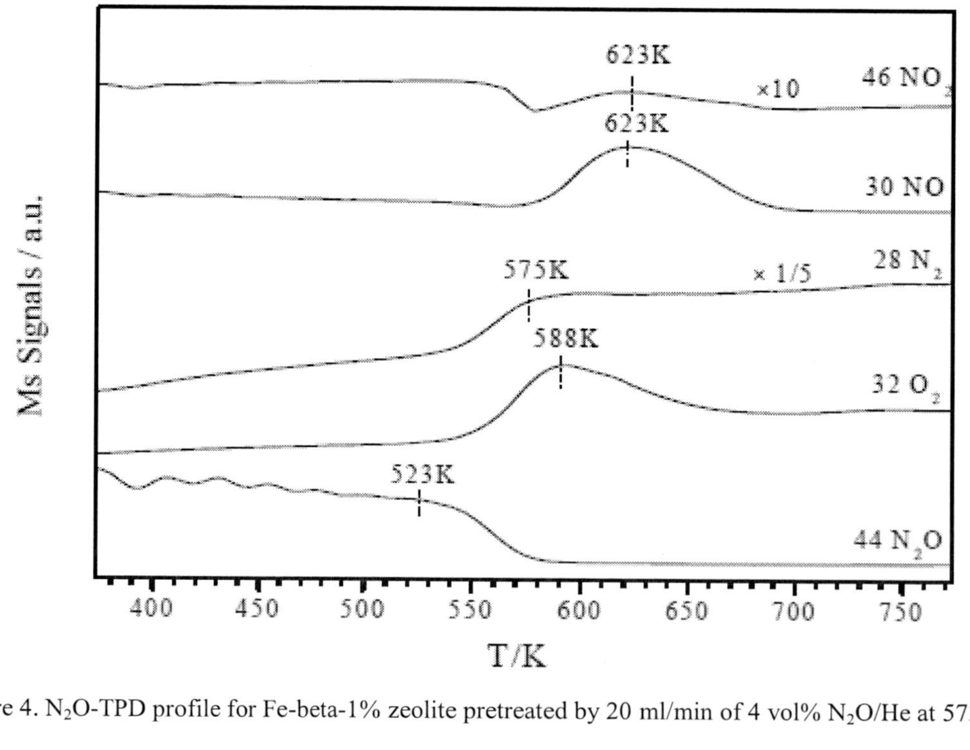

Figure 4. N_2O-TPD profile for Fe-beta-1% zeolite pretreated by 20 ml/min of 4 vol% N_2O/He at 573 K [2].

Based on the DRIFTS and N_2O-TPD results, the generation of N_2 and O_2 was realized via a dissociation of chemisorbed N_2O as formulated by Eqs.(2), (3) and (4).

$$\sigma N_2O \rightarrow N_2(g) + \sigma O \quad (2)$$

$$\sigma O \rightarrow N_2O(g) \rightarrow \sigma O_2 + N_2(g) \quad (3)$$

$$\sigma O_2 \rightarrow \sigma + O_2(g) \quad (4)$$

Eq.(2) formulated the decomposition of the first N_2O over Fe-Z with formation of nitrogen and Fe-O-Z (noted as σO). This oxygen species of Fe-O-Z was the so called α-O. Then, the second N_2O could be decomposed into N_2 and Fe-O_2-Z (noted as σO_2) over Fe-O-Z. It was worth noting that both N_2O participated in above two reactions with O end adsorbing upon the active center Fe of Fe-Z or Fe-O-Z. Finally O_2 was desorbed from Fe-O_2-Z as formulated by Eq.(4).

As shown in Figure 4 the desorption temperature of O_2 (588 K) was higher than that of N_2 (575 K), which revealed that desorption of O_2 would be the rate determining step for N_2O direct decomposition over Fe-beta zeolite.

NO and NO_2 appearing in the N_2O-TPD effluent were believed to be generated through the interaction between N_2O and α-O of Fe-O-Z as illustrated by Eqs.(5), (6) and (7). In Eq.(5) N_2O adsorbed upon the α-O with its N end and formed Fe-Z-$(NO)_2$ (noted as σON–NO) which could be dissociated into NO and Fe-Z-NO (noted as σNO) as shown in Eq.(6). Fe-Z-NO seems to be active enough to capture the O from N_2O to yield Fe-Z-NO_2 (noted as σNO_2) through a disproportionation process (Eq. (7)).

$$N_2O(g) + \sigma O \rightarrow \sigma ON\text{–}NO \quad (5)$$

$$\sigma ON\text{–}NO \rightarrow \sigma NO + NO(g) \quad (6)$$

$$\sigma NO + N_2O(g) \rightarrow \sigma NO_2 + N_2(g) \quad (7)$$

Subsequently, the adsorbed NO_2 was able to be desorbed from Fe-Z-NO_2 as shown in Figure 4 with a small signal of NO_2 and formulated by:

$$\sigma NO_2 \rightarrow NO_2(g) + \sigma \quad (8)$$

1.3.3. DFT Calculations on N_2O Decomposition over 5T-Fe-Beta Zeolite

Quantum calculation of density functional theory (DFT) [16] was used to investigate the mechanism of N_2O decomposition over Fe-beta zeolite that was simulated by a 5T model $[(SiH_3)_4AlO_4Fe]$ during DFT calculations. The N_2O decomposition mechanism simulated over 5T-Fe-beta zeolite model includes three parts and nine reaction steps as listed in Scheme 1. Part 1 formulates a decomposition of the first N_2O molecule over Fe-Z with the formation of N_2 and Fe-O-Z. Part 2 shows the subsequent decomposition of the second N_2O molecule over Fe-O-Z with another N_2 generation and Fe-O_2-Z. Part 3 describes the O_2 desorption from Fe-O_2-Z. The models used in these three parts were illustrated in Figure 5. Energy profiles of Part 1 and Part 2 represent in Figure 6, while the energy profile of Part 3 was not shown for the sake of brevity. For Step 1 with the formation of adsorbed N_2O (Fe-Z-N_2O) on Fe-Z, the adsorption energy was −13.18 kcal/mol and the enthalpy of reaction (ΔH_r) was −12.62 kcal/mol, which revealed that N_2O could be easily adsorbed on Fe-Z without energy barrier and with some heat releasing. Because of the low energy barrier the reverse reaction could also occur at the reaction temperature, which agrees well with the result obtained from the in situ N_2O-DRIFTS. Through the frequency calculation over 5T-Fe-Z-N_2O model, two frequencies associated with ν(N–O) and ν(N–N) were found with values of 1280 and 2240 cm^{-1}, respectively, which were in accordance with the former in situ N_2O-DRIFT study (Figure 3A and B).

Part 1

Fe-Z+N_2O → Fe-Z-N_2O	Step 1
Fe-Z-N_2O → Fe-Z-N_2O-TS	Step 2
Fe-Z-N_2O-TS → Fe-O-Z-N_2	Step 3
Fe-O-Z-N_2 → Fe-O-Z + N_2	Step 4
Fe-Z + N_2O → Fe-O-Z + N_2	Whole Reaction of Part 1

Part 2

Fe-O-Z+N_2O → Fe-O-Z-N_2O	Step 5
Fe-O-Z-N_2O → Fe-O-Z-N_2O-TS	Step 6
Fe-O-Z-N_2O-TS → Fe-O_2-Z-N_2	Step 7
Fe-O_2-Z-N_2 → Fe-O_2-Z + N_2	Step 8
Fe-O-Z + N_2O → Fe-O_2-Z + N_2	Whole Reaction of Part 2

Part 3

Fe-O_2-Z → Fe-Z + O_2	Step 9

Scheme 1. Reaction steps of mechanism with O_2 formation [2].

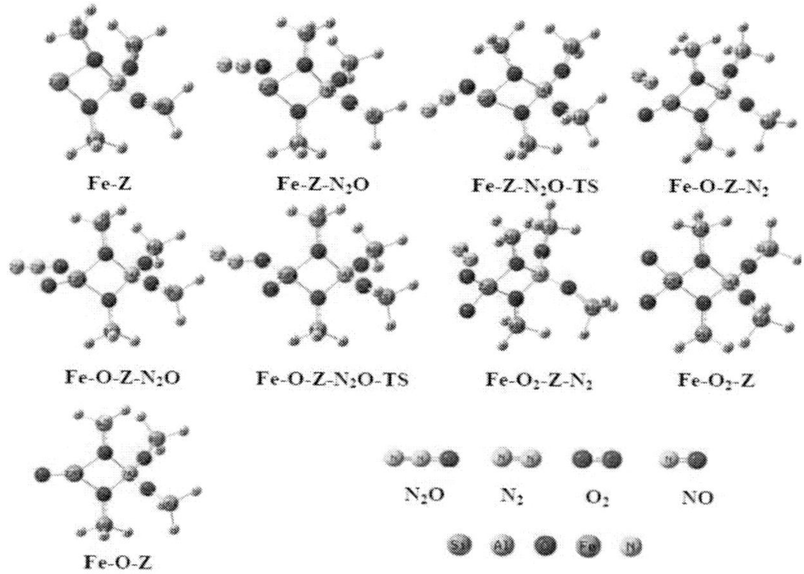

Figure 5. Optimized models used in DFT calculations [2].

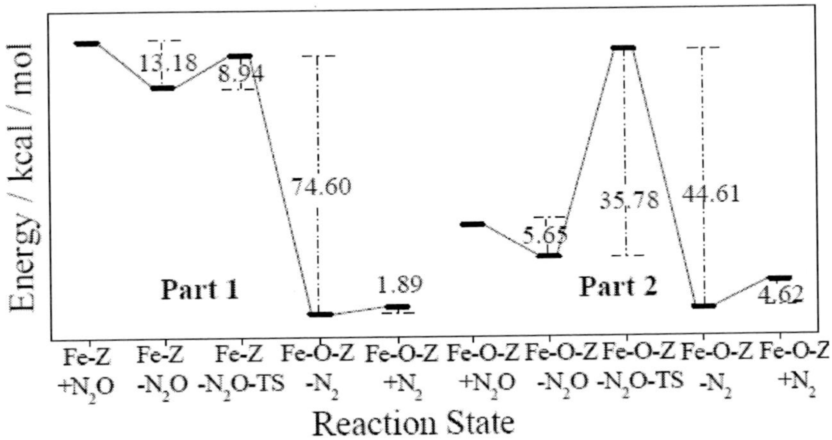

Figure 6. Energy profile for mechanism over 5T-Fe-beta model [2].

Step 2, showing the transform from the adsorbed state (Fe-Z-N_2O) to the transition state (Fe-Z-N_2O-TS), was the main reaction step which determined the activation barrier of Part 1. The transition state Fe-Z-N_2O-TS is characterized by a Fe–O (atom O of N_2O) bond distance of 2.259 Å, O–N′ (the middle N of N_2O) bond distance of 1.226 Å, N′–N″ bond distance of 1.137 Å and the bending of N_2O molecule from 180° in the adsorbed state to 163.6° in the transition state. The imaginary frequency associated with the transition-state mode is 489i cm^{-1}. This transition state was also testified by IRC calculations. As shown in Figure 6 an activation barrier of Step 2 was only 8.94 kcal/mol associated with a ΔH_r of 9.08 kcal/mol, which replied that decomposition of the first N_2O molecule on Fe-Z was endothermic with the easy formation of N_2 and α-O.

Step 3 involves a change from the transition state (Fe-Z-N_2O-TS) to the product state (Fe-O-Z-N_2), with energy barrier of −74.60 kcal/mol and the enthalpy of −73.65 kcal/mol, which revealed that this process released lots of heat with no energy barrier. The desorption of N_2 from Fe-O-Z-N_2 was formulated as Step A4 with a relatively small energy barrier of 1.89 kcal/mol and ΔH_r of 2.81 kcal/mol, suggesting that N_2 could be easily desorbed from Fe-O-Z when a small amount of energy was offered. This is consistent with the observation in N_2-TPD (not shown) that N_2 was able to be desorbed from Fe-beta zeolite at relatively low temperature (<373 K).

Energy calculations for the Overall Reaction of Part 1 were also studied. The total ΔH_r of Part 1 was −74.38 kcal/mol, indicating that the direct decomposition of the first N_2O molecule over Fe-Z was a strongly exothermic process.

Reaction Step 5 described an adsorption of the second N_2O molecule on Fe-O-Z formed in Step 4. The adsorption energy was −5.65 kcal/mol, while the reaction enthalpy ΔH_r was −6.74 kcal/mol. It revealed that no energy barrier existed in this reaction step and N_2O could be easily adsorbed on Fe-O-Z with a small amount of heat release. As noted, the reverse reaction could also occur at reaction temperature.

The transformation from the adsorbed state of Fe-O-Z-N_2O to the transition state of Fe-O-Z-N_2O-TS (formulated as Step 6) was regarded as the main reaction step of Part 2, because the general activation barrier of Part 2 was essentially determined by the activation barrier of 35.78 kcal/mol for Step 6 with a reaction enthalpy (ΔH_r) of 36.31 kcal/mol, implying that this step was endothermic. The transition state Fe-O-Z-N_2O-TS was characterized by a Fe-O′

(atom O of the first N_2O) bond distance of 1.624 Å, Fe-O″ (atom O of the second N_2O) bond distance of 1.839 Å, O–N′ (the middle N of N_2O) bond distance of 1.405 Å, N′–N″ bond distance of 1.159 Å and the bending of N_2O molecule from 180° in the adsorbed state to 129.1° in the transition state. The imaginary frequency associated with the transition-state mode is 800i cm^{-1}. This transition state was also testified by IRC calculations.

Step 7 showed the conversion from the transition state of Fe-O-Z-N_2O-TS to the product state of Fe-O_2-Z-N_2 with an energy barrier of −44.61 kcal/mol and a reaction enthalpy of −44.68 kcal/mol. It was similar to Step 3 of Part 1 that the reaction would quickly occur without an energy barrier and release large amount of heat as soon as the energy barrier of the former reaction step was crossed. In Step 8, N_2 was desorbed from Fe-O_2-Z-N_2 showing energy barrier of 4.62 kcal/mol and ΔH_r of 2.74 kcal/mol, suggesting that N_2 could be easily desorbed from Fe-O_2-Z.

For the Overall Reaction of Part 2 it could be found that the energy difference between initial reactants and final products was 9.86 kcal/mol and the general reaction enthalpy was −12.36 kcal/mol, implying that the second N_2O molecule decomposition over Fe-O-Z in Part 2 was exothermic.

Step 9 of Part 3 illustrates the process of O_2 desorption from Fe-O_2-Z. The energy barrier of this reaction step was as much as 63.20 kcal/mol, while the reaction enthalpy (ΔH_r) was 64.80 kcal/mol. Based on the energy calculations for these nine reaction steps it could be easily found that desorption of O_2 was the rate-limiting step for N_2O decomposition over Fe-beta zeolite. Reaction enthalpy calculations indicated that Part 1 and Part 2, processes of decomposition of N_2O molecule with O_2 formation, were exothermic processes. Part 3 for the desorption of O_2 was however a strongly endothermic process needing a large amount of energy.

Generally, N_2O direct decomposition over Fe-beta zeolite involves the following steps: N_2O was firstly chemisorbed upon the active site of single ionic Fe^{n+} (σ) (n<3, because of He pretreatment at 773K) on Fe-Z and formed Fe-Z-N_2O (σN_2O). This chemisorbed N_2O was subsequently dissociated into N_2 and Fe-O-Z (σO), accompanying with the rise of temperature. The second N_2O adsorbed on the Fe active center through its O end with formation of Fe-O-Z-N_2O (noted as Oσ-ON_2) which could be dissociated into N_2 and Fe-O_2-Z (O_2) as the temperature rose. Then O_2 could be released from Fe-O_2-Z to regenerate Fe-Z (σ) site, which was believe to be the rate-limiting step for N_2O decomposition over Fe-beta zeolite.

1.3.4. CT Analysis Based on Mulliken Population Analysis

As a continuation, the research aims at providing an in-depth insight into the N_2O decomposition mechanism by investigating the CT of the previously proposed reaction steps. The theoretical methods of Mulliken population analysis and FMO, which were established based on DFT, as well as the experimental DRIFTS were employed to investigate the CT during N_2O direct decomposition. During the calculations of Mulliken population the core electrons of Fe atom were represented by the LANL2DZ effective core potential (ECP), and its valence electrons were described by LANL2DZ basis set; while the common basis set of 6-31G(d) was used for atoms of Si, Al, O, N and H. Mulliken charges of the atoms in different models were listed in Table 2, in which the atoms of N_2O were labeled. Zeolite-framework atoms were considered as a whole part, with their Mulliken atomic charges being overlapped and noted as Z. The spin multiplicities (SM) of the modes were determined according to the following strategy. Before the geometry optimization, the single point energy (SPE) was calculated based on the possible SM numbers for each model. However, the right SM number was believed to be merely associated with the lowest SPE.

Table 2. Mulliken charges of the atoms (a.u.) in DFT calculation [17]

	Mulliken atomic charge/a.u.	N_2O+Fe-Z	N_2O-Fe-Z	TS1	N_2-Fe-O-Z	N_2+Fe-O-Z
Part 1	Q(Z)	-0.3383	-0.4132	-0.3549	-0.3538	-0.3438
	Q(Fe)	0.3383	0.3231	0.3485	0.7739	0.8446
	Q(O1)	-0.4990	-0.4416	-0.4244	-0.5150	-0.5008
	Q(N1)	0.6180	0.5852	0.4901	0.0868	0.0000
	Q(N2)	-0.1190	-0.0535	-0.0593	0.0081	0.0000
Part 2	Mulliken atomic charge/a.u.	N_2O+Fe-O-Z	N_2O-Fe-O-Z	TS2	N_2-Fe-O_2-Z	N_2+Fe-O_2-Z
	Q(Z)	-0.3438	-0.3530	-0.2837	-0.2594	-0.2477
	Q(Fe)	0.8446	0.7755	0.9852	0.8606	0.9336
	Q(O1)	-0.5008	-0.5448	-0.4095	-0.3450	-0.3444
	Q(O2)	-0.4990	-0.4712	-0.4931	-0.3482	-0.3415
	Q(N3)	0.6180	0.6192	0.2518	0.0891	0.0000
	Q(N4)	-0.1190	-0.0257	-0.0507	0.0029	0.0000

In order to clearly display the CT routes of each reaction step, atomic charge transfer (ACT) and Mulliken charge ratio (MCR) were cited and formulated as follows:

$$Q(ACT)_M = Q(M)_P - Q(M)_R \tag{9}$$

$$MCR_A = \frac{Q(ACT)_A}{\left|\sum Q(ACT)_A\right|} \times 100\% \tag{10}$$

$$MCR_D = \frac{Q(ACT)_D}{\sum Q(ACT)_D} \times 100\% \tag{11}$$

In Eq.(9) $Q(M)_P$ and $Q(M)_R$ respectively delegate the Mulliken atomic charges of M in a product state and reactant state; $Q(ACT)_M$ was the CT of M in each reaction step. The atomic M acting as a charge acceptor generally associates with a negative value of $Q(ACT)_A$, vice versa for $Q(ACT)_D$. As formulated by Eq.(10) and Eq.(11), which individually described the MCR calculations for the charge acceptor (MCR_A) and charge donor (MCR_D), $\sum Q(ACT)_A$ and $\sum Q(ACT)_D$ are respectively the total CT of all the charge acceptors and donors.

The CT routes of the nine reaction steps (as proposed in Scheme 1) were depicted in Figure 7a, b and c showing the CT routes originating from donor atoms to acceptor atoms and the CT amount represented by MCR of the charge donors.

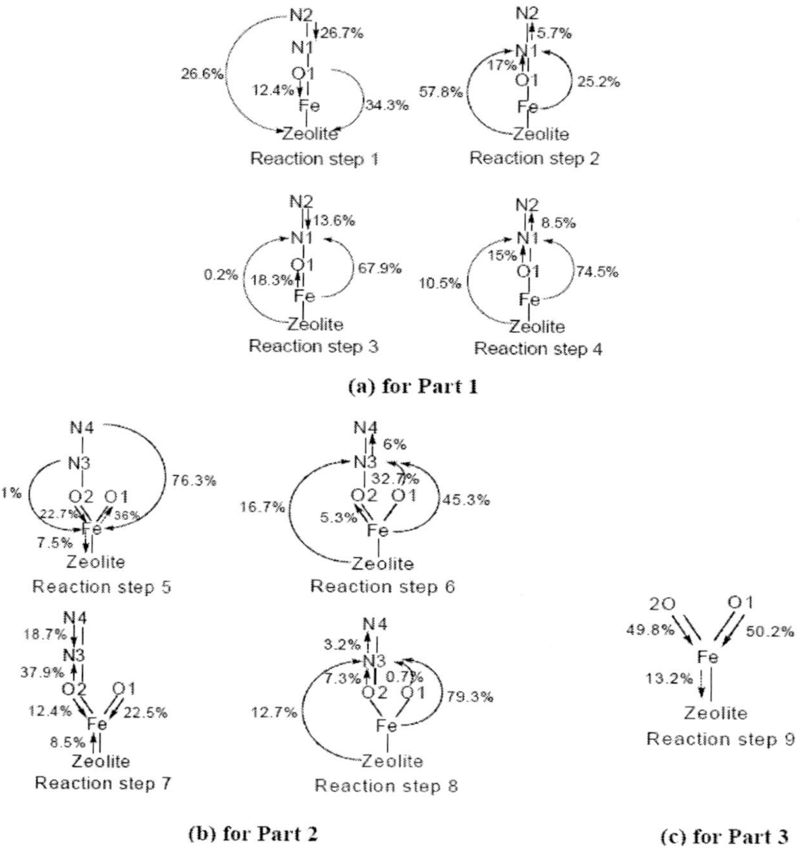

Figure 7. Charge-transfer diagram of each reaction step in Part 1 (a), Part 2 (b), and Part 3 (c) during N_2O direct decomposition over Fe-BEA zeolite based on Mulliken population analysis [17]/

Table 3. Optimized geometrical parameters (distances in Å, angles in o) of models used in DFT calculation [17]

	Model	Fe-O1	O1-N1	N1-N2	-	-	∠O1-N1-N2
Part 1	N_2O-Fe-Z	2.757	1.202	1.131	-	-	180.0
	TS1	2.259	1.226	1.137	-	-	163.6
	N_2-Fe-O-Z	1.671	3.143	1.103	-	-	-
	Fe-O-Z	1.664	-	-	-	-	-
	N2O	-	1.185	1.125	-	-	180.0
	N2	-	-	1.105	-	-	-
Part 2	Model	Fe-O1	Fe-O2	O1-O2	O2-N3	N3-N4	∠O2-N3-N4
	N_2O-Fe-O-Z	1.672	2.307	2.978	1.210	1.125	180.0
	TS2	1.624	1.839	2.770	1.405	1.159	129.1
	N_2-Fe-O2-Z	1.675	1.676	2.307	3.137	1.103	-
	Fe-O2-Z	1.673	1.675	2.311	-	-	-
	O2	-	-	1.216	-	-	-

The value of MCR could clearly display the role of atom playing in the reaction step, being important or normal. Because of the difficulty in judging the charge donors for some atoms, the dotted arrows were employed to show the CT routes; however, it does not significantly influence the CT analysis. On the basis of DFT calculations, the FMO of the adsorption states and transition states (TS) in the Part 1 and Part 2 were investigated.

1.3.4.1. For the Reaction Steps of Part 1

Part 1 including four steps was the decomposition of the first N_2O over Fe-BEA (noted as Fe-Z) (see Scheme 1). Firstly, N_2O adsorbed on the active center Fe through its O end and formed N_2O-Fe-Z complex. The CT routes of this step were profiled in Reaction step 1 of Figure 7a, which revealed that the charges were mainly transferred from N2 (26.6%) and O1 (34.3%) to the zeolite framework with part of charges immigrating from N2 (26.7%) to N1 and from O1 (12.4%) to Fe, respectively. N_2O obviously acted as a charge donor, while zeolite framework is the main charge acceptor. For the active center Fe, it accepted small amount of charges. After this step the charges of zeolite framework became more negative, whereas the charges of active center Fe hardly changed. The similar conclusion was made by Goursot et al.[18] who reported that during adsorption of NH_3 on TMI-zeolites the charges of zeolite framework became more negative, whereas the charges change for the metal is minor. Therefore, it could be found that zeolite framework acting as the main charge acceptor likely played an important role in Reaction step 1. In addition, it was also widely accepted that the structure modification was always attributed to the CT [19]. Seeing the structure parameters of N_2O and N_2O-Fe-Z (Table 3) it could be found that the bond lengths of O1-N1 and N1-N2 in N_2O-Fe-Z were both slightly elongated with respected to gaseous N_2O, and a weak Fe-O1 bond was formed with a bond length of 2.757Å.

The next step, involving a transformation from N_2O-Fe-Z to the first TS (TS1), was the rate-determining step of Part 1. The CT profile of this step (Reaction step 2 in Figure 7a) clearly depicted that almost all the charges were transferred to N1 (the middle N of N_2O). Due to the main CT to the atomic N1, the O-N_2 bond was greatly destabilized associated with the bond angle of N_2O strongly bent from 180° to 163.6° and the O-N_2 bond elongated from 1.202Å to 1.226Å. As noted, zeolite framework serving as the main charge donor played a vital role in this step. Based on CT study, the facilitation of N_2O dissociation by the zeolite framework was verified.

Reaction step 3 shows a pathway from the TS1 to the product state (N_2-Fe-O-Z) with a fracture of O-N_2 bond forming the α-O and N_2. The main CT route for this step was engaged in a transfer from active center Fe to N1 as shown in the Reaction step 3 of Figure 7a. Another part of charges (18.3%) were transferred from active center Fe to O1 beneficial for the formation of α-O, simultaneously with a significant shrinkage of Fe-O1 bond from 2.259Å to 1.671Å. These two kinds of CT greatly favored the dissociation of O-N_2 bond, resulting in the O-N_2 bond intensely elongated from 1.226Å to 3.143Å. Zeolite framework, however, donated little electrons (0.2%) in this step. Reaction step 3 was barrierless, which thanks to the deformation of N_2O in Reaction step 2, favoring the CT from the active center Fe to the N1 of N_2O. In the last step (Reaction step 4) of Part 1, the desorption of N_2 from N_2-Fe-O-Z, the CT routes were nearly the same as those of Reaction step 2, exhibiting that almost all the charges were transferred into N1 atom except that the main charge donor was the active center Fe rather than zeolite framework. Thus, it could be concluded that the active center Fe played a crucial role in Reaction steps 3 and 4 acting as the main charge donor with the electrons mainly transferred to the middle N of N_2O, which greatly facilitated the bond fracture of O-N_2 and the desorption of N_2, respectively.

1.3.4.2. For the Reaction Steps of Part 2

Part 2 described the decomposition of the second N_2O over Fe-O-Z. Similarly, the second N_2O could also be adsorbed upon Fe-O-Z with its O side bound to the active center Fe. The charges were accordingly transferred from the atoms of N_2O to the active center Fe and α-O (O1) with only a small fraction of charges (7.5%) being contributed to the zeolite framework as shown in Reaction step 5 of Figure 7b. After this step the bond length of O-N_2 was slightly elongated from 1.185Å to 1.210Å, and a relatively weaker Fe-O2 bond was formed showing a bond length of 2.307Å. Obviously, the active center Fe and α-O were crucial for this reaction step. Reaction step 6 indicating a change from the adsorbed state to the second TS (TS2) was another important step for N_2O direct decomposition displaying an energy barrier determining that of Part 2. The main CT route of this step was similar to that of Reaction step 2, with the charges being mostly transferred to the middle N of N_2O as shown in Reaction step 6 of Figure 7b. However, in this step the active center Fe and α-O, existing as the main charge donors, dominated the *de*N_2O behaviors. The primary CT to the middle N of N_2O combining with the CT from the active center Fe to O2 (5.3%) coordinately destabilized the O-N_2 bond, which resulted in the bond angle of N_2O strongly bent from 180° to 129.1^o and the bond length of O-N_2 obviously elongated from 1.210Å to 1.405Å. During the O-N_2 bond fracture of the second N_2O, O2 served as the dominant charge donor atom, whereas N3 and Fe acted as the main charge acceptors as shown in Reaction step 7 of Figure 7b. Due to the main CT from O2 to N3 and active center Fe, the O2-N3 bond was remarkably extended from 1.405Å to 3.137Å, and the O2-Fe bond was shrank from 1.839Å to 1.676Å. Therefore, it could be deduced that atomic O2 of the second N_2O made a great contribution to the facture of O-N_2 bond. The CT routes of the last step of Part 2 (Reaction step 8), were similar to those of Reaction step 4 with the charges being mainly transferred from the active center Fe to the middle N of N_2O, in which the active center Fe played the most important role.

1.3.4.3. For the Reaction Step of Part 3

Reaction step 9 illustrated in Part 3 is the last step for N_2O decomposition over Fe-BEA zeolite with the formation of O_2 and the regeneration of Fe-Z, which was believed as a rate-limiting step of N_2O decomposition [8]. The CT routes of this step (Figure 7c) clearly displayed that the charges were mainly transferred from O1 and O2 to Fe with small amount of charges being transferred to the zeolite framework. The catalytic cycle of N_2O direct decomposition over Fe-BEA zeolite was accomplished via this final reaction step.

1.3.4.4. Electronic Properties

Based on above CT analyses the electronic properties of zeolite framework, the active center Fe, α-O and N_2O were further discussed in this section. Firstly, as stated above zeolite framework made a great contribution in Part 1, respectively acting as a main charge acceptor in Reaction step 1 and a main charge donor in Reaction step 2. As shown in Table 2, the Mulliken atomic charges of zeolite framework did not change a lot after undergoing all the steps of Part 1, which indicated a charge reservoir property of the zeolite framework. It accepted large amount of charges from N_2O molecule in the initial step of Part 1 and subsequently donated them to the middle N (N1) of N_2O in the following steps, which was consistent with the theoretical study reported by Zasada et al. [20]. Moreover, Goursot et al. [18] once announced that the framework worked as a charge reservoir during adsorption and desorption of NH_3 on Cu-FAU. The charge reservoir property of zeolite framework during the N_2O direct decomposition may, to a certain degree, explain the diversely catalytic behaviors of zeolites with different topologies. The active center Fe played important roles in the last two steps of Part 1 and became crucial in Part 2 and Part 3. Especially in the Part 3, the rate-determining step for the whole decomposition process, the active center Fe acting as the main charge acceptor favored the releasing of the O_2 molecule and completed the whole catalytic cycle.

α-O was firstly found and named by Panov et al. [21] during the catalytic oxidation of benzene to phenol with N_2O as an oxidant over Fe-ZSM-5. This α-O was confirmed to be generated through N_2O adsorption upon Fe^{2+} ion of the reduced zeolite, in which N_2O is the sole source for α-O [21]. After that, the satisfactory reactivity of α-oxygen in N_2O direct decomposition [7] and organic molecule hydroxylation [22] was pointed out. Although α-O properties have been studied in many works and discussed in reviews [23], some points still subjected to debate, especially for the electronic property. Guesmi et al. [24] recently computed electronic properties, CT and vibrational frequencies of α-O, the isolated and binuclear irons in Fe-ZSM-5 models, showing Fe^{II} for the isolated model and Fe^{IV} for the binuclear one. Sojka et al. [25] reported that the high reactivity of O^- species (α-O) formed due to the dissociation of adsorbed N_2O on the catalyst surface arose from its charge donor property. In this study, the α-O reactivity could also be verified through analyzing its electronic property based on CT analyses. During the TS2 process (Reaction step 6) the α-O acting as another main charge donor donated the charges (32.7%) to N3 (the middle N of N_2O), which greatly favored the bond fracture of O-N_2. Therefore, the charge donor property of the α-O in Reaction step 6 gave a direct evidence of the high activity of α-O for N_2O direct decomposition. Differently, α-O could also attract charges from N_2O being another main charge acceptor (MCR of 36%) as for Reaction step 5, which was favorable for the adsorption of N_2O on Fe-O-Z. As previously reported [26] the adsorption energy of N_2O on TMI-zeolite (M=Fe, Co, Rh) seriously affected the energy barrier of the successive reaction step (from the adsorbed state to the TS): the higher the adsorption energy was, the lower the activation energy needed. Thus, the charge acceptor property of α-O during adsorption of the second N_2O facilitated the N_2O activation, which could also give us another clue of the high activity of α-O.

Additionally, similar behaviors were also observed in our early study on O_2-TPD over $LaCo_{1-x}Cu_xO_3$ perovskites [27] that the presence of α-O facilitated the adsorption/desorption of adjacently molecular oxygen upon its neighbor vacancies. It was also interestingly found that the N_2O acted as the charge donor molecule in the first step of both Part 1 and Part 2. Contrarily, during the decomposition steps of these two parts the middle N atom of N_2O worked as the main charge acceptor.

As listed in Table 1 the charge distributions of the corresponding N atoms of the first and the second N_2O (the middle N atoms N1 and N3 or the terminal N atoms N2 and N4) along with reaction steps were nearly the same, which indicated the similar electronic properties of the N atoms in these two N_2O molecules during direct decomposition of N_2O over Fe-BEA zeolite. The charge distributions of N atoms were also profiled in Figure 8 for a clear description.

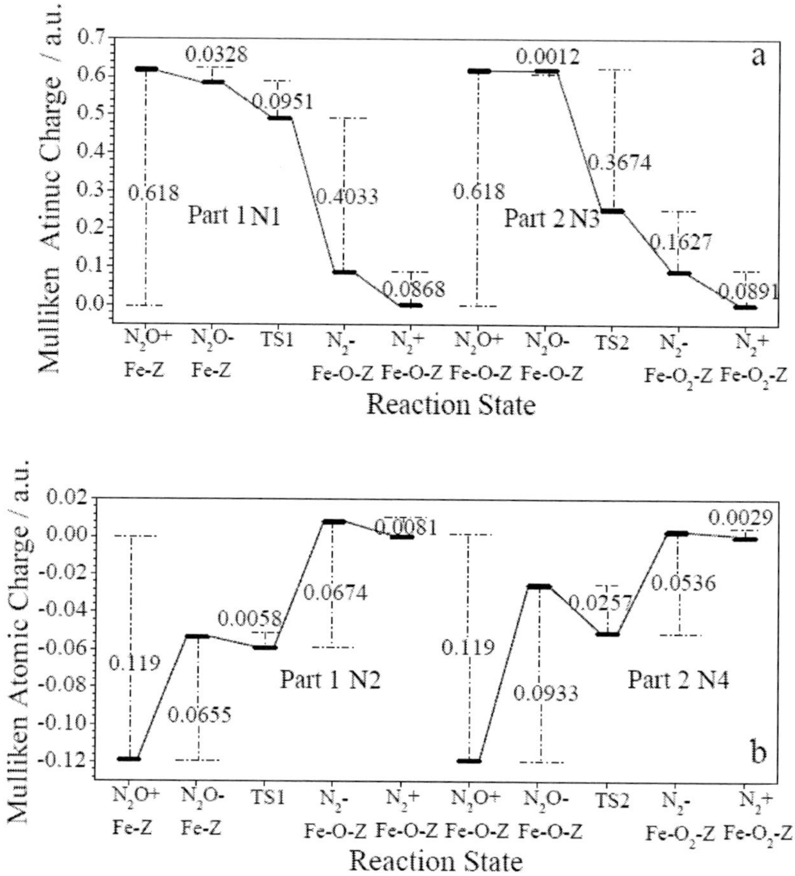

Figure 8. Mulliken atomic charge distributions of N atoms in N_2O along with reaction steps of Part 1 and Part 2: (a) charge distributions of the middle N atoms of the firstly and the secondly adsorbed N_2O molecules; (b) charge distributions of the terminal N atoms of the firstly and the secondly adsorbed N_2O molecules [17].

1.3.5. FMO Analysis

In this section, FMO analysis was applied to investigate the CT during the adsorption states and the TS involved in Part 1 and Part 2. The corresponding FMO diagrams were profiled in Figure 9a-9d.

As reported in the literature [28], the N_2O HOMO is a π-orbital associated with the O-N antibond and N-N bond, whereas the LUMO of N_2O is a π*-orbital with the antibonding character for both O-N bond and N-N bond. In Figure 9a, a CT occurring at the initial stage of the reaction (Reaction step 1) from the in-plan HOMO of N_2O to the $3d_z^2$-orbital of Fe by a π-bond, suggesting that the charges would be transferred from the N_2O molecule to the active center Fe during the adsorption of N_2O on Fe-Z. As the reaction further proceeded (Reaction step 2), the in-plane LUMO of N_2O (3π) started to populate and strongly interacted with 4s-orbital of the active center Fe by σ-bond (shown in Figure 9b), which provided a pathway for the CT from the active center Fe to the N_2O molecule and led to a severe bending of the transient N_2O^- intermediate. Thereafter, the Fe-O-Z and N_2 were generated through an asymmetric stretching oscillation.

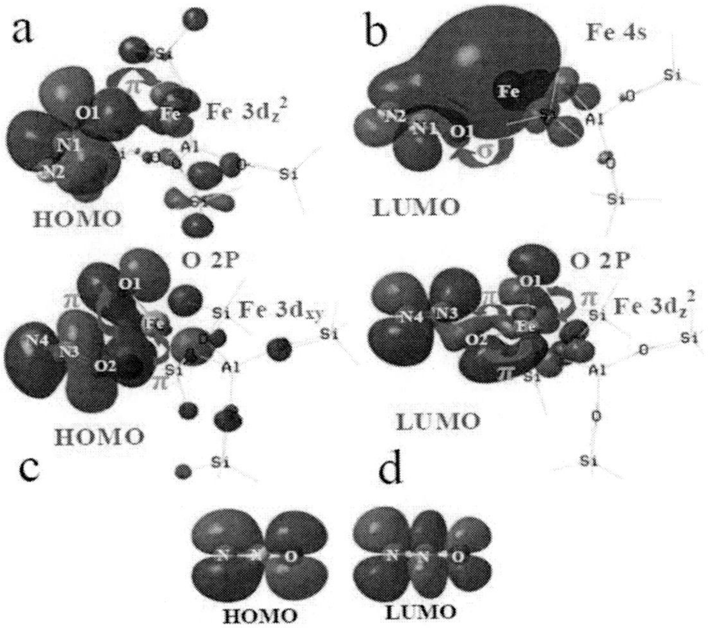

Figure 9. Diagrams of frontier molecular orbital analysis: (a) π-interaction between the N_2O in-plane HOMO and the Fe $3d_z^2$ in the first adsorbed state (N_2O-Fe-Z); (b) σ-interaction between Fe 4s and N_2O in-plane LUMO in the first transition state (TS1); (c) π-interactions of N_2O in-plane HOMO with Fe $3d_{xy}$ and O 2p in the second adsorbed state (N_2O-Fe-O-Z); and (d) π-interactions of Fe $3d_z^2$, O 2p and the N_2O in-plane LUMO in the second transition state (TS2) [17].

During the first step of Part 2 (Reaction step 5) the N_2O HOMO interacted with the orbitals of Fe $3d_{xy}$ and O 2p of α-O by the π-bond as shown in Figure 9c, giving a clue that the second N_2O molecule is a charge donor and adsorbed on Fe-O-Z with Fe and α-O being the charge acceptors. At the TS2 stage the strong interactions of the orbitals of Fe $3d_z^2$, O 2p (α-O) and the N_2O LUMO by the π-bond were observed in Figure 9d, indicating that the charges would be transferred from the active center Fe and the α-O to molecular N_2O. Generally speaking, the CT data referred from FMO analyses are in a good agreement with the results obtained from the CT analyses in the previous section (section 1.3.4).

1.3.6. DRIFTS Study

An in-situ DRIFTS system was employed in this section for the CT analysis. As shown in Figure 10, the vibration region of (2105-2280 cm^{-1}) was profiled after N_2O adsorption on Fe-Z at 573 K and then desorbed at different temperatures from 323 K to 573 K. The bands appearing in above wavenumber range was generally attributed to the asymmetric stretching motion [ν_{as}(N-N)] of N_2O according to literature assignment [29]. The band at 2235 cm^{-1} in Figure 4 was accordingly assigned to the ν(N-N) of the chemisorbed N_2O on Fe [12] as well as the bands at 2197-2218 cm^{-1} were ascribed to the weakly chemisorbed or some physisorbed N_2O [2]. In comparison with the literaturely verified vibrations of the gas phase N_2O [2224 cm^{-1} for ν(N-N)] [29], both our previous DFT frequency calculation for N_2O-Fe-Z (2240 cm^{-1}) [2] and the present low temperature (323 K) DRIFTS result (2235 cm^{-1}) showed a blue shift of ν(N-N) (2224 cm^{-1} → 2240 or 2235 cm^{-1}).

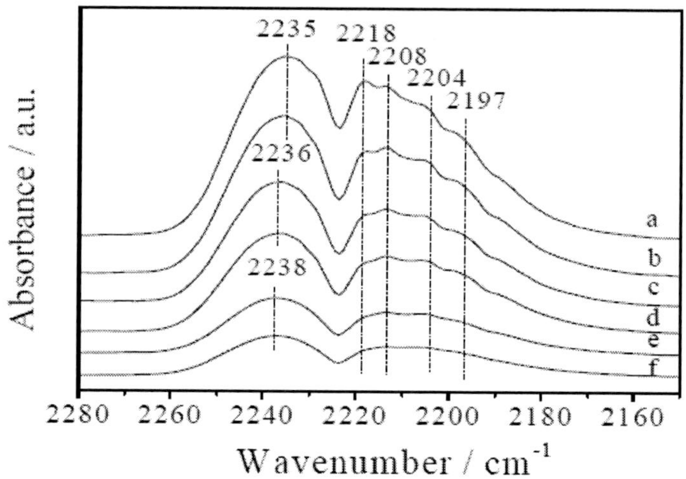

Figure 10. IR spectra of N_2O adsorbed on Fe-BEA-1% at 573K and desorbed at: (a) 323 K, (b) 373 K, (c) 423 K, (d) 473 K, (e) 523 K, (f) 573 K.[17].

It was recently announced that the CT between TMI-zeolite and adsorbed molecule could influence the vibration frequency of the adsorbate [19], resulting in the corresponding band shifts. Therefore, an attempt to correlate the present DRIFTS results with the CT analyses of section 1.3.4 has been done. Firstly, as stated above both the theoretical and experimental results indicated a blueshift of ν(N-N) as compared to gaseous N_2O during the adsorption of N_2O on Fe-Z. The band shifts of this process can be correlated with the CT of Reaction step 1 (N_2O + Fe-Z → N_2O-Fe-Z), in which the terminal N of N_2O molecule donated chargers to the middle N of N_2O, enhancing the N-N bond with the stretching vibration of ν(N-N) being increased and leading to a blue shift of ν(N-N). Accompanying with the increase of temperature from 323 to 573 K during the thermal desorption of N_2O-Fe-Z, the ν(N-N) shifted to the higher frequencies (2235 → 2238 cm^{-1}) (blue shift, Figure 10). The adsorbed complex (N_2O-Fe-Z) was known to be easily excited to the TS1, which was corresponding to the Reaction step 2 (N_2O-Fe-Z → TS1) with a small energy barrier of 8.94 kcal/mol [2].

As stated in the section 1.3.4 for Reaction step 2, the charges were mainly transferred to the middle N atom of N_2O together with small amount of charges immigrating to the end N atom. Due to these CT the ν(N-N) was increased, which was in accordance with the observation of blue shift of ν(N-N) in Figure 10. Generally speaking, the elevated temperature was able to promote CT, leading to an enhancement in the IR shifts. Finally, it could be concluded that the CT analyses of section 1.3.4 are totally consistent with the experimental DRIFTS data.

2. The Reaction of No + C_3H_6 + O_2 over the Mesoporous SBA-15 Supported Transition Metal Catalysts

The use of hydrocarbons as reducing agents (commonly with propene as a model delegate) for the catalytic reduction of hazardous nitrogen oxides, which are emitting from the automobile engines and can cause acid rain and photochemical smog, currently attracts much attention. Great effort has been made in order to improve the three-way performances of catalytic converters aiming at satisfying the requirement of increasingly strict regulations for environmental protection.

Noble metal-supported oxides have already been commercialized as three-way catalysts; however, the scarcity and relatively high price of noble metals (e.g. Pt, Rh, Pd, etc.) limit their practical application. Attempts were subsequently made to use molecular sieves mainly focusing on ZSM-5, Y, and Beta-type zeolites doped with cheap transition metals for the NO reduction [30-32]. Unfortunately, the pore blockage of these microporous zeolites due to the deposition of carbonaceous and sulfurous species in the exhaust effluent represents a fatal problem.

Mesostructured zeolites constitute a new state of association of molecular sieves having regularly spatial arrangement of pores with monodimensional size distribution tunable in the range of 2–50 nm. These nanoscaled materials promise spectacular improvement of catalytic activity in many reactions owing to their high specific surface areas and precise control of their surface chemistry. Their relatively larger pore sizes as compared to those of microporous zeolites facilitate the transportation of reactants and might lead to a better resistance to SO_2 and steam poisoning as well as deactivation caused by pore blockage.

Among the mesoporous zeolites, newly achieved SBA-15 exhibits excellent thermal stability and was thought as promising candidate acting as a support of some industrial catalysts [33]. Nevertheless, reports related to the utilization of mesostructured SBA-15 for the catalytic reduction of NO with C_3H_6 are so far scarce in the literature.

A series of SBA-15-based catalysts including parent SBA-15, mono-functional M (M = Cu, Cr, Fe, and Al)/SBA-15, and bifunctional M′ (M′ = Cu–Al, Cr–Al, and Fe–Al)/SBA-15 was accordingly prepared and characterized by N_2 adsorption, XRD, TEM, H_2-TPR, nuclear magnetic resonance (NMR), X-ray photoelectron spectroscopy (XPS) as well as the activity tests for catalytic reduction of NO with C_3H_6 in the presence of O_2 [34]. Emphasis will be placed on clarifying the relationship between the physicochemical properties of prepared catalysts and their depollution performances.

2.1. Structural Characterization

Figure 11 shows the small-angle XRD patterns of as-synthesized SBA-15-based samples without and with metal (Fe, Cu, Cr, and/or Al) doping. Three diffraction peaks at $2\theta = 0.94$, 1.6, and 1.8° were clearly discernable from the corresponding profile for the parent SBA-15 in Figure 11 which could be respectively indexed as (1 0 0), (1 1 0) and (2 0 0) *h k l* reflections of *P6mm* symmetry having a d-spacing of the (1 0 0) plane of 9.4 nm and cell parameter of $a_0 = 10.8$ nm, as listed in Table 4.

The formation of typical SBA-15 silica with well-ordered and hexagonally structured mesoporous channels was thus verified, according to the earlier study [33]. With the introduction of various metals (Fe, Cu, Cr and Al), this mesoporous structure of SBA-15 could be basically retained showing those characteristic peaks for (1 0 0), (1 1 0) and (2 0 0) reflections in the case of all investigated samples.

However, a drop in intensity of the main diffraction peaks in the small-angle region occurred along with Fe–Al, Cr–Al, Cu–Al, and Cr addition, reflecting the appearance of some deterioration of the ordered pore structures. This decline in the intensity of XRD peaks at around 0.94° became serious for the Cu/SBA-15, Fe/SBA-15, and Al/SBA-15 samples.

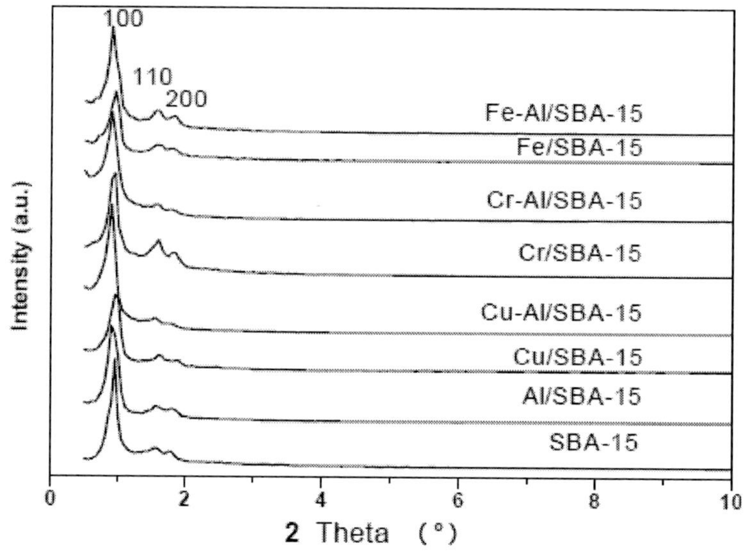

Figure 11. Small-angle XRD patterns of SBA-15, (Al, Cu, Cr, Fe)/SBA-15, and (Cu, Cr, Fe)-Al/SBA-15 [34].

Table 4. Characteristics of the mesoporous SBA-15-based catalysts [34]

Sample	BET area (m^2/g)	Pore volume (cm^3/g)			Pore diameter (nm)	d_{100} (nm)	Cell parameter a_0 (nm)	Wall thickness (nm)
		Micropore	Mesopore	Total				
SBA-15	645	0.10	0.71	0.81	8.0	9.4	10.8	2.8
Al/SBA-15	403	0.18	0.55	0.73	7.6	9.8	11.3	3.7
Cu/SBA-15	593	0.13	0.71	0.84	7.4	9.3	10.8	3.4
Cu-Al/SBA-15	441	0.07	0.53	0.60	6.2	9.8	11.3	5.1
Cr/SBA-15	568	0.13	0.48	0.61	6.8	9.3	10.8	4.0
Cr-Al/SBA-15	575	0.08	0.61	0.69	6.9	9.7	11.2	4.3
Fe/SBA-15	614	0.11	0.74	0.85	7.5	9.3	10.8	3.3
Fe-Al/SBA-15	513	0.19	0.61	0.80	5.2	9.7	11.2	6.0

As inferred from Table 4 data, the d-spacing of the (1 0 0) plane is almost constant accompanying with the introduction of Cu, Cr, and Fe components. It is therefore believed that these metal species were finely distributed and mainly confined in the mesoporous channel as the extra-framework species in the forms of either cations or their oxides without obviously modifying the structural parameters of SBA-15. In contrast, an obvious shift of the main diffraction peak to the lower angle was observed along with aluminum addition, hinting that the incorporation of Al into the SBA-15 skeleton. The relatively longer Al–O bond length as compared to Si–O bond was thought to be the reason causing a downward shift of the small-angle diffraction peak.

The textural properties of the as-synthesized catalysts were studied by N_2 adsorption representing a typical IV isotherm and a clear H1-type hysteresis loop (not shown), in accordance with the characteristic feature of mesoporous materials. BET specific surface areas and pore structures of as-synthesized SBA-15-based catalysts are also shown in Table 4. For the parent SBA-15, its BET surface area and total pore volume (V_t) were 645 m^2/g and 0.81 cm^3/g, respectively. It was noted that the surface area of SBA-15 support was reduced to some degree along with the active metals doping (especially for the Al-containing samples), following a decreasing order of SBA-15 > Fe/SBA-15 > Cu/SBA-15 > Cr-Al/SBA-15 ≈ Cr/SBA-15 > Fe-Al/SBA-15 > Cu-Al/SBA-15 > Al-SBA-15. The surface areas of Cu/SBA-15 and Fe/SBA-15 were comparable to that of bare SBA-15, indicating that the Cu and Fe components were well dispersed over the mesostructured silica without an obvious agglomeration.

In the other cases, the loss of surface areas of (Cr, Cr-Al, Fe-Al, and Cu-Al)/SBA-15 is associated with the diminution of mesopore volumes likely caused by some pore blockage (Table 4). Pore diameters determined by the BJH method fall into the mesoporous range of 5–8 nm. It has been observed that a drop in the pore size and a rise in the wall thickness occur simultaneously with the addition of active components, indicating that some metal species were confined inside the SBA-15 channels.

In order to clarify the pore architecture of catalysts, TEM measurements have been carried out for the bifunctional Cu–Al/SBA-15 sample with the image of (1 0 0) and (1 1 0) planes, being respectively shown in Figure 12a and b, exhibiting the orderly 2D hexagonal arrays with uniform pore sizes of about 6.3 nm and the parallel-arranged channels having wall thickness of about 4.9 nm, in good agreement with the corresponding values determined by N_2 adsorption (see Table 4). The long-range mesoporous-ordering characteristic reflects a good-quality of this SBA-15 structured material, without being remarkably disturbed due to the Al and Cu introduction.

Moreover, inside the mesopore channel metal clusters were hardly observed, suggesting these metals were mainly distributed over the mesopore intra-surface and even incorporated into the framework of SBA-15. In addition, a minor impurity phase other than that of SBA-15 was found and is speculated to be amorphous alumina.

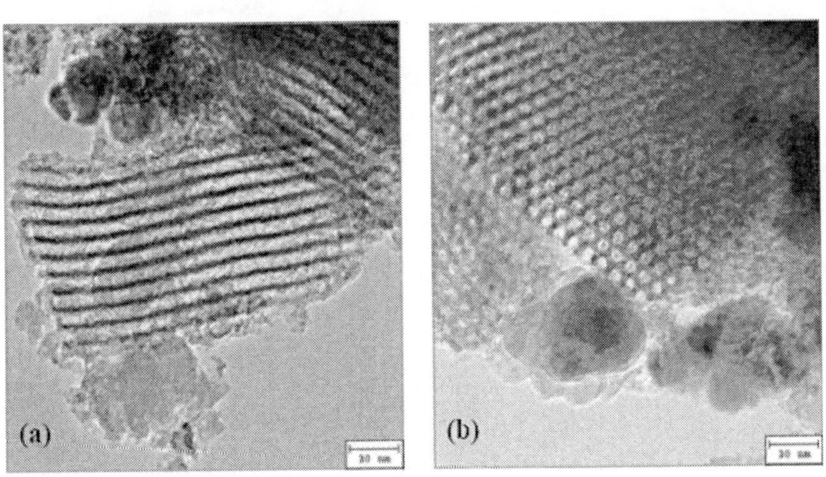

Figure 12. TEM images of Cu–Al/SBA-15 sample from (a) (1 0 0) and (b) (1 1 0) reflections [34].

2.2. Redox Properties

The reducibility of catalyst plays an important role in various reactions involving a redox mechanism. Figure 13 represents the H_2-TPR profiles of SBA-15-based catalysts. The high stability of SBA-15 structure was confirmed due to the absence of a reduction peak in the investigated temperature region under a 5% H_2/Ar atmosphere. Since SBA-15 is nonreducible under the conditions for H_2-TPR, the slight H_2 consumption peak appearing at 621 °C in the TPR profile of Al/SBA-15 was accordingly ascribed to Al^{3+} reduction. A multiple-step reduction was observed over Fe/SBA-15, showing a peak centered at 464 °C followed by a broad one at >794 °C. The former was attributed to the Fe^{3+} to Fe^{2+} reduction, the latter to the partial reduction of Fe^{2+} to metallic iron according to our previous investigation [35]. Two peaks at 399 and 538 °C were discerned from the TPR profile of Fe–Al/SBA-15, corresponding to the reduction of Fe^{3+} cations and iron oxides, respectively. The FeO_x reduction occurring in a similar temperature region was previously reported [10]. In addition, the peak appearing on the TPR plot of Fe/SBA-15 for $Fe^{2+} \rightarrow Fe^0$ reduction totally vanished in the TPR plot of Fe–Al/SBA-15. As for the Cr/SBA-15 sample, the corresponding TPR profile showed an intense peak at 509 °C and a shoulder at its rising part (443 °C). These reducing peaks were believed to be related to the partial reduction of Cr^{6+} ions and Cr_2O_3. It seems that Al incorporation is beneficial for Cr_2O_3 formation, displaying three overlapping peaks at 329, 437 and 543 °C.

The latter one was again attributed to the $Cr^{6+} \rightarrow Cr^{3+}$ reduction, while the former two were likely caused by the reduction of Cr_2O_3 with different crystalline domains. It is believed that the bigger metal oxide is more readily reduced than the smaller one [36]. The sharp peak at 288 °C observed with the Cu/SBA-15 sample was ascribed to the reduction of Cu^{2+} cations, as was done for Cu-ZSM-5 [37] and $LaFe_{0.8}Cu_{0.2}O_3$ [35]. It was noted that this copper reduction peak was shifted to lower temperature (247 °C) along with the Al incorporation, implying a better redox properties of Cu–Al/SBA-15.

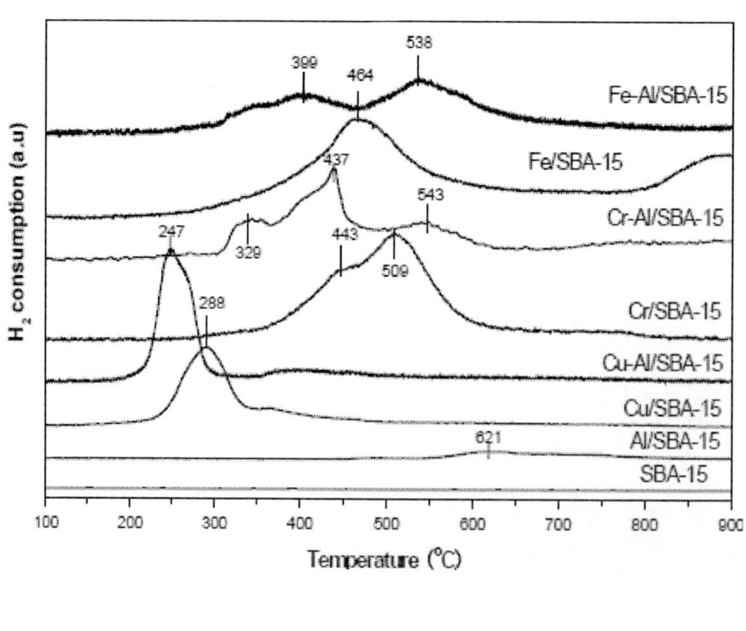

Figure 13. H_2-TPR profiles of SBA-15, (Al, Cu, Cr, Fe)/SBA-15 and (Cu, Cr, Fe)-Al/SBA-15 [34].

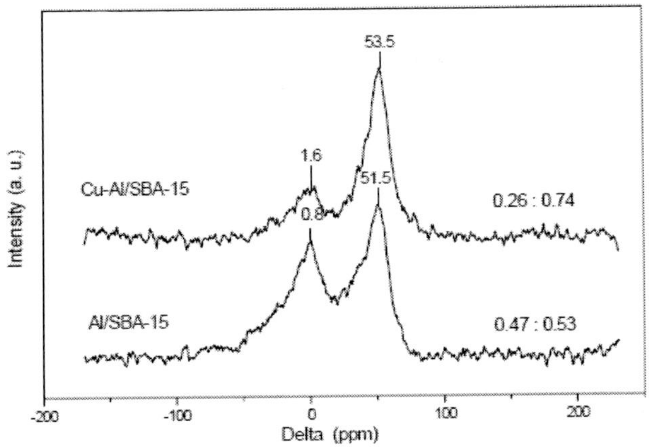

Figure 14. ^{27}Al MAS NMR of Al/SBA-15 and Cu–Al/SBA-15 [34].

2.3. Al Status Characterization

The type of Al coordination in both Al/SBA-15 and Cu–Al/SBA-15 was probed using ^{27}Al MAS NMR with spectra being depicted in Figure 14. The Al/SBA-15 showed a NMR peak at ca. 54 ppm attributable to the tetrahedral framework aluminum (Al^{IV}) intra-wall formed along with another resonance peak at around 0 ppm belonging to the octahedral coordinated aluminum (Al^{VI}) outside the SBA-15 skeleton, according to the literature assignment [37]. The ratio of Al(tetrahedral)/Al(octahedral) is 1.13 in the case of Al/SBA-15, suggesting that almost half of the aluminum in this sample existed in a tetrahedral state inside the SBA-15 wall while the other half likely remained mainly as alumina.

These extra-framework aluminas might be the cause of a serious drop in the corresponding surface areas (see Table 4). Interestingly, the NMR peak at 0 ppm (associated with octahedral aluminum) for Cu–Al/SBA-15 was totally suppressed and the opposite occurred in the resonance peak at ~54 ppm (assigned to tetrahedral aluminum). This strongly suggested that Al nuclei have been mainly incorporated into the framework with the assistance of copper loading.

2.4. Analysis of Surface Metallic Components

For the sake of getting insight into the surface composition and clarifying the nature of active sites, XPS was employed in this study as shown in Figure 15a–d. As regards the Cu 2p, the main peaks observed in Figure 15a were at BE ($2p_{1/2}$) = 953.1 eV and BE ($2p_{3/2}$) = 933.0 eV with its shake-up satellite. The valence of (divalent and/or reduced) copper species can be further differentiated according to their characteristic binding energies because only the Cu^{2+} species ($3d^9$) shows a shake-up satellite peak located about 10 eV higher than the Cu 2p3/2 transition. As verified by Figure 15a, Cu^{2+} species in both the bulk CuO and those dispersed as metallic ions were observed for Cu/SBA-15 sample displaying BE satellite peaks at 941.3 and 945.0 eV in accordance with the literature report [38]. In contrast, only the highly isolated Cu^{2+} ions were found for Cu–Al/SBA-15 sample.

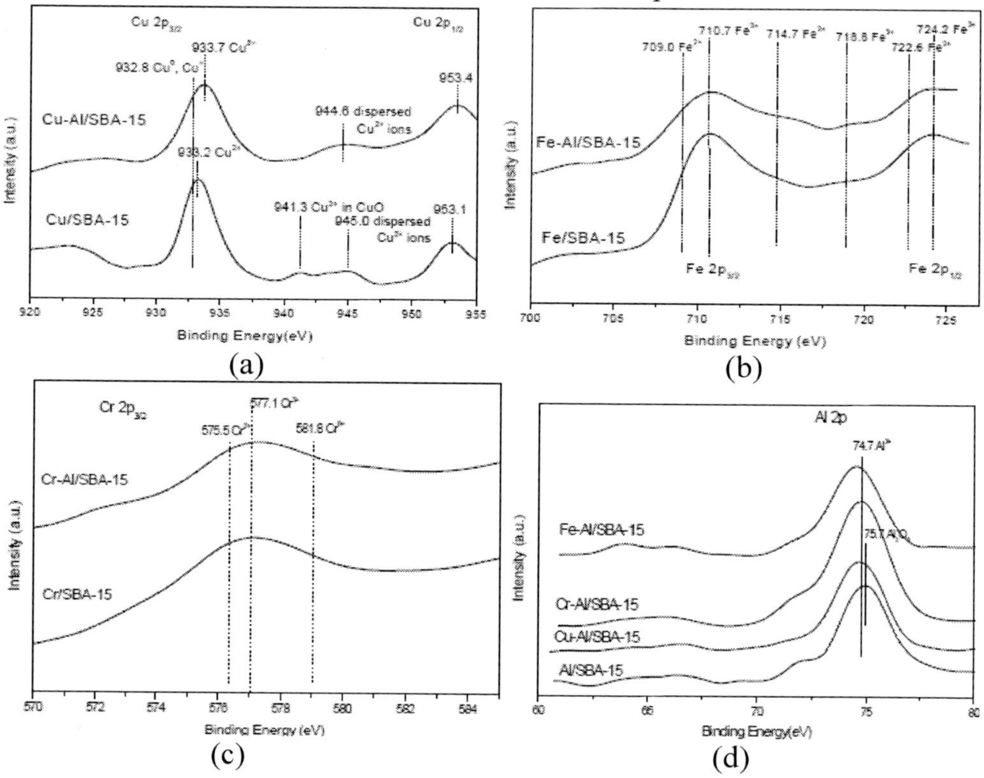

Figure 15. XPS spectra of (a) Cu 2p for Cu/SBA-15 and Cu–Al/SBA-15; (b) Fe 2p for Fe/SBA-15 and Fe–Al/SBA-15; (c) Cr 2p for Cr/SBA-15 and Cr–Al/SBA-15; (d) Al 2p for Al/SBA-15 and (Fe, Cr, Cu)-Al/SBA-15 [34].

The comparison of Fe 2p3/2 binding energy appearing in Figure 15b with the reference data [39] allows us to assign the Fe 2p3/2 line at 710.7 eV and its satellite at 718.8 eV to the Fe^{3+} ions. Thereafter, the shoulder line at 709.0 eV and its satellite at 714.7 eV were attributed to the partially reduced Fe^{2+} ions.

It is clearly seen from Figure 15b that the surface iron species of the Fe-loaded samples mainly existed in the form of Fe^{3+} species together with minor Fe^{2+} ones. In the light of the literature assignment [37], the Cr $2p_{3/2}$ signal approximately at BE = 577.1 eV was assigned to Cr^{3+} and its shoulder at BE = 575.5 eV was ascribed to Cr^{2+} (Figure 15c). X-ray photoelectron bands appearing at BE = 579 eV confirmed the presence of Cr^{6+} species [37].

It is clear that most of the chromium existed as trivalent cations, co-present with a small amount of Cr^{6+} and Cr^{2+} cations.

Three binding energies of Al 2p with values of 73.0, 74.5, and 75.4 eV were observed by Pashutski et al. [40] and thereafter assigned to pure metallic Al, AlO_x in zeolite framework, and anhydrous Al_2O_3, respectively. XPS results in Figure 15d suggest that most of the aluminum was successfully incorporated into the SBA-15 framework for all the Al-containing samples.

The slight shift of the Al 2p peak to lower binding energy in the case of Cu–Al/SBA-15 with respect to Al/SBA-15 is likely due to the diminution of Al_2O_3 phase formation.

2.5. Activity Test

The temperature dependence of NO and C_3H_6 conversions in the $C_3H_6 + NO + O_2$ reaction over the tested catalysts is shown in (Figure 16) and (Figure 17). NO conversion over the parent SBA-15 is marginal (below 5% at T < 600 °C).

Nevertheless, the conversion was significantly improved by the introduction of active metallic components (Figure 16a). Among monofunctional SBA-15 samples, a progressively increased NO conversion upon the rising temperature was found for Cr/SBA-15 catalyst.

Moreover, highly dispersed copper and iron could lead to a further enhancement giving NO conversions of 83% and 90% at 600 °C, respectively. The promotion of C_3H_6 conversion of SBA-15 was also verified via metal dopings (Figure 16b).

Al insertion in SBA-15 resulted in a moderated development of NO conversion (from 6% at 200 °C to 48% at 600 °C) with respect to the bare SBA-15 as shown in Figure 17a. With a comparison between Figure 16 and 17, it was found that the Al doping played a distinct role on affecting three-way performances of co-loaded (Cu, Cr, and Fe) metallic components.

A synergetic effect was achieved between copper and aluminum, resulting in an outstanding NO conversion over Cu–Al/SBA-15 starting at 200 °C and reaching almost 100% at a temperature as low as 450 °C together with 85% C_3H_6 conversion. Cr–Al/SBA-15, however, shows a NO conversion curve similar to that of Cr/SBA-15 except for a moderate promotion at temperatures <400 °C.

Conversely, the Al loading led to an obvious inhibition in NO reduction for Fe/SBA-15, yielding values below 25% at the tested temperature range for Fe–Al/SBA-15 sample. C_3H_6 conversions in Figure 17b roughly followed a decreasing order of Cu–Al/SBA-15 > Cr–Al/SBA-15 > Fe–Al/SBA-15 > Al/SBA-15.

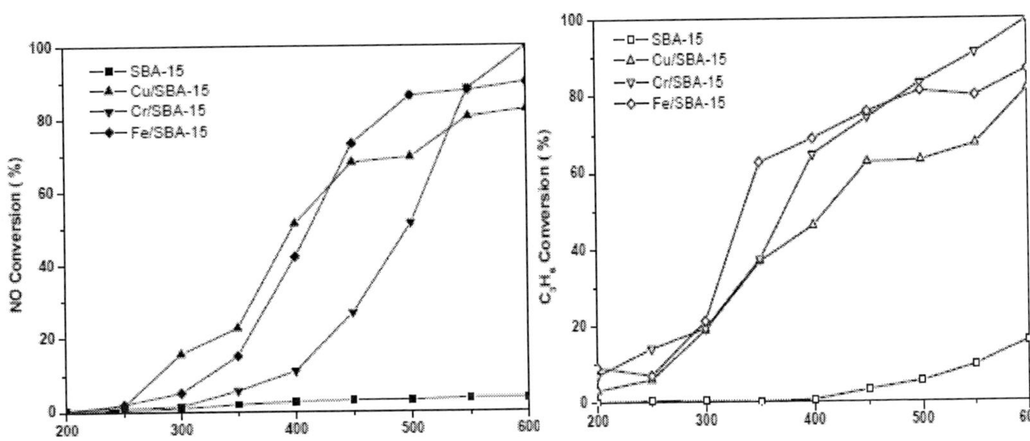

Figure 16. (a) NO conversion and (b) C_3H_6 conversion in $C_3H_6 + NO + O_2$ reaction over SBA-15 and (Cu, Cr, Fe)/SBA-15, Conditions: 3000 ppm NO, 3000 ppm C_3H_6, 1% O_2, 60,000 h^{-1} [34].

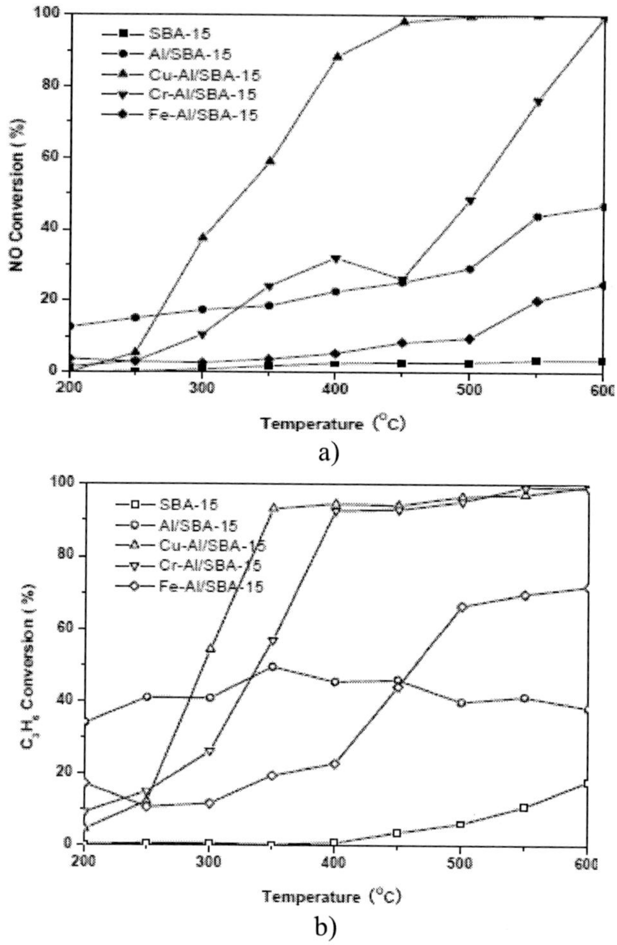

Figure 17. (a) NO conversion and (b) C_3H_6 conversion in C_3H_6 + NO + O_2 reaction over Al-SBA-15 and (Cu, Cr, Fe)-Al/SBA-15, Conditions: 3000 ppm NO, 3000 ppm C_3H_6, 1% O_2, 60,000 h^{-1} [34].

2.6. Correlation between Physicochemical Properties and Catalytic Performance

Based on the mechanism established in the literature [35, 41], an attempt to correlate the catalytic performances of metal-promoted SBA-15 catalysts in NO catalytic reduction with the local structure of metal cluster and SBA-15 support has been made. Since the parent SBA-15 was proven to be inactive for catalytic reduction of NO with C_3H_6 in the presence of oxygen, the metallic components loaded were believed to act as the active sites, resulting in an obvious enhancement in NO conversion (Figure 16a).

The role of SBA-15 is accordingly supposed to be beneficial for the metals dispersion. In this study, the successful formation of SBA-15 mesoporous structure processing long-distance ordering and satisfactory surface areas (around 400–650 m^2/g) for the prepared materials, whose structural characteristics remained even after metal doping and further calcination at 500 °C.

The characterization of local structure and coordination environment of active sites was further conducted by XRD (Figure 11), TPR (Figure 13), NMR (Figure 14) and XPS (Figure 15a–e). It was found that metals of Cu, Fe, and Cr were mainly distributed over the SBA-15 substrate as isolated metallic cations although a minor CuO phase in Cu/SBA-15 is evidenced by XPS pattern in Figure 15a and iron oxides in Fe–Al/SBA-15 are verified by H_2-TPR in Fig.13. In Figure 16a, the diverse catalytic performances for NO reduction of the loaded metals are essentially ascribed to their chemical nature. Copper- and iron-containing catalysts are of special interest because there are active in the transition of nitrogen oxides [42], with the low-coordination isolated ions as active sites [43].

According to Figure 15a and b, XPS analysis indicates that copper and iron components were fully distributed and mostly existing as Cu^{2+} and Fe^{3+} ions confined in the mesopore of SBA-15, whose mesoporous channel and superhigh surface favor metal dispersion and resistance to high-temperature sintering.

An obvious improvement was achieved in the case of Cu/SBA-15 and Fe/SBA-15 samples, respectively yielding NO conversions of 83% and 90% at 600 °C, as compared to SBA-15 support (Figure 16a). Furthermore, the satisfactory mobility of lattice oxygen of Cu/SBA-15, as evidenced by H_2-TPR in Figure 13, is also responsible for its good *de*NO_x activity.

An effort to modify the local environment of SBA-15 by means of Al doping has been done here. Unlike the other elements (e.g. Cu, Fe, and Cr), the incorporation of some Al into the SBA-15 skeleton was confirmed based on NMR experiments (Figure 14). The downward shift of low-angle diffraction peak of Al-loaded SBA-15 as compared to the unloaded SBA-15 also reflected the successful insertion of Al into the zeolite lattice (Figure 11). The different role was played by Al on affecting the SCR performance of co-loaded metals of copper, chromium, and iron, as depicted in Figure 16 and 17. The *de*NO_x activity (especially at low temperature) of Cu/SBA-15 was obviously improved with the assistance of aluminum. Clearly, the presence of aluminum is beneficial for further developing the redox properties of Cu/SBA-15 catalyst showing a TPR peak at relatively lower temperature (247 °C) for Cu–Al/SBA-15 (Figure 13).

Moreover, copper species facilitate Al insertion into the framework of SBA-15 (Figure 14). Furthermore, the formation of bulk CuO was detectable in the case of Cu/SBA-15 while this species was completely inhibited in the Cu–Al/SBA-15 samples with evidence from XPS (Figure 15a). The isolated copper ions were verified to be more effective for adsorption and reduction of NO than CuO species, while the CuO species promoted the detrimental oxidation of propene leading to the lack of reducing agent [44]. The as-synthesized Cu–Al/SBA-15 sample with skeleton Al generally associating with the dispersed Cu^{2+} ions thus displayed *de*NO_x activity superior to that of Cu/SBA-15 with some CuO phases formed upon the extra-framework.

In addition, Cu–Al/SBA-15 catalyst seems to involve into the "bifuctional mechanism" proposed by Bethke and Kung [41], in which Cu^{2+} sites favorite the formation of nitrite or nitrate species while AlO_x phase is beneficial for the further reduction of NO_2 by C_3H_6 towards N_2. Similarly, at <450 °C Cr–Al/SBA-15 exhibited a better SCR activity as compared to Cr/SBA-15, in accordance with the improvement of redox properties of Cr/SBA-15 via co-doping of Al; while the NO conversion at elevated temperatures essentially dependent on the concentration of isolated chromium species, thought as active centers for reactants adsorption, displaying a similar *de*NO_x curve for Cr/SBA-15 and Cr–Al/SBA-15 at $T > 450$ °C. In contrast, the decline in NO conversion was found for Fe–Al/SBA-15 as compared to Fe/SBA-15, which could be explained by the possible coverage of active sites (Fe^{3+}) by the additional Al component based on the relatively weaker XPS peak of Fe $2p_{3/2}$ for Fe–Al/SBA-15. Otherwise, the appearance of iron oxides was also believed to be harmful for NO catalytic reduction.

In short, the redox properties is crucially responsible for the low-temperature NO reduction by C_3H_6 likely via facilitating the corresponding reduction and/or oxidation process (e.g. pre-oxidation of C_3H_6, nitrite or nitrate formation) even at relatively lower temperatures. As the reaction temperatures became high enough, the reaction rate was somewhat correlated to the essential nature and the surface concentration of supported metals over SBA-15, which generally works as active sites for this SCR process. As a consequence, the synergistic effect was achieved between loaded copper and aluminum over high-surface-area SBA-15 support, resulting in a satisfactory *de*NO_x activity of ~88% at 400 °C and almost complete at 450 °C in the case of Cu–Al/SBA-15 catalyst.

3. Mesoporous SBA-15 Promoted by Transition Metals for Selective Catalytic Combustion of Acetonitrile

Acetonitrile (CH_3CN), acrylonitrile (C_2H_3CN), and hydrocyanic acid (HCN) commonly classified as very toxic nitrogen-containing voltaic organic compounds (VOCs) could lead to some seriously environmental problems because of their hazardous properties. These pollutants in the gas phase are able to affect human beings and even be lethal at ppm levels in air. The relatively lower operation temperature for catalytic combustion associated with a less NO_x formation made this technology especially suitable for gaseous nitrile elimination. Much attention was recently paid on the catalytic elimination of harmful acetonitrile (CH_3CN), coming from the tail gas of industrial acrylonitrile (C_2H_3CN) plant in significant amounts, because of its carcinogenicity.

Effort for searching the effective catalytic materials never ceased. Mesoporous SBA-15 was introduced as a support for the well dispersion of series of transition metals for CH_3CN selective combustion due to its superhigh surface area, ordered pore arrangement, adjustable pore sizes from 3 to 30 nm, and high hydrothermal and thermal stability [45].

Their relatively larger pore sizes as compared to those of microporous zeolites (e.g. ZSM-5) facilitate the transportation of reactants and might lead to a better resistance to SO_2 and steam poisonings as well as deactivation caused by pore blockage [38, 46]. Herein, a series of SBA-15-based catalysts including parent SBA-15, and mono-functional M'' (M'' = Cu, Co, Cr and Mn)/SBA-15 was prepared and characterized by N_2 physical sorption, X-ray diffraction (XRD), temperature programmed reduction by hydrogen (H_2-TPR), X-ray photoelectron spectroscopy (XPS) as well as the activity tests for selectively catalytic combustion of CH_3CN in the presence of O_2. Emphasis will be placed on both promoting the low-temperature activity and enhancing the N_2 selectivity via a modification of SBA-15 by transition metals. The relationship between the physicochemical properties of prepared catalysts and their depollution performances was also addressed.

3.1. Physicochemical Properties

The crystal structure of M'' (M'' = Cu, Co, Cr and Mn)/SBA-15 was identified by small-angle XRD showing characteristic peaks of 0.9°, 1.6° and 1.8° for classical meosporous silica with *P6mm* symmetry [33]. The effect of metal loading on the modification of SBA-15 structure was slight. Their structural parameters were further listed in Table 4 and 5, respectively. The constant of d_{100} value implied the well dispersion of metal over SBA-15 surface (mainly as cations or tiny metal oxides). This coincides with the absence of diffraction peaks for the relatively larger crystal of metal oxides. Specific surface area and pore structure of the prepared SBA-15-based catalysts were determined by N_2 sorption with the results being reported in Table 4 in Section 2.1 and Table 5. The BET surface area and pore volume for parent SBA-15 are 645 m^2/g and 0.81 cm^3/g, respectively. Both of them would drop to some degree with the introduction of transition metals displaying a sequence of SBA-15 > Cu/SBA-15 > Co/SBA-15 > Mn/SBA-15 > Cr/SBA-15. This was explained by the occurring of partial pore blockage due to some remaining metal oxides, leading to a decline in pore volume, pore diameter and specific surface area. It was clear that metal doping could cause an increase in the wall thickness along with a diminution of pore diameter. This change might be beneficial for improving of the hydrothermal stabilities of mesoporous catalysts. The redox properties of prepared samples were investigated by H_2-TPR, with the result for Cu/SBA-15 and Cr/SBA-15 being shown in the previous Figure 13 and detailedly discussed in Section 2.2. In Figure 18, the TPR plot for Mn/SBA-15 and Co/SBA-15 were shown. Only one peak for Mn/SBA-15 was observed, which was assigned to the reduction of MnO_2 according to Tang's report [47]. As for the Co/SBA-15 sample, three reducing peaks at 326, 453 and 671°C were discerned. In the light of literature assignments [48], the former two peaks were attributed to CoO_x reduction, while the latter one was related to the reduction of Co^{n+} ions inside the SBA-15 skeleton. Generally speaking, the fact that transition metals were dispersed over SBA-15 support mainly as cations or highly isolated oxides was verified. The excellent redox properties of Cu/SBA-15 could also be confirmed according to its first TPR peak appearing at relatively lower temperatures (around 288 °C). In order to know the chemical status of loaded metals and their local environment, XPS analyses for the prepared catalysts were conducted, with the results being respectively depicted in Figure 15 for Cr/SBA-15 and Cu/SBA-15 (see Section 2.4) and Figure 19 for Mn/SBA-15 and Co/SBA-15. The XPS for Cr/SBA-15 suggest that the chromium species were mostly existed as Cr^{3+} ions together with minor Cr^{2+} and Cr^{6+}. The divalent copper was clearly observed over Cu/SBA-15 showing its peaks at BE=941.3ev for CuO and BE=945ev for dispersed Cu^{2+} ions.

Table 5. Structural parameters of (Mn, Co)/SBA-15

Sample	BET area (m^2/g)	Pore volume (cm^3/g)			Pore diameter (nm)	d_{100} (nm)	Cell parameter a_0 (nm)	Wall thickness (nm)
		Micropore	Mesopore	Total				
Mn/SBA-15	579	0.09	0.75	0.84	6.1	9.8	11.3	5.2
Co/SBA-15	588	0.11	0.74	0.85	5.4	9.8	11.3	5.9

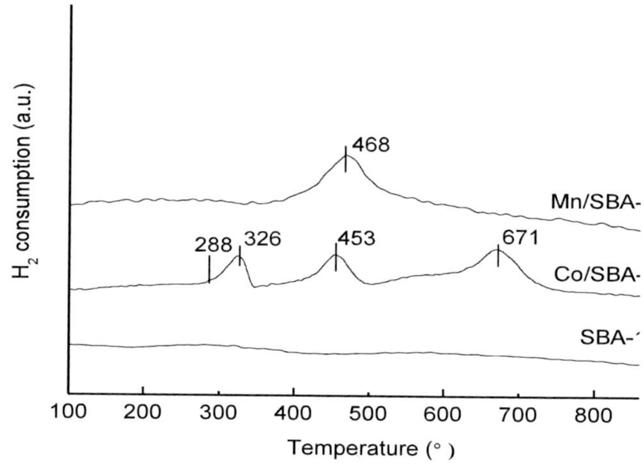

Figuure18. H_2-TPR profiles of SBA-15 and (Cr, Mn)/SBA-15.

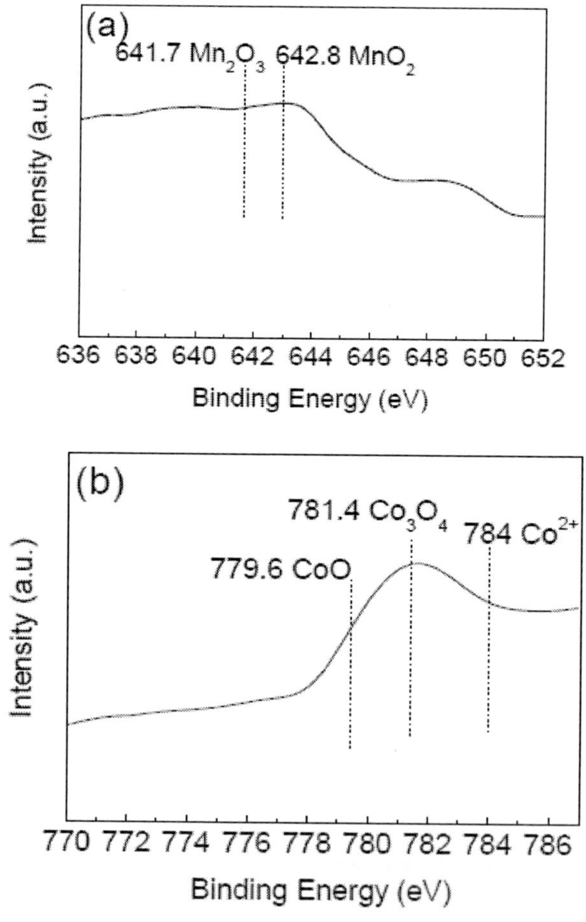

Figure 19. XPS spectra of (a) Mn_{2p}; (b)Co_{2p} for (Mn, Co)/SBA-15.

As seen from Figure 19a for Mn/SBA-15, Mn $2p_{3/2}$ peak at BE=641.7ev was ascribed to Mn_2O_3 as well as the one at BE=642.8ev was attributed to MnO_2. The XPS pattern for Co/SBA-15 was illustrated in Figure 19b, representing peaks at BE=779.6ev and 781.4ev assigned to CoO and Co_3O_4 [49] associated with one at BE=784ev ascribed to Co^{2+} in the mesoporous channel [45].

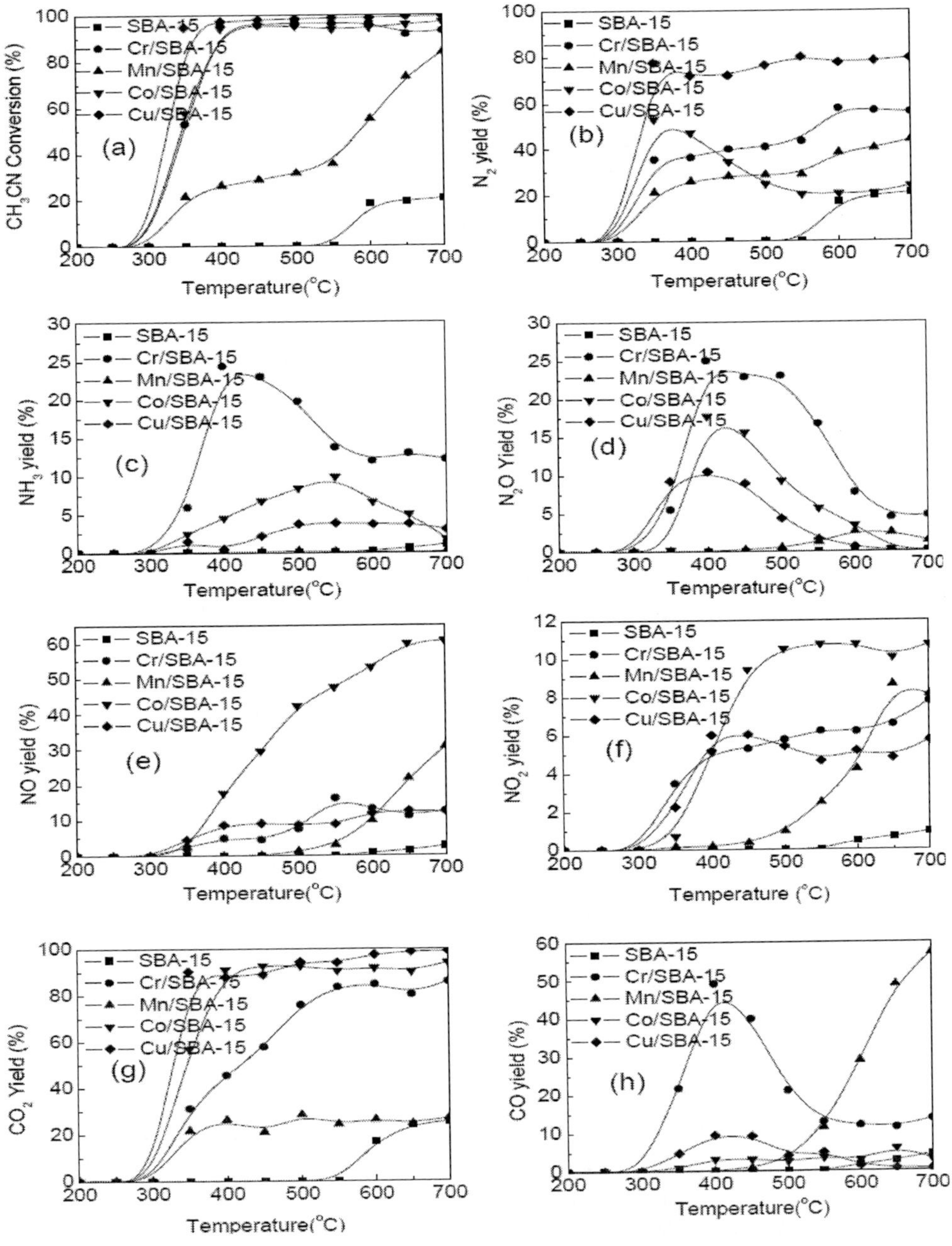

Figure 20. Catalytic performance as a function of temperature during CH_3CN combustion over SBA-15 and (Cr, Mn, Co, Cu)/SBA-15. (a)CH_3CN conversion; (b)N_2 yield; (c)NH_3 yield; (d)N_2O yield; (e)NO yield; (f)NO_2 yield; (g)CO_2 yield; (h)CO yield; Conditions: 1% CH_3CN, 5% O_2, $20000h^{-1}$.

3.2. Performance for Selective Catalytic Combustion of CH_3CN

The selective catalytic combustion of acetonitrile over M''/SBA-15 was investigated under a space velocity of 20000 h^{-1}, as illustrated in Figure 16a-h. Apart from the target products of N_2 and CO_2, the generation of series of harmful by-products of NH_3, NO, N_2O, NO_2, and CO are inevitable with their selectivities strongly dependent on the chemical properties of active components. CH_3CN conversion as a function of temperature represents in Figure 20a. It was seen that SBA-15 merely had a marginal conversion for CH_3CN at temperatures below 600°C. Among the M''/SBA-15 samples, the highest acetonitrle conversion was achieved over Cu/SBA-15, yielding an almost complete conversion at >350°C. Co/SBA-15 and Cr/SBA-15 showed the moderate conversions (~95 at T>350°C), while the worst combustion activity was found in the case of Mn/SBA-15, achieving a conversion of only 84% at temperature as high as 700°C. The yields of N_2 and the other N-containing byproducts with respect to temperature were depicted in Figure 20b-f. As indicated by Figure 20b, the highest N_2 yield was observed again in the case of Cu/SBA-15, with a value reaching approximately 80% as temperature higher that 350°C. N_2 yield for Co/SBA-15 was higher than those for Cr/SBA-15 and Mn/SBA-15 at T<400°C. However, the conversion for $CH_3CN \rightarrow N_2$ followed an order of Cr/SBA-15 > Mn/SBA-15 > Co/SBA-15 at the elevated temperatures. The easy formation of ammonia over Cr/SBA-15 was true, while N_2O yields were obtained over Cr/SBA-15 (25%), Co/SBA-15 (18%) and Cu/SBA-15 (25%) at 400°C, as clearly seen from Figure 20d. NO yield over Co/SBA-15 sample was fully developed along with the increasing temperatures and reached almost 60% at 650°C. This value is even higher than that for N_2 yield. The similar tendency for NO_2 yield over the tested samples was also observed, apart from an inhibition of NO_2 generation at high temperature, as described in Figure 20f. The highest NO_2 yield was again achieved over Co/SBA-15. Figure 20g and 20h respectively show the CO_2 and CO yields over M''/SBA-15 catalysts. Due to the excellent oxidation ability of Cu/SBA-15 and Co/SBA-15 samples, the CO_2 is the only carbonaceous product due to organic carbon combustion. As a result, Cu/SBA-15 and Co/SBA-15 displayed combustion activities superior to Cr/SBA-15. Additionally, Mn/SBA-15 showed a rather poor performance for combustion, with CO_2 yields maintaining at about 25%, simultaneously with an increasing CO yield of 58% at 700°C. Overall, Cu/SBA-15 has a much better activity for CH_3CN combustion towards N_2 as compared with the other metal-modified SBA-15 catalysts. The thermal stabilities and resistance to SO_2 poisoning were evaluated based on the best-performing Cu/SBA-15 (see Figure 21). This sample was tested at 400°C under the usual reaction atmosphere to check its stability. The ideal initial activity with a CH_3CN conversion of 97% and a N_2 yield of 71% was well maintained even after 48 h reaction, indicating a good stability of this SBA-15-based catalyst. Moreover, the Cu/SBA-15 was also calcined at 800°C for 5 h before reaction to further identify its thermal stability. This thermally treated sample represented 85% CH_3CN conversion and 65% N_2 yield. This deactivation with respect to the fresh sample was accordingly ascribed to the sintering of Cu species during higher-temperature calcination.

The resistance to SO_2 poisoning was investigated via an exposure to 10 ppm SO_2 for 20 h after 10 h usual reaction, and then switching to 50 ppm for 4 h and 100 ppm for 4 h, finally releasing SO_2 from the stream for an other 10 h. Interestingly, it was seen that 10-100 ppm SO_2 addition hardly affected the combustion performance of Cu/SBA-15, suggesting its good resistance to SO_2 poisoning. It might correspond to the relatively larger pore size of SBA-15 in comparison with microporous zeolites, which effectively avoids the pore blockage caused by sulfurous compounds deposition.

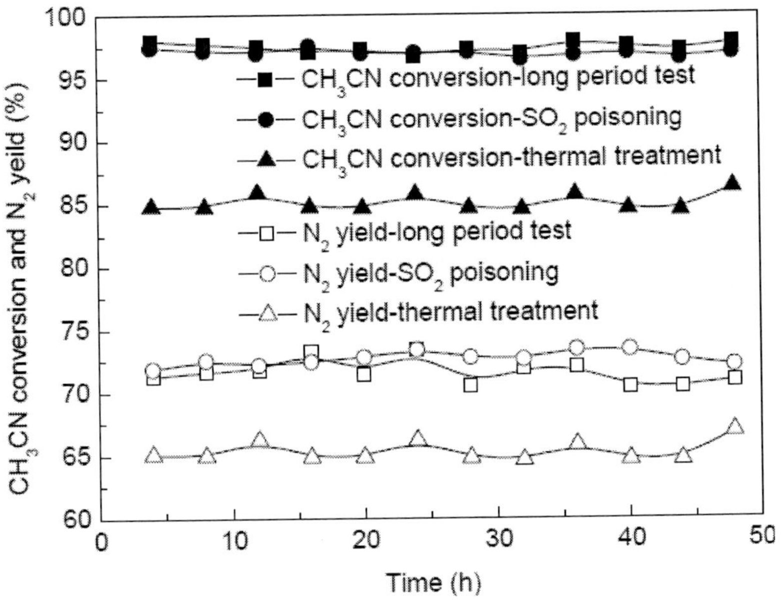

Figure 21. Catalytic performance as a function of time during CH_3CN combustion over Cu/SBA-15 sample, Conditions: 1% CH_3CN, 5% O_2, 0-100 ppm SO_2, 20000h^{-1}.

3.3. Correlation between Physicochemical Properties and Catalytic Performance

No obvious CH_3CN combustion was observed over the bare SBA-15 at T<500°C, while the SBA-15 activity was fully developed by means of transition metals doping. It thus deduces that the loaded metals act as active centers and play a crucial role for catalytic combustion. Series of catalysts with a long-range-ordering pore and high specific surface area (about 568-645m^2/g) were presently prepared. Thanks to the superhigh surface area of SBA-15 support, the well dispersion of transition metal mainly as cations or oxide crystallites was ensured.

The optimal performance for CH_3CN selective combustion was achieved in the case of Cu/SBA-15 with the copper species mainly existing in a Cu^{2+} ionic form. According to Figure 16a, the activity of Cu/SBA-15 was much higher with respect to that of the parent SBA-15. This indicates that copper serves as the active center to accelerate the combustion process. The small-angle XRD peaks and the high surface area suggest that Cu loading did not destroy the SBA-15 structure. Similarly, the minor change in surface area could also be ascribed to the high dispersion of copper species. Moreover, the obviously superior combustion activity of Cu/SBA-15, especially at low temperature, with a comparison to the other M''/SBA-15 samples was observed, which was thought to be correlated to their redox properties. According to the temperature of the earliest reduction peak, Cu/SBA-15 possesses of the best redox ability and can be readily reduced at 288°C. On the contrary, Mn/SBA-15 shows the worst redox properties. This terrible redox ability consists well with its bad capability for CH_3CN conversion. Additionally, the highest N_2 yield was achieved over Cu/SBA-15. The generation of NH_3 and NO_x was effectively suppressed. The essentially chemical nature of copper could also be responsible for this excellent performance. As reviewed by Centi and Perathoner for the excellent NO_x transformation over Cu species, the low coordinated Cu^{2+} ions were deemed as the cores for NO reduction [42-44]. Hence, the abundant Cu^{2+} ions are in favor of the coupling the interaction between NH_3 and NO_x, yielding the environmental-friendship N_2. As a result, the selectivity towards N_2 is significantly enhanced. Conversely, the byproducts of NO and NO_2 due to a deep oxidation become significant (especially at high temperatures) over Co/SBA-15, which likely corresponds to its good oxidation ability. For the Mn/SBA-15 sample with poor oxidation ability, the CH_3CN combustion was totally inhibited. The nitrile conversion merely took place at T>500°C, where the lattice oxygen become available. The participation of lattice oxygen would be beneficial for the formation of NO and NO_2. In addition, the CO_2 and CO yields obtained over the investigated samples match well with their redox properties. Briefly, catalytic performance for CH_3CN combustion strongly depends on both the redox properties of M''/SBA-15 catalysts and the chemical nature of the loaded transition metals. Redox properties mainly lead to the destruction of methyl group (CH_3-), while the chemical nature might play a crucial role for the selective oxidation of cyanide group (-CN).

CONCLUSION

In this chapter, our continued work on the application of catalytic materials for the purification of exhaust gases including N_2O from the industrial plants of adipic acid or nitric acid as well as NO and HCs from automobile, CH_3CN from acrylonitrile plant was depicted. The following conclusions can be made on the experimental results.

- The catalytic activity for N_2O decomposition of bare beta zeolite was obviously enhanced by Fe loading through an ion-exchange process. As the ion-exchange content of iron was relatively lower (<1%), the iron mainly existed in ionic Fe^{n+}, which was active for N_2O decomposition, on Fe-beta zeolite; while the high Fe loading (>1%) would lead to the aggregation with the formation of FeO_x species, which is less active. Energy calculations revealed that O_2 desorption process was the rate-determining step in N_2O decomposition and an energy barrier of 63.2 kcal/mol.
- The CT analysis on each reaction step of the "O_2-formation" mechanism for N_2O direct decomposition over Fe-BEA zeolite was thereafter studied based on a Mulliken population analysis. Zeolite framework generally acting as a charge reservoir played important role on reaction steps involved in Part 1, whereas the active center Fe and α-O seem to be crucial for reaction steps involved in Part 2, in which the high activity of α-O was also confirmed by means of the CT analysis. The N_2O commonly working as a charge donor was adsorbed on the Fe-BEA zeolite. During the decomposition reaction steps the middle N atom of N_2O acting as the main charge acceptor received large amount of charges, which is beneficial for the fracture of O-N_2 bond and N_2 formation. It was found that the electronic properties of

the corresponding N atoms in the firstly adsorbed N_2O molecule and the secondly adsorbed N_2O molecule were nearly the same. In Part 3 both two O atoms acting as the charge donors mainly donated the charges to the active center Fe and then were desorbed from Fe-O_2-Z, completing a catalytic cycle.

- Mesostructured SBA-15 was obtained via hydrothermal synthesis with a highly structural ordering and a specific surface area of 645 m^2/g. Due to good dispersion, metal-promoted mesoporous silica exhibits remarkable *de*NO_x activity, which strongly depends on the essentially chemical nature of metals doped and their dispersities. An optimal performance for NO catalytic reduction with C_3H_6 was achieved in the case of bifunctional Cu–Al/SBA-15. This synergistic effect was accordingly ascribed to the promotion of redox properties of active copper species with the assistance of aluminum. Additionally, the Al presence is beneficial for suppressing the formation of CuO cluster, which is thought to be less active for NO-SCR, however, could facilitate the reductants consumption due to an unfavourable oxidation process as compared to isolated Cu^{2+} ions.
- Series of metal-modified SBA-15 were used for CH_3CN combustion. Besides the target N_2 and CO_2, the generation of series of harmful by-products of NH_3, NO, N_2O, NO_2, and CO was detected. CH_3CN conversion follows a sequence of Cu/SBA-15 > Co/SBA-15 and Cr/SBA-15 > Mn/SBA-15 > SBA-15. Among the M''/SBA-15 samples, the highest acetonitrile conversion and N_2 yield were achieved in the case of Cu/SBA-15, which is likely attributed to its excellent redox properties and ideal ability of copper cores facilitating the coupling of NH_3 and NO_x to yield the unharmful N_2.

REFERENCES

[1] Centi, G; Galli, A; Montanari, B; Perathoner, S and Vaccaria, A. Catalytic decomposition of N_2O over noble and transition metal containing oxides and zeolites. Role of some variables on reactivity, *Catal. Today*, 1997, Volume 35, Issues 1-2, 113-120.

[2] Chen, BH; Liu, N; Liu, XY; Zhang, R; Li, YP; Li, YX; Sun, XL. Study on the direct decomposition of nitrous oxide over Fe-beta zeolites: From experiment to theory, *Catal. Today*, 2011, Volume 175, Issue 1, 245-255.

[3] Øygarden, AH and Pérez-Ramírez, J. Activity of commercial zeolites with iron impurities in direct N_2O decomposition. *Appl. Catal. B*, 2006, Volume 65, Issues 1-2, 163-167.

[4] Pieterse, JAZ; Pringruber, GD; van Bokhoven, JA and Booneveld, S. Hydrothermal stability of Fe-ZSM-5 and Fe-BEA prepared by wet ion-exchange for N_2O decomposition. *Appl. Catal. B*, 2007, Volume 71, Issues 1-2, 16-22.

[5] Delahay, G; Mauvezin, M; Coq, B and Kieger, S. Selective catalytic reduction of nitrous oxide by ammonia on iron zeolite beta catalysts in an oxygen rich atmosphere: effect of iron contents. *J. Catal.*, 2011, Volume 202, Issue 1, 156-162.

[6] Lobree, LJ; Wang, IC; Reimer, JA and Bell, AT. Investigations of the State of Fe in H-ZSM-5. *J. Catal.*, 1999, Volume 186, Issue 2, 242-253.

[7] Pannov, GI; Sobolev, VI and Kharitonov, AS. The role of iron in N_2O decomposition on ZSM-5 zeolite and reactivity of the surface oxygen formed. *J. Mol. Catal.*, 1990, Volume 61, Issue 1, 85-97.

[8] Sun, KQ; Xia, HA; Hensen, E; van Santen; R and Li, C. Chemistry of N_2O decomposition on active sites with different nature: Effect of high-temperature treatment of Fe/ZSM-5. *J. Catal.*, 2006, Volume 238, Issue 1, 186-195.

[9] Guzmán-Vargas, A; Delahay, G and Coq, B. Catalytic decomposition of N_2O and catalytic reduction of N_2O and N_2O + NO by NH_3 in the presence of O_2 over Fe-zeolite. *Appl. Catal. B*, 2003, Volume 42, Issue 4, 369-379.

[10] Park, JH; Choung, JH; Nam, IS and Ham, SW. N_2O decomposition over wet- and solid-exchanged Fe-ZSM-5 catalysts. *Appl. Catal. B*, 2008, Volume 78, Issues 3-4, 342-354.

[11] Munteanu, G; Ilieva, L and Andreeva, D. Kinetic parameters obtained from TPR data for α-Fe_2O_3 and Au/α-Fe_2O_3 systems. *Therm. Acta*, 1997, Volume 291, Issues 1-2, 171-177.

[12] Berlier, G; Spoto, G; Bordiga, S; Ricchiardi, G; Fisicaro, P; Zecchina, A; Rossetti, I; Selli, E; Forni, L; Giamello, E and Lamberti, C. Evolution of extraframework iron species in Fe silicalite: 1. effect of Fe content, activation temperature, and interaction with redox agents. *J. Catal.*, 2002, Volume 208, Issue 1, 64-82.

[13] Debeila, MA; Covile, NJ; Surrell, MS and Hearne, GR. DRIFTS studies of the interaction of nitric oxide and carbon monoxide on Au–TiO_2. *Catal. Today*, 2002, Volume 72, Issues 1-2, 79-87.

[14] Wood, BR; Reimer, JA and Bell, AT. Studies of N_2O adsorption and decomposition on Fe-ZSM-5. *J. Catal.*, 2002, Volume 209, Issue 1, 151-158.

[15] Pirngruber, GD. and Pieterse, JAZ. The positive effect of NO on the N_2O decomposition activity of Fe-ZSM-5: A combined kinetic and in situ IR spectroscopic study. *J. Catal.*, 2006, Volume 237, Issue 2, 237-247.

[16] Kohn, W. and Sham, LJ. Self-Consistent Equations Including Exchange and Correlation Effects. *Phys. Rev.*, 1965, Volume 140, A1133-A1138.

[17] Liu, N; Chen, BH; Li, YP; Zhang, RD; Liang, X; Li, YX; Lei, ZG. Charge transfer analysis on the direct decomposition of nitrous oxide over Fe-BEA zeolites: An experimental and density functional study, *J. Phys. Chem. C*, 2011, Volume 115, Issue 1-2, 12883-12890.

[18] Goursot, A; Coq, B; Fajula, F. Toward a molecular description of heterogeneous catalysis: transition metal ions in zeolites. *J. Catal.* 2003, Volume 216, Issues 1-2, 324-332.

[19] Ajay, M; Joshi, W; Nicholas, D; Kendall, T; Thomson. Interaction of carbon monoxide with small gold clusters inside TS-1 pores. *J. Phys. Chem. C*, 2007, Volume 111, Issue 30, 11424-11436.

[20] Pietrzyk, P; Zasada, F; Piskorz, W; Kotarba, A; Sojka, Z. Computational spectroscopy and DFT investigations into nitrogen and oxygen bond breaking and bond making processes in model *de*NO_x and *de*N_2O reactions. *Catal. Today*, 2007, Volume 119, Issues 1-4, 219-227.

[21] Panov, GI; Kharitonov, AS; Sobolev, VI. Oxidative hydroxylation using dinitrogen monoxide: a possible route for organic synthesis over zeolites. *Appl. Catal., A*, 1993, Volume 98, Issue 1, 1-20.

[22] Rodkin, MA; Sobolev, VI; Dubkov, KA; Watkins, NH; Panov, GI. Room-temperature oxidation of hydrocarbons over FeZSM-5 zeolite. *Stud. Surf. Sci. Catal.*, 2000, Volume 130, 875-880.

[23] Panov, GI; Uriarte, AK; Rodkin, MA; Sobolev, VI. Generation of active oxygen species on solid surfaces. Opportunity for novel oxidation technologies over zeolites. *Catal. Today*, 1998, Volume 41, Issue 4, 365-385.

[24] Guesmi, H; Berthomieu, D; Kiwi-Minsker, L. Reactivity of oxygen species formed upon N_2O dissociation over Fe-ZSM-5 zeolite: CO oxidation as a model. *Catal. Commun.*, 2010, Volume 11, Issue 12, 1026-1031.

[25] Sojka, Z; Che, M. EPR Investigation of the Activation of N_2O on Mo/SiO_2 Catalysts via Electron Transfer: From N_2O as a Ligand to Adsorbed O^- Ion. *J. Phys. Chem.*, 1996, Volume 100, Issue 35, 14776-14785.

[26] Yakovlev, AL; Zhidomirov, GM; van Santen, RA. N_2O decomposition catalysed by transition metal Ions. *Catal. Lett.*, 2001, Volume 75, Numbers 1-2, 45-48,.

[27] Zhang, RD; Villanueva, A; Alamdari, H; Kaliaguine, S. Catalytic reduction of NO by propene over $LaCo_{1-x}Cu_xO_3$ perovskites synthesized by reactive grinding. *Appl. Catal. B*, 2006, Volume 62, Issues 3-4, 220-233.

[28] Delabie, A; Vinckier, C; Flock, M; Pierloot, K. Evaluating the activation barriers for transition metal N_2O reactions. *J. Phys. Chem. A*, 2001, Volume 105, Issue 22, 5479-5485.

[29] Zecchina, A; Cerruti, L; Borello, E. An infrared study of nitrous oxide adsorption on á-chromia. *J. Catal.*, 1972, Volume 25, Issue 1, 55-64.

[30] Hoost, TE; Laframboise, KA and Otto, K. Co-adsorption of propene and nitrogen oxides on Cu-ZSM-5: An FTIR study. *Appl. Catal. B*, 1995, Volume 7, Issues 1-2, 79-93.

[31] Shen, Q; Li, L; Hao, Z. and Xu, Z. Highly active and stable bimetallic Ir/Fe-USY catalysts for direct and NO-assisted N_2O decomposition. *Appl. Catal. B*, 2008, Volume 84, Issues 3-4, 734-741.

[32] Delahay, G; Coq, B and Broussous, L. Selective catalytic reduction of nitrogen monoxide by decane on copper-exchanged beta zeolites. *Appl. Catal. B*, 1997, Volume 12, Issue 1, 49-59.

[33] Zhao, D; Feng, J; Huo, Q; Melosh, N; Fredrickson, GH; Chmelka, BF and Stucky, GD. Triblock Copolymer Syntheses of Mesoporous Silica with Periodic 50 to 300 Angstrom Pores. *Science*, 1998, Volume 279, Number 5350, 548-552.

[34] Zhang, R; Shi, D; Zhao, Y; Chen, B; Xue, J; Liang, X; Lei, Z. The reaction of NO + C_3H_6 + O_2 over the mesoporous SBA-15 supported transition metal catalysts. *Catal. Today*, 2011, Volume 175, Issue 1, 26-33.

[35] Zhang, R; Villanueva, A; Alamdari, H and Kaliaguine, S. Cu- and Pd-substituted nanoscale Fe-based perovskites for selective catalytic reduction of NO by propene. *J. Catal.*, 2006, Volume 237, Issue 2, 368-380.

[36] Zhang, R. and Kaliaguine, S. Lean reduction of NO by C_3H_6 over Ag/alumina derived from Al_2O_3, AlOOH and $Al(OH)_3$. *Appl. Catal.* B, 2009, Volume 78, Issues 3-4, 275-287.

[37] Zhang, L; Zhao, Y; Dai, H; He, H and Au, CT. A comparative investigation on the properties of Cr-SBA-15 and CrO_x/SBA-15. *Catal. Today*, 2008, Volume 131, Issues 1-4, 42-54.

[38] Batista, J; Pintar, A; Mandrino, D; Jenko, M and Martin, V. XPS and TPR examinations of γ-alumina-supported Pd-Cu catalysts. *Appl. Catal. A*, 2001, Volume 206, Issue 1, 113-124.

[39] Yamashita, T and Hayes, P. Analysis of XPS spectra of Fe^{2+} and Fe^{3+} ions in oxide materials. *Appl. Surf. Sci.*, 2008, Volume 254, Issue 8, 2441-2449.

[40] Pashutski, A; Hoffman, A and Folman, M. Low temperature XPS and AES studies of O_2 adsorption on Al(100). *Surf. Sci. lett.*, 1989, Volume 208, Issue 3, L91-L97.

[41] Bethke, KA and Kung, HH. Supported Ag Catalysts for the Lean Reduction of NO with C_3H_6. *J. Catal.*, 1997, Volume 172, Issue 1, 93-102.

[42] Centi, G and Perathoner, S. Nature of active species in copper-based catalysts and their chemistry of transformation of nitrogen oxides. *Appl. Catal. A*, 1995, Volume 132, Issue 2, 179-259.

[43] Iwamoto, M. Heterogeneous catalysis for removal of NO in excess oxygen. Progress in 1994. *Catal. Today*, 1996, Volume 29, Issues 1-4, 29-35.

[44] [44] Wan, Y; Ma, J; Wang, Z; Zhou, W and Kaliaguine, S. Selective catalytic reduction of NO over Cu-Al-MCM-41. *J. Catal.*, 2004, Volume 227, Issue 1, 242-252.

[45] da Cruz, RS; Mascarenhas, AJS; Andrade, HMC. Co-ZSM-5 catalysts for N_2O decomposition. *Appl.Catal. B*, 1998, Volume 18, Issues 3-4, 223-231.

[46] Xu, JF; Ji, W; Shen, ZX; Tang, SH; Ye, XR; Jia, DZ; Xin, XQ. Preparation and Characterization of CuO Nanocrystals. *J. Solid State Chem.*, 1997, Volume 147, Issue 2, 516-519.

[47] Tang, X; Li, Y; Huang, X; Xu, Y; Zhu, H; Wang, J; Shen, W. MnO_x-CeO_2 mixed oxide catalysts for complete oxidation of formaldehyde: Effect of preparation method and calcination temperature. *Appl. Catal. B*, 2006, Volume 62, Issues 3-4, 265-273.

[48] Smeets, PJ; Meng, Q; Corthals, S; Leeman, H; Schoonheydt, RA. Co-ZSM-5 catalysts in the decomposition of N_2O and the SCR of NO with CH_4: Influence of preparation method and cobalt loading. *Appl. Catal. B*, 2008, Volume 84, Issues 3-4, 505-513.

[49] Lishka, MA; Barr. TL. ESCA studies of the surface chemistry of zeolites. *J. Am. Chem. Soc.*, 1986, Volume 108, Issue 12, 3178–3186.

In: Zeolites: Synthesis, Chemistry and Applications
Editors: Moisey K. Andreyev and Olya L. Zubkov
ISBN: 978-1-61942-861-4

Chapter 3

ZEOLITE FROM FLY ASH-IRON OXIDE MAGNETIC NANOCOMPOSITE: SYNTHESIS AND APPLICATION AS AN ADSORBENT FOR REMOVAL OF CONTAMINANTS FROM AQUEOUS SOLUTION

Denise A. Fungaro**, Mitiko Yamaura,
Terezinha E. M. Carvalho and José E. A. Graciano
Environment and Chemistry Centre, Nuclear and Energy Research Institute
(IPEN – CNEN/SP), São Paulo, Brazil

ABSTRACT

This chapter presents and discusses current research on the study of the adsorption efficiency of zeolite from fly ash-iron oxide nanocomposites towards removal of hazardous pollutants from aqueous solutions. The magnetic properties of iron oxides have been combined with the adsorption features of zeolites synthesized from fly ash to produce a magnetic nanocomposite adsorbent. Magnetite nanoparticles in suspension were synthesized by precipitating iron ions in a NaOH solution. Zeolite was synthesized from coal fly ash by alkaline hydrothermal treatment. Zeolite-iron oxide composite was prepared by mixing zeolite with magnetite suspension with weight ratios of 3:1, 2:1 and 1:1. The resulting materials were characterized by X-ray diffraction, surface area, magnetization measurements, Fourier transform infrared and X-ray fluorescence analyses. Batch tests were carried out to investigate the adsorption mechanism of dyes and metal ions from aqueous solution onto magnetic nanocomposite. The composite adsorbents after be used to adsorb contaminants in water were removed from the medium by a simple magnetic process. The effect of various experimental parameters such as contact time, pH, adsorbent dose and temperature were investigated. The experimental data were analyzed using the pseudo-first-order, pseudo-second-order and intraparticle diffusion equations adsorption kinetic models. The Langmuir and Freundlich isotherm models were tested for their applicability. The results indicated that zeolite from fly ash-iron oxide magnetic nanocomposite, a low cost adsorbent, exhibited considerable

* E-mail address: dfungaro@ipen.br.

potential for application in removal of contaminants from aqueous solution. Future prospects for research are discussed.

1. INTRODUCTION

Magnetic separation technology as an efficient, fast and economical method for separating magnetic materials has been widely used in textile, biology and environmental protection [1–4]. Some examples of this technology are the use of magnetic poly(oxy-2,6-dimethyl1,4-phenylene) for the adsorption of organic dyes [5], polymer-coated magnetic particles for oil spill remediation [6], polyacrylic acid-bound iron oxide magnetic nanoparticles for the adsorption of methylene blue [7] and the use of montmorillonite–Cu(II)/Fe(III) oxide magnetic materials for the removal of humic acid [8].

Adsorption is an important technique for metals ion, color and any other pollutant removal from industrial wastewaters and radioactive wastes. The prediction of batch adsorption kinetics is necessary for the design of industrial adsorption columns.

The adsorbents combining magnetic separation technology with adsorption process have been widely used in environmental purification [9–11]. The main advantage of this technology is that it can dispose a mass of wastewater in a very short period of time and produce no contaminants such as flocculants [12]. After the adsorption is carried out, the adsorbent can be separated from the medium by a simple magnetic process.

Magnetic composites based on activated carbon/iron oxide, clay/iron oxide and NaY commercial zeolite/iron oxide have been prepared via a simple precipitation method and used for the removal of contaminants from aqueous effluents [13-15]. These materials possessed a high surface area and a high adsorption capacity.

The magnetized adsorbents have recently been used for removal of dyes from water [16].

Most commercially available magnetic particles are rather expensive and cannot be used large-scale processes. Magnetic modification of low cost adsorbents could lead to materials for biotechnology and environmental applications.

Fly ash is formed by combustion of coal in coal-fired power station as a waste product. The generation rate of coal fly ash in Brazil is approximately 4 million tons per year and is predicted to increase.

Great part of this fly ash is applied as raw material for cement and concrete production (~ 30% of the total), while a small part is used in road-stabilizer, glass composites and asphalt mixtures. The remaining solid wastes are disposed in on-site ponds, nearby abandoned or active mine sites, or landfills.

There is a big concern related to the non-use fly ash since is endangering surrounding surface and ground waters as it can release different and potentially toxic trace and major elements during and after its transport to the dump. Further on, leached elements are entering soils and sediments and through biosphere can enter food-chain. The coal ash in the environment is exposed to various influences, and solubilization processes take place, sometimes transforming almost insoluble metals to soluble species, and leading to pollution of water and soil [17–20].

Efficient disposal of coal fly ash is a worldwide issue due to its massive volume and harmful risks to the environment. Furthermore, the conversion of a low-cost waste product into a higher level product would make the disposal of the remaining unused ash far more economically viable. Thus continuous research is needed to develop an alternative technology for utilizing coal ashes. As a technique for recycling coal fly ash, synthesis of zeolites from coal fly ash has attracted a great deal of attention.

Holler and Wirsching [21] and Henmi [22] were the first to report the modification of fly ash by hydrothermal treatment in alkaline solutions at elevated temperatures and pressures to produce zeolites. Since them, the synthetic approach has been modified to include microwave assisted synthesis [23] and the fusion of ash with base prior to hydrothermal treatment [24].

Zeolitic materials have been widely used as low-cost adsorbents as reviewed by Querol et al. and Rayalu et al. [25, 26].

In our group we have been developing zeolites synthesized from Brazilian coal fly ashes to remove metals ions from aqueous effluents and soil. The efficiency of zeolitic materials as adsorbent allowed investigations focused on the removal of dyes [27-31].

In this study, the development of a new kind of magnetic adsorbent which can be effective for the removal of contaminants from aqueous solution was examined. For this purpose, the adsorption properties of zeolite synthesized from coal fly ashes were combined with the magnetic properties of iron oxides to produce zeolite-iron oxide magnetic composite. The adsorption studies such as effect of contact time, pH, adsorbent dose and temperature were explored in batch experiments and thermodynamic, kinetic and adsorption isotherm analyses were used to elucidate the adsorption mechanism.

2. Materials and Methods

2.1. Materials

The sample of coal fly ashes from baghouse filter was obtained from a coal-fired power plant located at Figueira City, in Paraná State, Brazil. All chemicals used in this study were of analytical grade. Reactive Orange 16 (RO16; C.I. 17757; $C_{20}H_{17}N_3O_{10}S_3Na_2$; molar mass = 601.54 g mol^{-1}) was purchased from Sigma–Aldrich (50% purity) and Indigo Carmine (IC; C.I. 73015; $C_{16}H_8N_2Na_2O_8S_2$; molar mass = 466.35 g mol^{-1}) was supplied by Vetec Química Fina Ltda (100% purity). The standard solutions of metal ions Cd^{2+}, Pb^{2+} and Zn^{2+} were prepared from their salts (Merck). Stock solutions were prepared in deionized water (Millipore Milli-Q) and the solutions for adsorption tests were prepared by diluting. For pH adjustment, 0.01 mol L^{-1} HNO_3 and 0.5 mol L^{-1}NaOH were used.

2.2. Preparation of the Zeolite from Fly Ash

Coal fly ash (CFA) was used as starting material for zeolite synthesis by means of hydrothermal treatment. In synthesis experiment, 20 g of fly ash was heated to 100 °C in an oven for 24 h with 160 mL of 3.5 mol L^{-1} NaOH solution. The zeolitic material was repeatedly washed with deionized water until pH 11 and dried at 50 °C for 12 h [22]. The zeolitic product obtained was labeled as ZFA.

2.3. Preparation of the Zeolite-Iron Oxide Magnetic Nanocomposites

The composites were prepared using two different methods. In the first method, magnetite particles were prepared by adding of 2 mol L^{-1} NaOH solution drop by drop in a 100 mL solution of ferrous sulfate (1.8 g) with agitation until the pH reached 11. The slurry was heated on a water bath. After that, the magnetite was washed with distilled water and dried at room temperature. One part of the magnetite powder was separated and stored (MAG). Other part of magnetite particles was redispersed in aqueous solution and zeolite synthesized from fly ash (ZFA) was added slowly with agitation. The zeolite/magnetite ratio was 3:1 (w/w). The obtained zeolite-iron oxide magnetic composite was washed with distilled water, dried at room temperature and milled [30]. The magnetite composite obtained was labeled as ZM3.

In the second method, the composites were prepared by adding a 5 mol L^{-1} NaOH solution into a mixed solution of 0.25 mol L^{-1} ferrous chloride and 0.5 mol L^{-1} ferric chloride (molar ratio 1:2) until obtaining pH 11 at room temperature. The slurry was filtrated and washed repeatedly with distilled water until obtaining pH 7. Then particles were redispersed in aqueous solution and zeolite synthesized from fly ash (ZFA) was added. The amount of zeolite was adjusted in order to obtain zeolite/iron oxide weight ratios of 1:1, 2:1 and 3:1. The composites were filtrated and the obtained materials were dried in an oven at 50 °C [27]. The magnetic composites were named ZC1, ZC2 and ZC3 for the zeolite/iron oxide weight ratios of 1:1, 2:1 and 3:1, respectively.

Figure 1 show a photograph of the zeolite-iron oxide magnetic nanocomposites attracted toward the magnet.

Figure 1. Magnetic responses of the ZM3 (left) and ZC2 (right) in contact with a magnet.

2.4. Characterization Techniques

The chemical composition of zeolite was determined by a RIX-3000 RIGAKU X-ray fluorescence spectrometer (XRF) equipped with a Rh X-ray tube (operated at 50 kV-60 mA). The surface area was determined by N_2 adsorption isotherm using NOVA 1200 (Quantachrome Corp.). Before adsorption experiments, the samples were degassed at 150 °C for 12 h. The specific surface area was obtained by five points at p/p^0 between 0.05 and 0.20 applying the Brunauer–Emmet–Teller equation to the adsorption data. Bulk density was determined by helium pycnometer (Micromeritics – Accupyc 1330). The phases of the zeolite were determined by X-ray diffraction analyses (XRD) with an automated Rigaku multiflex diffractometer with Cu anode using Co Kα radiation at 40 kV and 20mA over the range (2θ) of 5–80° with a scan time of 1°/min. The images of scanning electron microscopy (SEM) were obtained using the scanning electron microscope Philips XL-30. Magnetization measurements were obtained at room temperature in magnetic fields up to 20 kOe using a vibrating sample magnetometer (Princeton Applied Research, model 4500). Fourier transform infrared (FTIR) spectra were recorded using KBr pellet technique on FTIR spectrometer Nicolet, Nesus 670.

2.5. Adsorption Experiments

The adsorption capacity (mg g^{-1}) of adsorbents was calculated using the Eq. (1):

$$q_t = \frac{(C_0 - C_t)\,V}{M} \tag{1}$$

where q_t is the adsorbed amount of adsorbate per gram of adsorbent at any time t, C_0 and C_t the concentrations of the adsorbate in the initial solution and at any time t, respectively (mg L^{-1}); V the volume of the adsorbate solution added (L) and M the amount of the adsorbent used (g).

Through the following equation (Eq. (2)), it was possible to determine the extraction efficiency in the different experiments performed:

$$R\,(\%) = \left(\frac{C_o - C_t}{C_o}\right)100 \tag{2}$$

where R is the efficiency of extraction or retention percentage, C_0 (mg L^{-1}) is the initial concentration of each adsorbate and C_t (mg L^{-1}) represents the concentration of the adsorbate at time t.

All the experiments, except the study of the influence of temperature, were carried out at laboratory ambient temperature, 298 ± 2 K.

2.5.1. Dye Adsorption

The adsorption was performed using the batch procedure. The experiments were carried out by shaking 50 mg of adsorbent (ZM3) with 5 mL of dye solutions with initial concentration of 10.0 mg L^{-1} in a shaker operated at 360 rpm for 30-420 min in glass bottles. Samples were withdrawn at appropriate time intervals and placed on top of a magnet for 60 min to settle the particles and 4 ml of supernatant was taken for the measurement. The concentration of dye in the supernatant solution was analyzed using a UV spectrophotometer (Cary 1E – Varian) by measuring wavelength of maximum absorption at 493 nm for RO16 and 610 nm for IC. The same experiment was repeated with variations in some of the parameters such as: pH (2-10), mass of adsorbent (25-150 mg) and temperature (293-313 K). Each experiment was duplicated under identical conditions with maximum deviation was found to be 3%. For adsorption isotherms, samples with concentrations ranging from 1.2 to 20.0 mg L^{-1} for RO16 and 2.7 to 19.5 mg L^{-1} for IC were agitated till the equilibrium was achieved.

2.5.2. Heavy Metal Ion Adsorption

Batch adsorption experiments were carried out by shaking 1 g of composite (ZC1, ZC2 or ZC3) with 100 mL of solutions containing metal ions. After shaking, the suspensions were placed on top of a magnet for 60 min to settle the particles and 25 mL of supernatant was taken for the measurement. The supernatant concentrations were determined by complexometric titration with EDTA. In kinetic studies, the contact time was 2, 4, 8, 15 and 24 h with concentrations in the range 260–520 mg L^{-1}. Adsorption isotherms of metal ions at initial concentrations in the range 260–1660 mg L^{-1} were determined after shaking the suspensions for 24 h. Each experiment was duplicated under identical conditions with maximum deviation was found to be 5%.

2.6. Kinetic Models

A study of adsorption kinetics is desirable as it provides information about the mechanism, which is important for the efficiency of the process. It is important to be able to predict the rate at which contamination is removed from aqueous solutions in order to design an adsorption treatment plant.

The rate constants of the adsorbate removal from the solution by zeolite-iron oxide magnetic nanocomposites were determined using Lagergren pseudo-first-order (Eq. 3) and pseudo-second-order (Eq. 4) with linear form of equations as follows:

$$\log(q_e - q_t) = log - \frac{k_1 t}{2.303} \tag{3}$$

$$\frac{t}{q_t} = \frac{1}{k_2 q_e^2} + \frac{1}{q_e} t \tag{4}$$

where q_e is the amount of dye adsorbed at equilibrium (mg g^{-1}), qt is the amount of dye adsorbed at time t (mg g^{-1}), k_1 is the rate constant of the pseudo-first-order adsorption (min^{-1}), and k_2 is the rate constant of the pseudo-second-order kinetics (g mg^{-1} min^{-1}) [32, 33].

The possibility of intraparticle diffusion resistance affecting adsorption was explored using the appropriate intraparticle diffusion model as [34]:

$$q_t = k_{id} t^{\frac{1}{2}} + C \tag{5}$$

where k_{id} is the intraparticle diffusion rate constant (mg g^{-1} $min^{-0.5}$). According to Eq. (5), a plot of qt versus $t^{0.5}$ should be a straight line with a slope k_{id} and intercept C when adsorption mechanism follows the intraparticular diffusion process. Values of the C give an idea about the thickness of boundary layer, i.e., the larger the intercept is the greater the boundary layer effect.

2.7. Adsorption Isotherms

The equilibrium adsorption isotherm is of importance in the design of adsorption systems. Several adsorption isotherm equations are available and two important isotherms are selected in this study, the Langmuir and Freundlich isotherms [35, 36].

The equilibrium data obtained in the present study were analyzed using the linear forms of the expressions of Langmuir (Eq (6)) and Freundlich (Eq (7)) isotherm models:

$$\frac{C_e}{q_e} = \frac{1}{Q_0 b} + \frac{C_e}{Q_0} \tag{6}$$

$$\log q_e = logK_F + \frac{1}{n} \, logC_e \tag{7}$$

where C_e is the equilibrium concentration (mg L^{-1}), q_e the amount adsorbed at equilibrium (mg g^{-1}), Q_0 is the maximum amount of adsorbate per unit weight of adsorbent to form a complete monolayer on the surface (mg g^{-1}); b is the Langmuir isotherm constant (L mg^{-1}), related to the affinity of the adsorption sites; K_F [(mg g^{-1}) (L $mg^{-1})^{1/n}$] and n are the Freundlich constants related to adsorption capacity and adsorption intensity of adsorbents, respectively.

The Langmuir isotherm can also be represented in terms of a dimensionless constant separation factor or an equilibrium parameter:

$$R_L = \frac{1}{1 + bC_o} \tag{8}$$

where b is the Langmuir constant (L mg^{-1}) and C_o the initial concentration of adsorbate (mg L^{-1}). R_L values between 0 and 1 indicate favorable adsorption [37]. The R_L is equal to the ratio of the unused adsorbent capacity to the maximum adsorbent capacity and thus it can be measurement of the adsorbent capacity used and the affinity between the adsorbate and adsorbent.

The non-linear regression Chi-square (χ^2) test was employed as a criterion for the fitting quality due to the inherent bias resulting from linearization of isotherm models. This statistical analysis is based on the sum of the squares of the differences between the experimental and model calculated data, of which each squared difference was divided by the corresponding data obtained by calculating from models [38]. The Chi-square can be represented by Eq. (9).

$$\chi^2 = \sum \frac{(q_e \exp - q_e calc)^2}{q_e calc} \tag{9}$$

where $q_e$exp is the equilibrium capacity of the adsorbent obtained from experiment (mg g^{-1}) and $q_e calc$ is the equilibrium capacity obtained by calculating from the model (mg g^{-1}). A low value of χ^2 indicates that experimental data fit better to the value from the model. Therefore, it is necessary to analyze the data set using the Chi-square test to confirm the best fit isotherm for the adsorption system combined with the values of the correlation coefficient.

3. Results and Discussion

3.1. Dye Adsorption Studies

3.1.1. Kinetic Studies

Fig. 2 shows the effect of contact time on the adsorption capacities of RO16 and IC by ZM3. It is clear that the adsorption capacity of ZM3 increased rapidly in the initial stages of contact time and gradually increased with prolonging the contact time until equilibrium. It can be seen that the adsorption equilibrium of RO16 and IC on ZM3 were reached at 300 and 360 min, respectively.

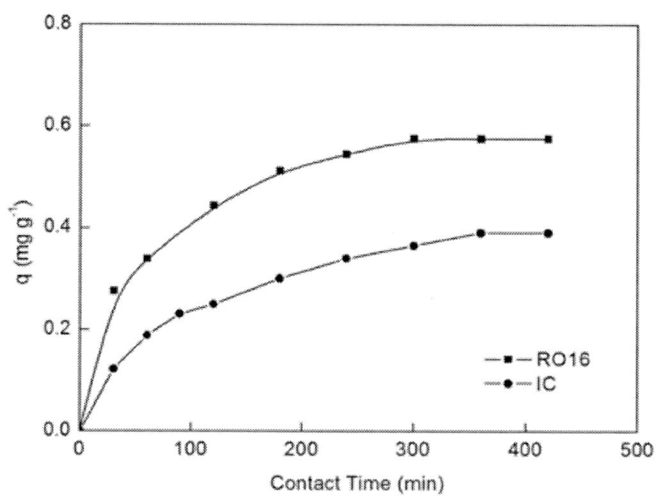

Figure 2. Effect of contact time on the adsorption of RO16 and IC by ZM3.

In order to investigate the adsorption processes of RO16 and IC by ZM3 pseudo-first-order (Eq. 3), pseudo-second-order kinetic (Eq. 4) and intraparticle diffusion (Eq. 5) models were used.

The calculated kinetic constants values and the corresponding linear regression correlation constants are given in Table 1. The pseudo second-order kinetic model better describes all of the dyes adsorption according to the correlation coefficients ($R_2 \geq 0.99$) than those of pseudo first-order model. In addition, the experimental adsorption capacity values were very close to the model-calculated adsorption capacity data (not shown), verifying the high correlation of adsorption to the pseudo-second-order model. The fitted linear regression plots of pseudo-second-order model are shown in Fig. 3.

The plots of intraparticle diffusion of RO16 and IC onto the ZM3 are illustrated in Fig. 4. The linearity of fitting lines in plots points to the presence of intraparticle diffusion. However, the deviation of the lines from the origin indicates that the pore diffusion is not the only rate controlling step [34, 39]. As the plot is often multilinear for many adsorption systems, it is common to segment it into two or more straight lines and to assume that different adsorption mechanisms control the step represented by each straight line [40]. Referring to RO16/ZM3 system, two distinct regions are observed (Fig. 4a). The first section of plots indicates that boundary layer diffusion probably limited dye adsorption. The second section shows the occurrence of intraparticle diffusion as the adsorption limiting step [41, 42].

Table 1. Kinetic parameters for the adsorption of dyes onto ZM3

Dyes	Pseudo-first-order			
	k_1 (min^{-1})	R_1		
RO16	1.15 x 10^{-2}	0.995		
IC	0.94 x 10^{-2}	0.973		
	Pseudo-second-order			
	k_2 ($g\ mg^{-1}\ min^{-1}$)	h ($mg\ g^{-1}\ min^{-1}$)	q_e ($mg\ g^{-1}$)	R_2
RO16	3.90 x 10^{-2}	1.29 x 10^{-2}	0.659	0.999
IC	2.08 x 10^{-2}	0.470 x 10^{-2}	0.477	0.989
	Intraparticle diffusion			
	C ($mg\ g^{-1}$)	k_i ($mg\ g^{-1}\ min^{-0.5}$)	R_i	
RO16	0.251	2.83 x 10^{-3}	0.956	
IC	0.037	1.87 x 10^{-2}	0.998	

Table 2. Langmuir and Freundlich parameters for the adsorption of RO16 and IC onto ZM3

Dye	Langmuir			
	Q_o ($mg\ g^{-1}$) b ($L\ mg^{-1}$) R χ^2			
RO16	1.06	0.582	0.984	0.169
IC	0.583	0.359	0.988	0.0382
	Freundlich			
	K_f n R χ^2 [($mg\ g^{-1}$) ($L\ mg^{-1}$)$^{1/n}$]			
RO16	0.340	2.00	0.874	0.452
IC	0.191	2.68	0.972	0.0161

The absence of such features in the plot of IC/ZM3 system (Fig. 4b) indicated that the steps were indistinguishable from one another and that the intraparticle diffusion was a prominent process right from the beginning of dye-solid interaction [43].

3.1.2. Adsorption Isotherms

The isotherm parameters for RO16 and IC of two models are presented in Table 2. The isotherms of the two dyes with the experimental data points and the two theoretical isotherms plotted on the same graph are given in Fig. 5.

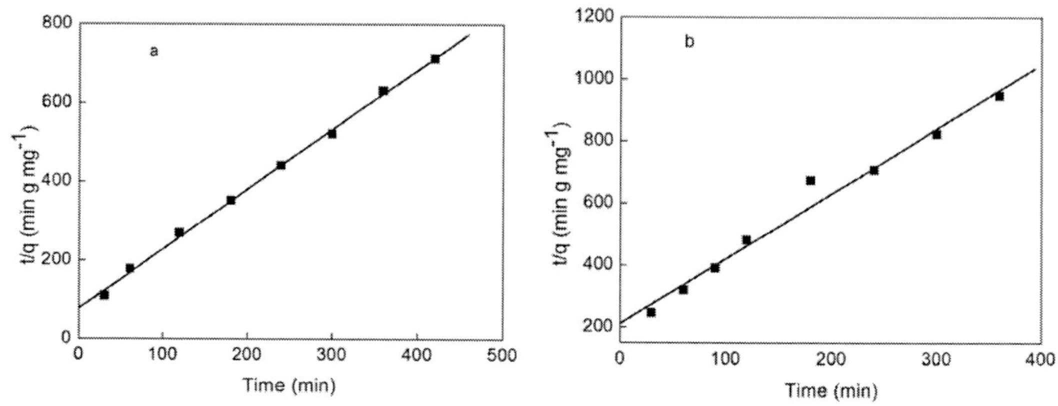

Figure 3. Pseudo-second-order adsorption kinetics of (a) RO16, (b) IC on ZM3.

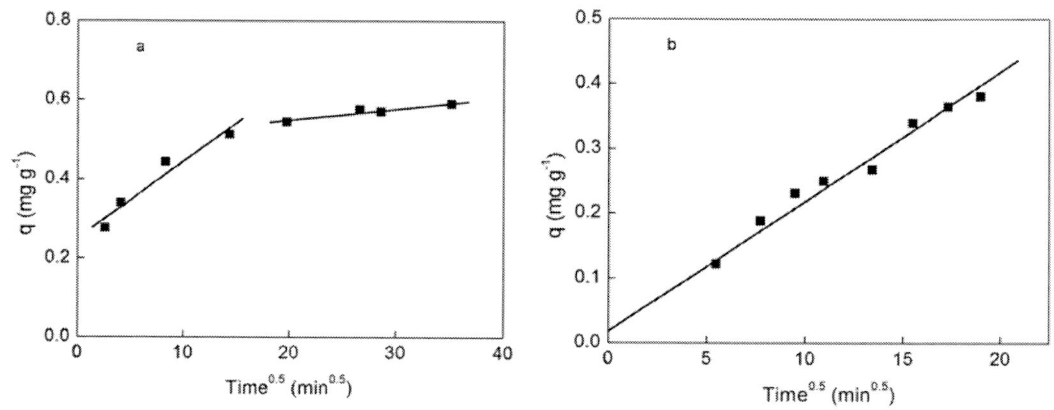

Figure 4. Intraparticle diffusion model of (a) RO16, (b) IC on ZM3.

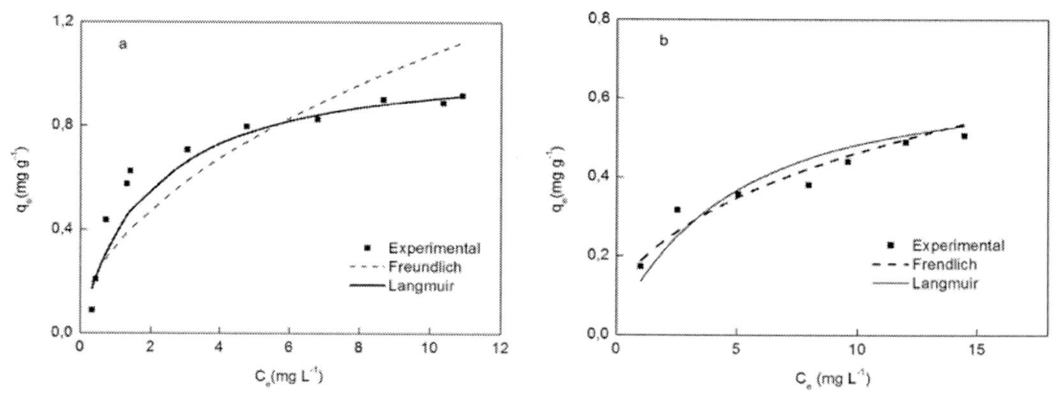

Figure 5. Adsorption isotherms of dyes onto ZM3: (a) RO16; (b) IC.

The Langmuir isotherm gave a better fit than the Freundlich isotherm by giving greater R value closer to unity (0.98) and smaller value of χ^2 for RO16 adsorption by ZM3. For IC, the correlation coefficients for the Freundlich model and the Langmuir model are much similar but according to the χ^2 values, the best-fit isotherm model was the Freundlich isotherm model

[38]. The value of n, which is significantly higher than unity, indicated that both RO16 and IC dyes are favourably adsorbed by zeolitic material for under the experimental condition [44].

In general for the adsorption process, the degree of adsorption depends on both the active sites of the adsorbent and the adsorbate properties. An interaction occurs between active sites on zeolite-iron oxide magnetic nanocomposite (Si-O and O-H) and SO_3^-, –N=N– and -N=C– -C=C– groups and aromatic skeletal vibrations of dyes [41, 45, 46].

3.1.3. Effect of Solution pH

The pH was shown to be an important parameter in terms of the adsorption capacity, influencing not only the surface charge of the adsorbent but also the level of ionization of the material present in the solution. The effect of initial pH values for the adsorption of RO16 and IC dyes in solution onto ZM3 was investigated in the range of pH 2.0–10.0 (Fig. 6).

The adsorption of RO16 dyes on ZM3 is controlled by a pH-independent adsorption mechanism. The fact that the adsorption of dye onto zeolite from fly ash composite is not significantly affected with changing of initial pH of the dye solution may be attributed to the weak electrostatic interaction between the dye molecules and the sites on the surface of the zeolite particles [47] and neither the chemical species present in solution nor the effects of pH on the material superficies are important parameters for the adsorption processes.

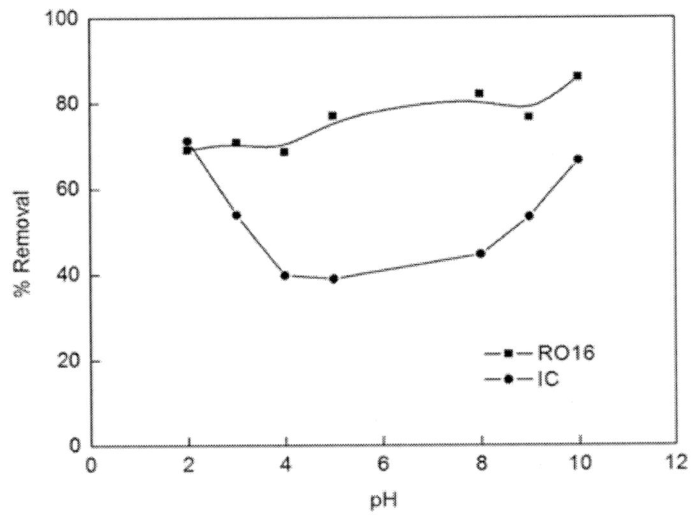

Figure 6. Effect of initial solution pH on dyes removal.

It can be seen that the adsorption of IC was pH-dependent (Fig. 6). The amount adsorbed decreased as pH decreased from 2 to 4 and from 10 to 8 and remained constant in the pH interval between 4 to 8. The chemical species has an important effect on the adsorption of indigo carmine on the magnetic composite. Indigo carmine may undergo a redox process involving the uptake of one proton per electron transferred or two protons per two electrons transferred. Transfer of the second electron results in formation of the hydroquinone [48]. These results may indicate that the species present interacts with the adsorbent and the interactions between IC and composite surface can occurs due electrostatic force and weak interaction of van der Walls forces. Previous studies have demonstrated that another possible interaction mechanism of the IC with magnetic zeolitic material composite is catalytic degradation [48, 49].

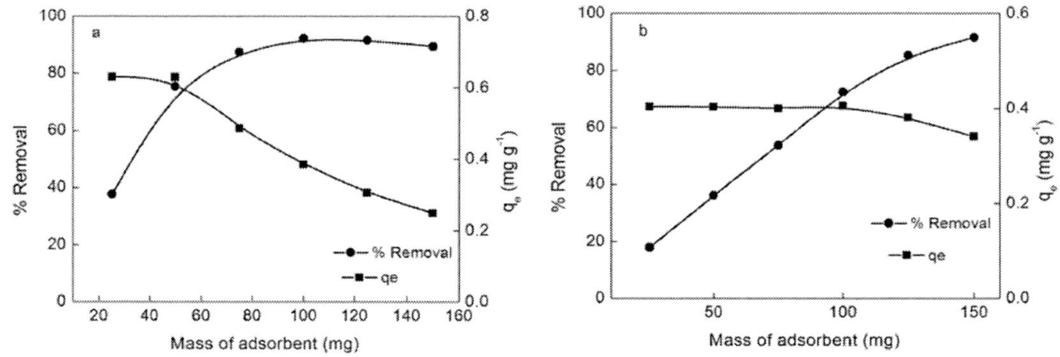

Figure 7. Effect of adsorbent mass on the percentage of removal and amount of dye adsorbed on ZM3: (a) RO16; (b) IC.

3.1.4. Effect of Adsorbent Dosage

The study of adsorbent mass for the removal of dyes from aqueous solution was carried out using adsorbent masses ranging from 25 to 150 mg of ZM3 and fixing the initial dye concentration (Fig. 7).

The highest amount of dye removal was attained for adsorbent mass of 100 mg for RO16 (Fig. 7a). For adsorbent dosages higher than this value, the percentage of dye removal remained almost constant. The increases in the removal efficiency with the adsorbent dose can be attributed to increase surface area and the availability of more adsorption sites as already reported in several papers [50, 51].

On the other hand, the increase in the adsorbent masses promoted a remarkable decrease in the amount of RO16 uptake per gram of adsorbent (q_e). The drop in adsorption capacity is basically due to sites remaining unsaturated during the adsorption and particle aggregation [52, 53]. This effect can be also mathematically explained by combining of equations [54].

For IC, along with the increase of adsorbent dosage from 25 to 150 mg, the percentage of dye removal increased and the adsorption capacity (q_e) decreased after 100 mg (Fig. 7b).

3.1.5. Effect of Temperature

As the temperature can influence the final value of adsorption, the study of its effects on the adsorption process becomes an important aspect in waste water treatment. The variation was explained on the basis of thermodynamic approach. The thermodynamic parameters, ΔG^o, ΔH^o and ΔS^o, were calculated with the help of following Eqs. (10)-(13):

$$K_C = \frac{C_A}{C_S} \tag{10}$$

$$\Delta G° = -2{,}303RT\ logK_C \tag{11}$$

$$\Delta H° = 2{,}303R\left(\frac{T_1T_2}{T_2 - T_1}\right) log\ \frac{K_{C2}}{K_{C1}} \tag{12}$$

$$\Delta S^\circ = \frac{\Delta H^\circ - \Delta G^\circ}{T} \qquad (13)$$

where C_A is the equilibrium concentration of the adsorbate ions on adsorbent (mg L^{-1}); C_S is the equilibrium concentration of the adsorbate ions in the solution (mg L^{-1}); R is the gas constant, T is temperature on the absolute scale and K_C, K_{C1}, and K_{C2} are the equilibrium constants at temperature T, T_1 and T_2, respectively.

The values of these parameters were calculated and are presented in Table 3. The values of ΔG^o of dye/ZM3 adsorption systems are all negative, which indicates the spontaneous adsorption processes. Moreover, the increase in the absolute value of ΔG^o with increasing temperature indicates that higher temperatures facilitated the adsorption. The positive value of ΔH^o indicates that the adsorptions of both dyes onto ZM3 are endothermic. The positive value of ΔS^o suggests that in the dye/ZM3 adsorption systems, although the adsorption process caused an entropy decrease for the adsorbed dye molecules, the entropy increase of dye molecules in dissolved state due to rising temperature was much greater, and consequently the ΔS^o of the whole adsorption system increased.

Table 3. Thermodynamic parameters for adsorption of both dyes onto zeolite from fly ash-iron oxide magnetic nanocomposite

Dye	T (K)	ΔG^o (kJ mol^{-1})	ΔH^o (kJ mol^{-1})	ΔS^o (J $K^{-1}mol^{-1}$)
RL16	298	-0.71	74.0	251
	303	-1.96	13.2	50.2
	313	-2.46		
IC	298	-0.44	1.83	7.60
	303	-0,47	22.7	76.4
	313	-1.24		

3.2. Adsorption Studies of Zn^{2+} Ions onto Zeolite-Iron Oxide Magnetic Nanocomposites

3.2.1. Adsorption Isotherms

Adsorption isotherms were determined for the adsorption of Zn^{2+} ions onto pure zeolite (ZFA) and zeolite/Fe oxide composites (ZC1, ZC2 and ZC3). The results are shown in Fig. 8. The adsorption capacities increased with increasing equilibrium Zn^{2+} concentrations and eventually attained a plateau value.

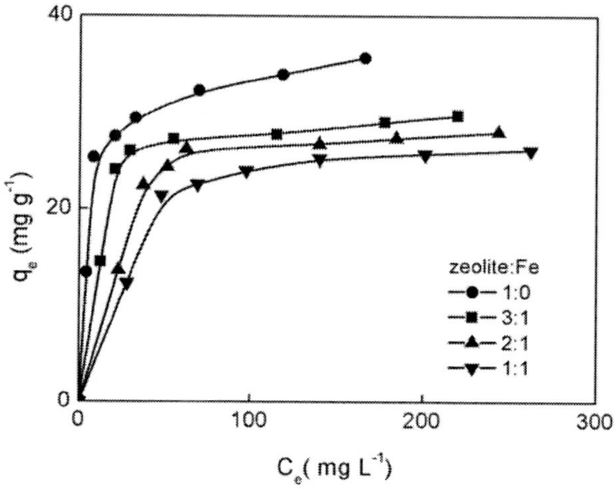

Figure 8. Equilibrium adsorption isotherms for Zn^{2+} ions onto pure zeolite and zeolite/Fe oxide composites 3:1, 2:1 and 1:1.

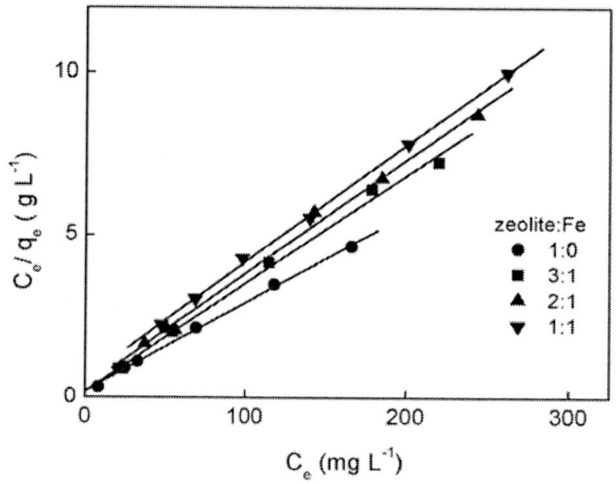

Figure 9. Langmuir plots for the adsorption of Zn^{2+} ions onto pure zeolite and zeolite/Fe oxide composites.

Table 4. Langmuir and Freundlich parameters for the adsorption of Zn^{2+} ions onto pure zeolite and zeolite/Fe oxide composites

Adsorbent	Langmuir Q_o (mg g^{-1}) b (L mg^{-1}) R_L R^2			
Zeolite:Fe 1:1	27.8	0.062	0.668	0.999
Zeolite:Fe 2:1	28.6	0.108	0.536	0.998
Zeolite:Fe 3:1	30.0	0.175	0.417	0.998
Zeolite	36.8	0.113	0.016	0.999
	Freundlich $K_f{}^*$ n R^2			
Zeolite:Fe 1:1	53.4	8.75	0.869	
Zeolite:Fe 2:1	17.8	12.4	0.737	
Zeolite:Fe 3:1	21.1	16.5	0.824	
Zeolite	19.3	8.42	0.983	

(*)[(mg g^{-1})(L mg^{-1})$^{1/n}$].

The adsorption capacity for the Zn^{2+} decreases in the order zeolite > 3:1 composite > 2:1 composite > 1:1 composite. It is interesting to note that the presence of the iron oxide does not significantly affect the adsorption capacity of the zeolite on the composites. The lower adsorption capacity of the composites is probably related to the decrease in surface area caused by the presence of the iron oxide. Similar results were obtained for the adsorption of Zn^{2+} on clay/iron oxide composites [15].

The isotherm shapes are largely determined by the adsorption mechanism and can therefore be used to diagnose the nature of the adsorption [55]. The adsorption isotherm for solution may be classified into four main classes relating to their shapes termed S, L, H and C and subgroups 1, 2, 3, 4 or max. Figure 8 clearly shows that adsorption isotherms are L2 type. In L2-type isotherms, adsorption of solute on the adsorbent proceeds until a monolayer is established, with the formation of more than one layer not being possible.

Figures 9 and 10 depict linear Langmuir and Freundlich plots for the adsorption of Zn^{2+} ions onto pure zeolite and zeolite/Fe oxide composites, respectively. The isotherm constants and correlation coefficients calculated for Freundlich and Langmuir equations are given in Table 4.

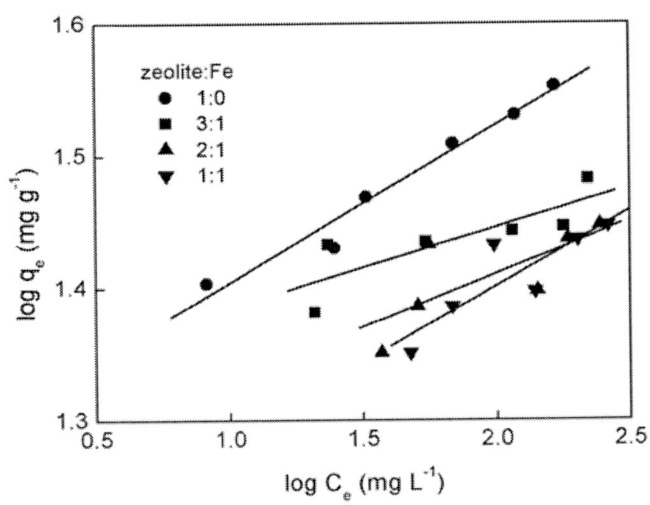

Figure 10. Freundlich plots for the adsorption of Zn^{2+} ions onto pure zeolite and zeolite/Fe oxide composites.

It can be observed from the Table 4 that the according the correlations coefficients criterion (R^2), the Langmuir model fits the experimental data better than the Freundlich model. The calculated R_L values were between 0 and 1 at all the adsorbents studied indicating the favorable condition for adsorption process.

The 2:1 zeolite-iron oxide nanocomposite (ZC2) was chosen for the studies of kinetic, thermodynamic characteristics and adsorption process of Cd^{2+} and Pb^{2+} from aqueous solution.

3.2.2. Effect of Contact Time and Initial Concentration

Figure 11 shows the effect of initial Zn^{2+} concentration at different contact time on adsorption of ZC2. The results showed that the extent of adsorption increased rapidly in the initial stages before 5 h and then approaches equilibrium after 10 h.

During the initial stage of adsorption, a large number of vacant surface sites are available for adsorption. After lapse of some time, the remaining vacant surface sites are difficult to be occupied due to repulsive forces between the solute molecules on the solid surface and the bulk phase. Besides, the metal ions are adsorbed into the pores that get almost saturated with metal ions during the initial stage of adsorption. Thereafter, the metal ions have to traverse farther and deeper into the pores encountering much larger resistance. This results in the slowing down of the adsorption during the later period of adsorption.

An increase in the initial Zn^{2+} concentration leads to an increase in the adsorption capacity of the Zn^{2+} on composite. The initial concentration provides an important driving force to overcome all mass transfer resistances of the ion between the aqueous solution and solid phases. Hence a higher initial concentration of ion enhances the sorption process.

The time to reach equilibrium was independent of initial Zn^{2+} concentration. The adsorption of Zn^{2+} decreased from 86 to 54% by increasing Zn^{2+} concentration from 260 to 520 mg L^{-1}. The increase in initial Zn^{2+} concentration results in the fast attainment of saturation of composite and higher residual Zn^{2+} in the equilibrium solution. Further, it was observed that removal curves are single, smooth and continuous suggesting the formation of monolayer of adsorbate on the surface of the adsorbent [56].

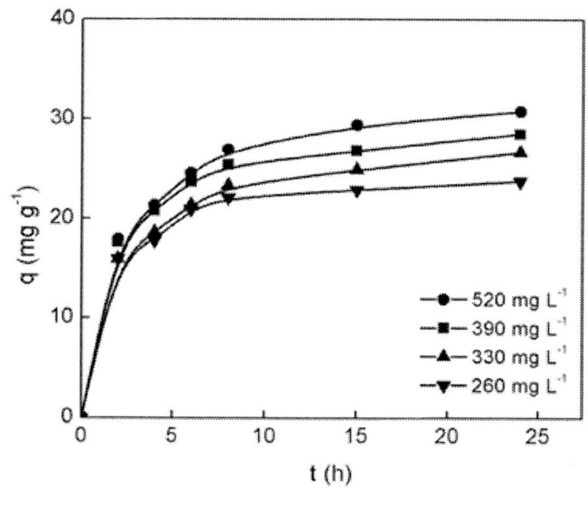

Figure 11. Adsorption kinetics for Zn $^{2+}$ onto zeolite/iron oxide (2:1) composite.

3.2.3. Kinetic Models

Figures 12-14 present the plots for the adsorption of Zn^{2+} on zeolite/iron oxide (2:1) using the pseudo-first-order kinetics, pseudo-second-order kinetics and intraparticle diffusion model. The calculated kinetic constants values and the corresponding linear regression correlation constants are given in Table 5.

As seen in Table 5, the correlation coefficients values obtained from pseudo-second-order kinetic were found to be higher than for the pseudo-first-order and intraparticle diffusion models. Kinetic studies on adsorption of composite for Zn^{2+} ions by using the pseudo-second-order rate expression provided the best fitting kinetic model.

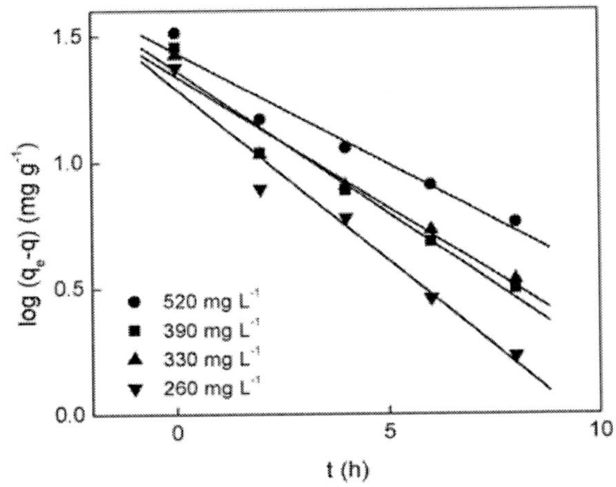

Figure 12. Pseudo-first-order kinetic plot for Zn^{2+} onto zeolite/iron oxide (2:1) composite.

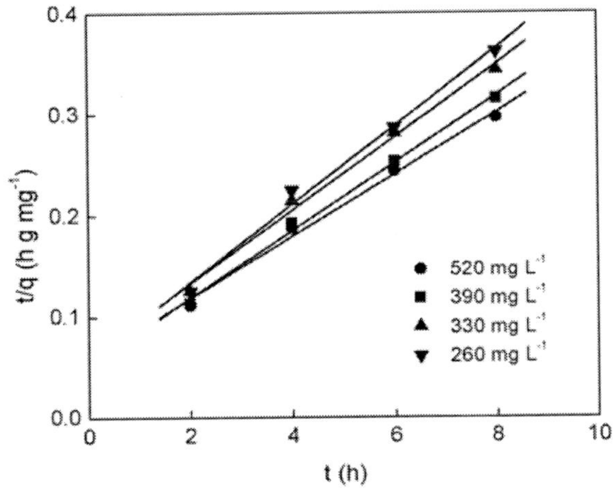

Figure 13. Pseudo-second-order kinetic plot for Zn^{2+} onto zeolite/iron oxide (2:1) composite.

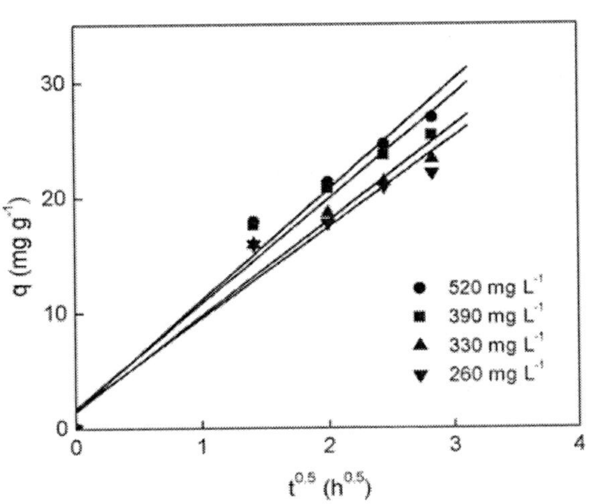

Figure 14. Intraparticle diffusion plot for Zn^{2+} onto zeolite/iron oxide (2:1) composite.

Table 5. Kinetic constants for Zn^{2+} onto zeolite/iron oxide (2:1) composite

[Zn] (mg L^{-1})	Pseudo- first-order			
	K_1 (h^{-1})	R_1^2		
260	3.14 x 10^{-1}	0.984		
330	3.34 x 10^{-1}	0.979		
390	2.61 x 10^{-1}	0.980		
520	2.04 x 10^{-1}	0.976		
	Pseudo- second-order			
	K_2 (g mg^{-1} h^{-1})	h (mg g^{-1} h^{-1})	q_e (mg g^{-1})	R_2^2
260	2.62 x 10^{-2}	0.176	25.9	0.995
330	2.14 x 10^{-2}	0.164	27.7	0.996
390	2.18 x 10^{-2}	0.193	29.8	0.998
520	0.164 x 10^{-2}	1.76	32.6	0.996
	Intraparticle diffusion			
	C (mg g^{-1})	K_i (mg g^{-1} $h^{-0.5}$)	R_i^2	
260	1.68	7.86	0.974	
330	1.53	8.23	0.981	
390	1.78	9.04	0.979	
520	1.58	9.53	0.985	

For the diffusion model plots (Fig. 14), the linearity of the fitting lines points the presence of intraparticle diffusion. However, the deviation of the lines from the origin indicates that although intraparticle diffusion was involved in the adsorption process, it was not the rate-controlling step [34].

3.2.4. Evaluation of Thermodynamic Parameters

The adsorption of Zn^{2+} from solutions aqueous on the zeolite/iron oxide (2:1) composite at different temperatures was studied. The values of standard free energy (ΔG^o), enthalpy (ΔH^o) and entropy (ΔS^o) were calculated (Eqs 10-13) and are presented in Table 6.

Table 6. Thermodynamic parameters for adsorption of Zn^{2+} on zeolite/iron oxide (2:1) at different temperatures

Temperature (K)	ΔG^o (kJ mol^{-1})	ΔH^o (kJ mol^{-1})	ΔS^o (J K^{-1} mol^{-1})
298	-3.90	6.42	34.6
303	-4.07	47.8	171
313	-5.78		

It is obvious from this table that the negative and small values of free energy change (ΔG^o) were an indication of spontaneous nature of the adsorption process. The positive values of standard enthalpy change (ΔH^o) at different temperatures were indicating the endothermic nature of the adsorption process. The positive values of (ΔS^o) suggested the increased randomness at the solid/solution interface during the adsorption of the zinc onto composite.

3.3. Adsorption Isotherms of Pb^{2+} and Cd^{2+} on 2:1 Zeolite/Iron Oxide Composite

The adsorption isotherms of Pb^{2+} and Cd^{2+} on 2:1 zeolite/iron oxide composite (ZC2) are shown in Fig. 15. The isotherm of Zn^{2+} is plotted on the same graph for comparison of results. The shape of isotherms are of ''L2'' type according to Giles classification for isotherms [55]. This shape indicated that the data have reached a maximum value, resulting in the presence of the plateau. L-isotherm type (or Langmuir isotherm type) is usually associated with ionic substrates (e.g., metal cations) adsorption with weak competition from the solvent molecules.

The Q_o, b, K_f, n values and the linear regression correlations (R^2) for Langmuir and Freundlich are listed in Table 7. Based on these data, it showed that the Langmuir isotherm fitted the data better with regression coefficients $R^2 > 0.99$. The R_L values shown were $0 < R_L < 1$ suggesting the favorable adsorption isotherm of metals onto ZC2.

Using the Langmuir model, the maximum adsorption capacity for the metals can be estimated as: $Pb^{2+} > Cd^{2+} > Zn^{2+}$. The same order of selectivity obtained with zeolite-from fly ash-iron oxide composite was determined with natural zeolites and is related to the characteristics of the metal ions [57, 58].

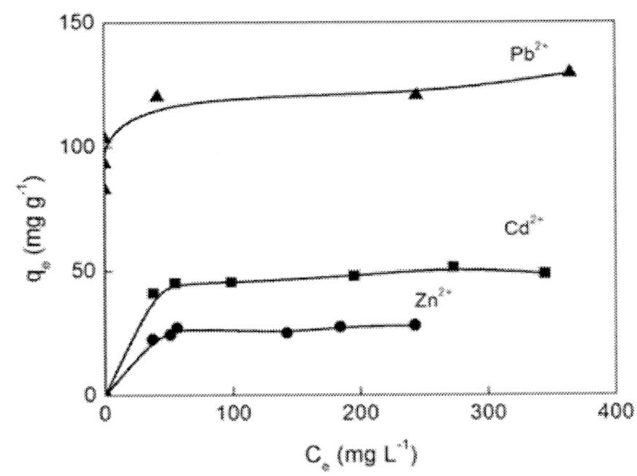

Figure 15. Adsorption isotherms of Pb^{2+}, Cd^{2+} and Zn^{2+} on 2:1 zeolite/iron oxide composite.

Table 7. Langmuir and Freundlich parameters for the adsorption of metal ions onto 2:1 zeolite/iron oxide composite

Adsorbate	Langmuir			
	Q_o (mg g^{-1})	b (L mg^{-1})	R_L	R^2
Zn^{2+}	28.5	0.108	0.536	0.998
Cd^{2+}	50.8	0.114	0.147	0.999
Pb^{2+}	127	0.657	0.917×10^{-3}	0.999
	Freundlich			
	K_f^*	n	R^2	
Zn^{2+}	17.8	12.4	0.737	
Cd^{2+}	31.7	12.7	0.918	
Pb^{2+}	109	41.7	0.674	

(*)[(mg g^{-1})(L mg^{-1})$^{1/n}$].

The selectivity could be the result of various factors which influence ion-exchange behavior in zeolites. The specificity of a surface containing fixed charges for metal ions can be accounted for in terms of ion hydration and electrostatic bond energies [59, 60]. The metal ions with higher electronegativity adsorb more readily and the obtained selectivity series is in agreement with the metal ions electronegativity, namely Pb^{2+} (2.33) > Cd^{2+} (1.69) > Zn^{2+} (1.65) [61]. Also, the differential adsorption of metal ions may be ascribed to the difference in their ionic radii. The smaller the ionic radius, the greater its tendency to hydrolyze leading to reduced adsorption. The ionic radii of Pb^{2+}, Cd^{2+} and Zn^{2+} are 1.21 Å, 0.97 Å and 0.74 Å, respectively [61]. The adsorption process of metal cations by zeolites has been extensively studied [62]. Most studies considered it as a cation exchange process between metal cations presence in water and cations in the zeolite framework. Researchers also proposed the adsorption of metal cations was due to the surface reaction with terminal hydroxyl groups on zeolites and the combination of positive charges of metal cations and negative charges on zeolite surfaces. *Metal* removal is often accomplished by *precipitating,* particularly at high initial *metal* ion concentrations.

4. Characterization of Materials

4.1. Physicochemical Properties

Table 8 shows some properties of coal fly ash (CFA), zeolite from fly ash (ZFA) and zeolite/Fe oxide composites (ZM3 and ZC2). The chemical compositions and BET results suggested that CFA was partly converted into zeolitic products. The chemical composition of materials is mainly silica, alumina and iron oxide. A significant amount of Na element is incorporated in the zeolitic product ZFA due the hydrothermal treatment with NaOH solution. The inclusion of magnetite during synthesis of composite ZM3 does not seem to affect the BET area of the zeolite (there is a perceptible decrease, from 53 to 51, but it must be taken into account that in the composite sample one quarter of the weight corresponds to magnetite). The fact that the surface area is roughly maintained indicates that most of the pore mouths of the zeolitic channels remain open after the inclusion of magnetite. The value of surface area of the composite prepared via iron chloride ions co-precipitation method (ZC2) was much higher than that of zeolitic material (ZFA). This is probably related to the porous texture of the formed iron oxide contributing for the increase of the surface area and microporosity. Colors of iron oxide are very dependent on grain size and chemical composition and this feature is a first auxiliary attribute in their identification. The ZM3 was black suggesting the presence of pure magnetite while the ZC2 was brownish product, indicating a mixture of two iron oxides, magnetite and maghemite.

4.2. X-Ray Diffraction Analysis

XRD patterns of samples are shown in Fig. 16. It can be seen that the peaks in the synthesized magnetite MAG, Fe_3O_4, are in agreement with the data of the face-centered cubic (fcc) of the reported pattern of reference magnetite in the ICDD nº 19-629. All MAG peaks showed broad diffraction lines and well defined peaks indicating that the crystallites were very small.

The XRD analyses of all zeolitic materials showed that hydroxysodalite (ICDD 00-011-0401) was the major zeolitic phase with peaks of quartz (ICDD 001-0649) and mullite (ICDD 002-0430) of fly ash that remained after the alkaline hydrothermal treatment.

The obtained zeolite from coal fly ash (ZFA) showed crystalline components no peaks of magnetite indicating the absence of same in accordance with the composition of the material.

Table 8. Chemical compositions and textural properties of materials

	CFA	ZFA	ZM3	ZC2
SiO_2	27.7	23.9	11.08	9.8
Al_2O_3	9.1	10.8	4.6	4.4
Fe_2O_3	39.8	48.3	77.8	80.2
K_2O	9.4	2.2	0.80	0.85
CaO	4.0	4.9	2.1	1.6
TiO_2	3.0	3.7	1.2	0.96
ZnO	1.7	1.6	0.43	0.44
As_2O_3	1.4	0.09	0.09	0.09
SO_3	1.3	1.1	0.44	0.31
MgO	0.46	0.51	0.17	0.20
PbO	0.44	0.51	0.16	0.16
U_3O_8	0.30	0.35	0.35	0.35
Na_2O	0.22	0.86	0.33	0.28
MnO	0.21	0.25	0.14	0.14
ZrO_2	0.15	0.19	0.04	0.03
SrO_2	0.13	0.16	0.16	0.02
P_2O_5	0.13	0.08	0.44	0.44
Cr_2O_3	0.13	0.10	0.10	0.03
Rb_2O	0.10	0.03	0.03	0.03
NiO	0.09	0.13	0.11	0.05
GeO_2	0.07	0.03	0.03	0.03
CuO	0.07	0.08	0.08	0.08
Ga_2O_3	0.05	0.02	0.02	0.02
S_{BET} (m^2/g)	10.6	53.0	51.0	78.2
Bulk density (g/cm^3)	2.40	2.54	2.96	2.96

The crystallite size (D) of magnetite particles for samples of MAG, ZM3 and ZC2 was estimated from the X-ray diffraction peak (311) using the Scherrer formula [63] represented by the Equation 14. The Scherrer equation is limited to nano-scale and is given for particles of spherical shape (shape factor of typical value 0.9). In this work in order to apply the Scherrer formula the spherical shape of nanoparticles was considered.

$$D = \frac{0.9\,\lambda}{\beta_{1/2} cos\theta} \tag{14}$$

where λ is the wavelength of the X-ray source ($Cu_{K\alpha 1}$), $\beta_{1/2}$ is the angular width at the half maximum intensity of the diffraction peak (rad) and θ is the Bragg angle.

The estimated crystallite size of magnetite of the samples are given in Table 9. As can be seen the crystallite is nanosize, which may be smaller or equal to the grain size. The nanoparticles are also evidenced in the XRD pattern by broad peaks and distinct. The magnetite prepared via iron chloride ions co-precipitation method present in the ZC2 composite showed crystallite size smaller than the ZM3 composite.

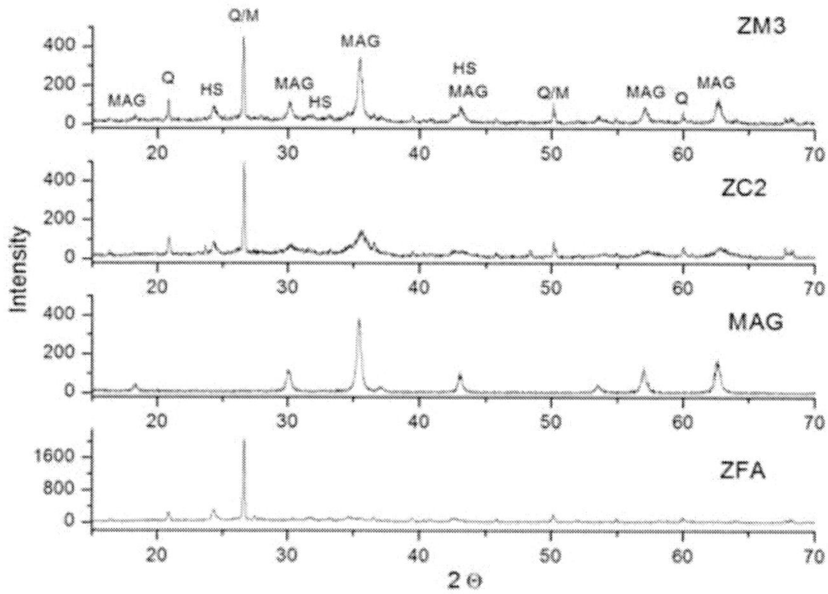

Figure 16. X-ray diffractograms for zeolite from fly ash (ZFA), magnetite (MAG) and zeolite/Fe oxide composites (ZM3 and ZC2). M = mulitte; Q = quartz; MAG = magnetite; HS = hydroxysodalite.

Table 9. XRD crystallite size of the samples of MAG, ZM3 and ZC2 for the (311) peak of magnetite

sample	MAG	ZM3	ZC2
crystallite size (nm)	17	18	8

4.3. SEM Studies of Materials

As can be seen in Fig. 17, the SEM images of ZFA, ZM3, ZC2 and MAG reveal non-uniformly sized particles. The particles form agglomerates which consist of irregularly shaped grains of various sizes, usually less than 500 nm. However some are as big as 1 μm and there is not a narrow distribution of particle sizes.

The original fly ash particles had a smooth surface because the surface is covered by an aluminosilicate glass phase [29]. By contrast, the surface of zeolitic materials is rough, indicating that zeolite crystals were deposited on the surface of underlying fly ash particles during the hydrothermal treatment.

Figure 17. SEM images of zeolite from fly ash (ZFA), zeolite/Fe oxide composites (ZM3 and ZC2) and magnetite (MAG).

4.4. Magnetization Measurements

The particle morphology and distribution of particle sizes can affect the magnetic properties such as residual magnetism. Among other effects, the response to magnetic force has been attributed to the size and shape of the particles.

Magnetic materials can retain a memory of an applied field once it is removed. This behavior is called hysteresis and a plot of the variation of magnetization with magnetic field is called a hysteresis loop.

The coercivity of fine particles has a strong dependence on the particle size. The decrease in coercivity with decreasing particle size in the single domain region is due to thermal effects. Below a critical diameter, the coercivity is zero because the thermal effects are strong enough to overcome the anisotropy energy and demagnetize the sample. Such particles are referred to as superparamagnetic and the hysteresis loop is not observed [64].

The magnetic properties of the zeolite/Fe oxide composites ZM3 and ZC2 were evaluated using a vibrating-sample magnetometer. The two powdered samples were subjected to magnetic measurements at 300 K cycling the field between ± 20 kOe and theirs magnetic properties were investigated on the obtained hysteresis loops shown in Fig. 18.

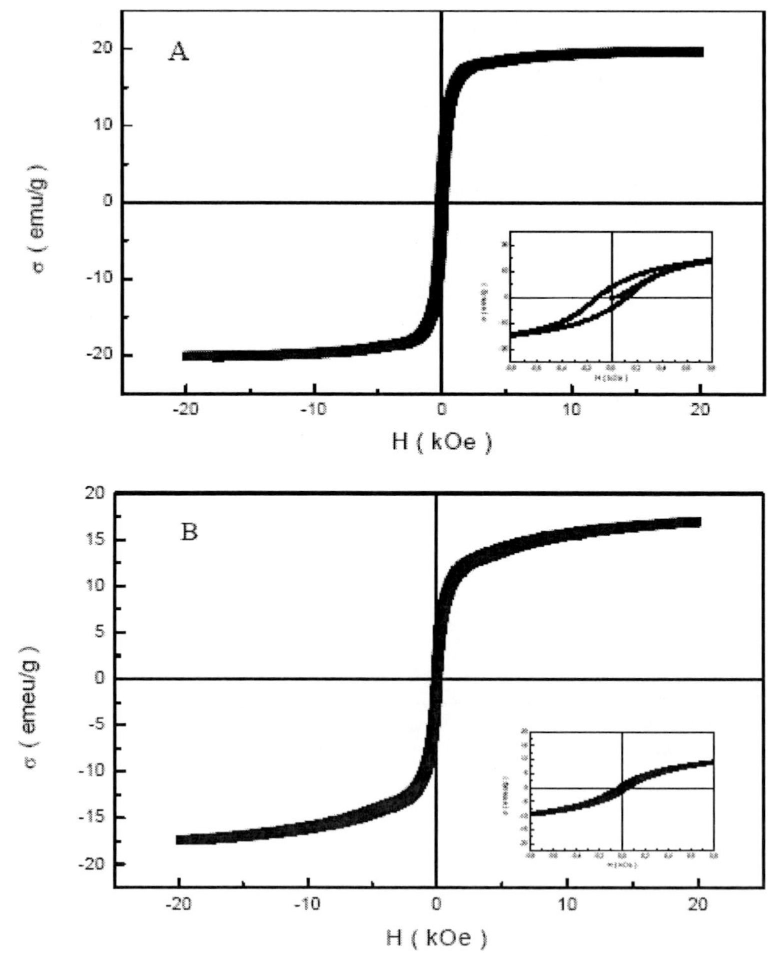

Figure 18. Magnetization loops measured at 300 K for the particles of (A) ZM3; (B) ZC2.

From the plot of magnetization (M) and magnetic field (H) and its enlargement near the origin, the very weak hysteresis revealed the resultant magnetic nanoparticles were nearly superparamagnetic with a saturation magnetization (Ms) of 19.9 and 19.7 emu g^{-1} for ZM3 and ZC2, respectively, at room temperature (300 K) and so that the average size of the particles is a few tens of nanometer which is in agreement with the XRD and SEM results.

As it was seen in the SEM images (Fig. 17), the samples are formed by grains of particles, so the result of the magnetic properties represents a total for all particles present in the cluster. The non-zero coercivity can be explained because the samples were more affected by the grain growth than by the reduction process and the formation of very fine grains.

This experiment proved that zeolite from fly ash-iron oxide nanocomposites possessed magnetism and could be potentially used as a magnetic adsorbents to remove compounds in liquid phase. As displayed in Fig. 19, the magnetic adsorbent can be removed from the liquid by applying an external magnetic field and re-dispersed (or resuspension) in another liquid medium after removal of the external magnetic field.

Figure 19. Separation of zeolite from fly ash-iron oxide composite from aqueous solution by a magnet after adsorption process.

4.5. FT-IR Analysis

The FTIR spectra of the materials are given in Fig. 20. For MAG it is possible to visualize the absorption band of strong intensity at 568 cm^{-1} which can be assigned to the Fe-O stretching mode of the tetrahedral and octahedral sites of the magnetite [65].

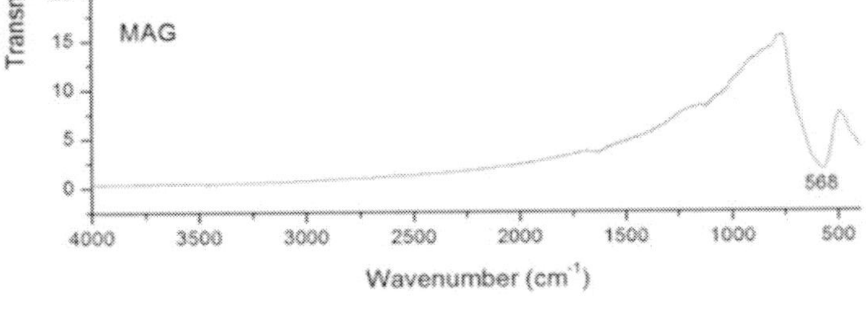

Figure 20. (Continued).

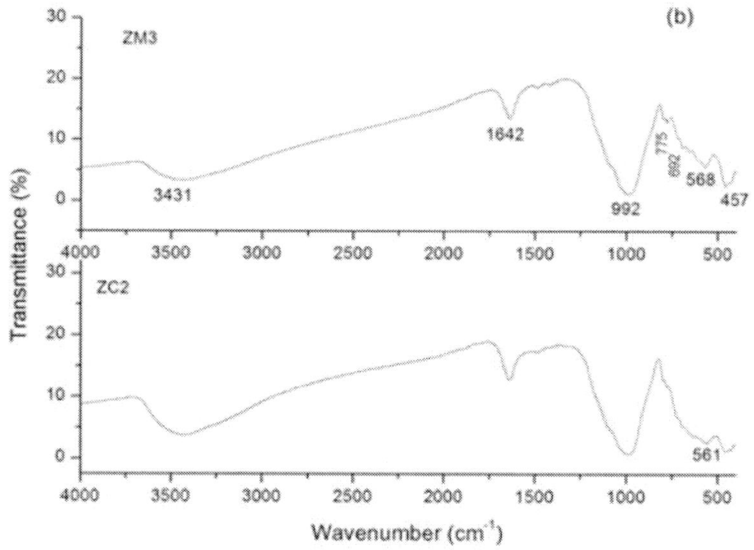

Figure 20. FTIR spectra of: (a) zeolite from fly ash (ZFA) and magnetite (MAG); (b) zeolite/Fe oxide composites (ZM3 and ZC2).

The absorption bands in the range 992-456 cm^{-1} for ZFA were characterized as the infrared absorption bands of the zeolite hydroxysodalite [66]. The broad band at 992 cm^{-1} arises from the asymmetric stretch, ν_{as} (T-O-T, T = Si, Al). The bands located in the 775-653 cm^{-1} region are attributed to the symmetric stretch ν_s (T-O-T). The band around at 456 cm^{-1} is assigned to the bending vibration of δ(O-T-O). The broad band at 3431 cm^{-1} and a band at 1642 cm^{-1} can be related to the water molecules located inside the channels of zeolites and/or associated with exchangeable cations [67].

The FT-IR spectra (Fig. 20b) of zeolite-iron oxide composites ZM3 and ZC2 show the overlapping of the absorption bands of the zeolite from coal fly ash and magnetite particles. The appearance of a band of the Fe-O vibrations of the magnetite particles can be seen around 568-561 cm^{-1}.

Conclusion and Future Perspectives

Zeolite synthesized from fly ash has a great potential in environmental applications and interesting alternative to replace natural and commercial zeolites for adsorption in water pollution treatment. Converting fly ash into zeolites not only alleviates the disposal problem but also converts a waste material into a marketable commodity.

Zeolite from fly ash-iron oxide nanocomposites were synthesized and investigated for the removal of dyes and toxic metals from aqueous solutions. A series of experiments were carried out in a batch adsorption technique to obtain the effect of process variables on adsorption.

The data obtained through this work support the view that the zeolite from fly ash-iron oxide nanocomposite is an effective low cost adsorbent for the removal of contaminants from aqueous solution.

If magnetic zeolite-iron oxide nanocomposite is used as the adsorbent for contaminants in water, magnetic separation will be applied and the clear solution could be easily decanted off or removed by pipette. Furthermore, supporting of magnetite nanoparticles on zeolitic material from coal fly ashes during the preparation process prevents the coaggregation of the iron oxide nanoparticles and is of help for their storage and pelletization.

Development of the adsorption process requires further investigation in the direction of testing zeolite from fly ash-iron oxide nanocomposites with real industrial effluents. Once batch conditions, characterization of adsorbents, adsorptive capacity and the uptake mechanism are investigated work should be done to design and carry out some pilot-plant scale studies to check their feasibility at industrial level.

The use of zeolite from fly ash-iron oxide nanocomposites for application in radioactive liquid waste treatment is also a promising prospect.

REFERENCES

[1] M. Yamaura, R. L. Camilo and M. C. F. C. Felinto, *J. Alloys Compd.* 344 (2002) 152.

[2] Z. H. Ai, Y. Cheng, L. Z. Zhang and J. R. Qiu, *Environ. Sci. Technol.* 42 (2008) 6955.

[3] S. Pal and E. C. Alocilja, *Biosens. Bioelectron.* 24 (2009) 1437.

[4] K. P. Singh, S. Gupta, A. K. Singh and S. Sinha, *J. Hazard Mater.* 186 (2011) 1462.

[5] I. Safarik, M. Safarikova and V. Buricova, *Collect. Czech. Chem. Commun.* 60 (1995) 1448.

[6] J. D. Orbell, L. Godhino, S. W. Bigger, T.M. Nguyen and L.N. Ngeh, *J. Chem. Educ.* 74 (1997) 1446.

[7] S. Y Mak, and D. H. Chen, *Dyes Pigm.* 61(2004) 93.

[8] P. Xianjia, L. Zhaokun, and Z. Hongmei, *Chemosphere* 63 (2006) 300.

[9] Q.L. Zhang, Y.C. Lin, X. Chen and N. Y. Gao, *J. Hazard. Mater.* 148 (2007) 671.

[10] P. Yuan, M. Fan, D. Yang, H. He, D. Liu, A. Yuan, J. Zhu and T. Chen, *J. Hazard. Mater.* 166 (2009) 821.

[11] G. S. Zhang, H. J. Liu, R. P. Liu and J. H. Qu, *J. Colloid Interface Sci.* 335 (2009) 168.

[12] Y Feng, J. L. Gonga, G. M. Zeng, Q. Y. Niu, H. Y. Zhang, C.G. Niu, J.H. Deng and M. Yan, *Chem. Eng. J.* 162 (2010) 487.

[13] L. C. A. Oliveira, D. I. Petkowics, A. Smaniott, and S. B. C. Pergher, *Water Res.* 38 (2004) 3699.

[14] L. C. A. Oliveira, R. V. R. A. Rios, J. D. Fabris, V. K. Garg, K. Sapag and R. M. Lago, *Carbon* 40 (2002) 2177.

[15] L. C. A. Oliveira, R. V. R. A. Rios, J. D. Fabris, K. Sapag, V. K. Garg and R. M. Lago, *Appl. Clay Sci.* (2003) 22 169.

[16] K. P. Singh, S. Gupta, A. K. Singh and S. Sinha, *J. Hazard. Mater.* 186 (2011) 1462.

[17] J. Levandowski and W. D. Kalkreuth, *Int. J. Coal Geol.* 77 (2009) 269.

[18] F. S. Depoi, D. Pozebon and W. D. Kalkreuth, *Int. J. Coal Geol.* 76 (2008) 227.

[19] A. S. Meawad, D. Y. Bojinova and Y. G. Pelovski, *Waste Manage* 30 (2010) 2548.

[20] P. S. Polic, M. R. Ilic and A. R. Popovic, *Handb. Environ. Chem.* (2005) Doi:10.1007/b11733

[21] H. Holler and U. Wirsching, *Fortschr. Miner.* 63 (1985) 21.
[22] T. Henmi, *Clay Sci.* 6 (1987) 277.
[23] X. Querol, A. Alastuey, A. Lopez-Soler, J. M. Andres, R. Juan, P. Ferrer and C. R. Ruiz, *Environ. Sci.Technol.* 31 (1997) 2527.
[24] N. Shigemoto, H. Hayashi and K. Miyaura, *J. Mater. Sci.* 28 (1993) 4781.
[25] X. Querol, N. Moreno, J. C. Umaña, A. Alastuey, E. Hernández, A. López-Soler and F. Plana, *Int. J. Coal Geol.* 50 (2002) 413.
[26] S. S. Rayalu, A. K. Bansiwal, S. U. Meshram, N. Labhsetwar and S. Devotta, *Catal. Sur. Asia* 10 (2006) 74.
[27] D. A. Fungaro and J. E. A. Graciano, *Adsorpt. Sci. Technol.* 25 (2007) 729.
[28] D. A. Fungaro, M. Bruno and L. C. Grosche, Desalin. *Water Treat.* 2 (2009) 231.
[29] D. A. Fungaro, L. C. Grosche, A. S. Pinheiro, J. C. Izidoro and S. I. Borrely, *Orbital Elec. J. Chem.* 2 (2010) 235.
[30] D. A. Fungaro, M. Yamaura and T. E.M. Carvalho, *J. At. Mol. Sci.* 2 (2011) 305.
[31] T. E. M. Carvalho, D. A. Fungaro, C. P. Magdalena and P. Cunico, *J. Radioanal. Nucl. Chem.* 289 (2011) 617.
[32] Y. S. Ho and G. McKay, *Can. J. Chem. Eng.* 76 (1998) 822.
[33] Y. S. Ho, D. A. J. Wase, and C. F. Forster, *Environ. Technol.* 17 (1996) 71.
[34] W. J. Weber and J. C. Morris, *J. Sanit. Engin. Div. ASCE* 89 (1963) 31.
[35] I. Langmuir, *J. Am.Chem. Soc.* 40 (1918) 1361.
[36] H. M. F. Freundlich, *J. Phys. Chem.* 57 (1906) 385.
[37] K. R. Hall, L. C. Eagleton, A. Acrivos and T. Vermeulen, *Ind. Eng. Chem. Fundam.* 5 (1966) 212.
[38] Y. S. Ho, *Carbon* 42 (2004) 2115.
[39] Y. S. Ho, *Water Res.* 37 (2003) 2323.
[40] S. J. Allen, G. McKay and K. Y. H. Khader, *Environ. Pollut.* 56 (1989) 39.
[41] N. Dizge, C. Aydiner, E. Demirbas, M. Kobya, and S. Kara, *J. Hazard. Mater.* 150 (2008) 737.
[42] W. H. Cheung, Y. S. Szeto and G. McKay, *Bioresour. Technol.* 98 (2007) 2897.
[43] K. G. Bhattacharyya and A. Sharma, *Dyes Pigm.* 65 (2005) 51.
[44] W. A. Helby, *Chem. Eng.* 59 (1952) 153.
[45] B. Acemioglu, *J. Colloid Int. Sci.* 274 (2004) 371.
[46] G. Atun, G. Hisarlý, A. E. Kurtoglu, and N. Ayar, *J. Hazard. Mater.* 187 (2011) 562.
[47] A. Sharma and K. G. Bhattacharyya, *Indian J. Chem. Technol.* 12 (2005) 285.
[48] E. Gutierrez-Segura, M. Solache-Rios, and A. Colin-Cruz, *J. Hazard. Mater.* 170 (2009) 1227.
[49] M. B. Kasiri, H. Aleboyeh, and A. Aleboyeh, *Appl. Catal.* 84 (2008) 9.
[50] C. Namasivayam and R. T. Yamuna, *Environ Pollut.* 89 (1995) 1.
[51] C. Namasivayam, N. Muniasamy, K. Gayathri, M. Rani and K. Ranganathan, *Biores Technol.* 57 (1996) 37.
[52] C. Namasivayam, D. Prabha, and M. Kumutha, *Bioresour. Technol.* 64 (1998) 77.
[53] A. Shukla, Y. H. Zhang, P. Dubey, J. L. Margrave and S. S. Shukla , *J. Hazard. Mater.* 95 (2002) 137.
[54] B. Royer, N. F. Cardoso, E. C. Lima, V. S. O. Ruiz, T. R. Macedo and C. Airoldi, *J. Colloid Interface Sci.* 336 (2009) 398.

[55] C. H. Giles, T. H. MacEwan, S. N. Nakhwa, and D. Smith, *J. Chemical Society* (1960) 3973.

[56] C. Namasivayam and D. J. S. E. Arasi, *Chemosphere,* 34, 401 (1997).

[57] M. J. Zamzow and J. E. Murphy, *Separ. Sci. Tech.* 27 (1992) 1969.

[58] S. Kesraoui-Ouki, C. Cheeseman and R. Perry, *Environ. Sci. Technol.* 6 (1993) 1108.

[59] H. W. Sherry, *Ion exchange*, J. A. Manisky, ed.; Marcel Dekker, New York, cap. 2. (1969).

[60] S. K. Ouki and M. Kavannagh, *Waste Manage. Res.* 15 (1997) 383.

[61] G. Wulfsberg, *Principles of Descriptive Chemistry*, Brooks/Cole Publishing, Monterey, CA, p.2, (1987).

[62] W. Qiu and Y. Zheng, *Chem. Eng. J.* 145 (2009) 483.

[63] B. D. Cullity, *Elements of X-rays diffraction,* Addison Wesley Publishing Company, (1972).

[64] S. Gangopadhyay, G. C. Hadjipanayis, B. Dale, C. M. Sorensen and K. J. Klabunde, *Nanostruct. Mater.* 1 (1992) 77.

[65] J. Giri, S. Thakurta, J. Bellare, A. Nigam and D. Bahadur, *J. Magn. Magn. Mater.* 293 (2005) 62.

[66] C. Covarrubias, R. García, R. Arriagada, J. Yánez and M. T. Garland, *Microp. Mesop. Mat.* 88 (2006) 220.

[67] Th. Mouhtaris, D. Charistos, N. Kantiranis, A. Filippidis, A. Kassoli-Fournaraki and A. Tsirambidis, *Microp. Mesop. Mat.* 61 (2003) 57.

In: Zeolites: Synthesis, Chemistry and Applications
Editors: Moisey K. Andreyev and Olya L. Zubkov
ISBN: 978-1-61942-861-4

Chapter 4

ALUMINOSILICATE/SILICOALUMINOPHOSPHATE ZEOLITE COMPOSITE: SYNTHESIS, CHARACTERIZATION AND CATALYTIC PERFORMANCE

Xin Zhang*
School of Chemical Engineering, Northwest University,
Xi'an, China

ABSTRACT

HZSM-5/SAPO-34 composite and HBeta/SAPO-11 composite were prepared by one-step hydrothermal crystallization, respectively. The structure, texture, morphology and acidity of these as-prepared samples were comparatively investigated by powder X-ray diffraction (XRD), infrared spectroscopy (IR), solid state nuclear magnetic resonances (NMR), N_2 isothermal adsorption-desorption, scanning electron microscope (SEM), temperature-programmed desorption of adsorbed ammonia (NH_3-TPD) and infrared spectroscopy of adsorbed pyridine (Py-IR), etc..

In addition, the catalytic performance of HZSM-5/SAPO-34 composite in the conversion of ethanol to propylene, and the catalytic performance of HBeta/SAPO-11 composite in the cracking of 2-butylene to propylene were studied.

Chemical interaction of aluminosilicate with silicoaluminophosphate zeolites in the composites occurred. Such interaction modified structure, texture, morphology and acidity of these composites, which in turn significantly improved their catalytic reactivity.

Keywords: Aluminosilicate zeolite; Silicoaluminophosphate zeolite; Hydrothermal synthesis; Propylene; 2-Butylene; Ethanol

---

* E-mail: zhangxinzhangcn@yahoo.com.cn.

1. INTRODUCTION

Aluminosilicate/silicoaluminophosphate zeolite composite prepared by chemical synthesis has been considered as a type of promising catalyst material, because it can not only combine the advantages of aluminosilicate and silicoaluminophosphate zeolites but also cause the composite distinctive properties. For example, aluminosilicate/silicoaluminophosphate zeolite composite exhibited obvious synergetic catalysis because of its structural and acidic features; however, the physical mixture might inhibit the close contacts of different zeolites and thereby weaken their synergetic catalysis [1-4]. Therefore, in recent years, the chemical synthesis, characterization and catalytic performance of zeolite composite have attracted great attention.

Comparing with abundant research works on aluminosilicate/ aluminosilicate zeolite composites, only limited information is available on the chemical synthesis, characterization and catalytic application of aluminosilicate/silicoaluminophosphate zeolite composites. Recently, some aluminosilicate/silicoaluminophosphate zeolite composites have been developed as catalysts for FCC gasoline upgrading [1-2], methanol to olefins (MTO) [3], the synthesis of high-octane hydrocarbon from syngas [4], etc.. The preparation method and the composition of the zeolite composites modified their physicochemical properties, which in turn affected its catalytic performance [1-4].

It is widely recognized that HZSM-5 has excellent reactivity for the hydrodesulfurization (HDS) and aromatization of light hydrocarbons. On the other hand, SAPO-11 has been successfully applied for the isomerization of hydrocarbons. Therefore, it is reasonable to conceive that HZSM-5/SAPO-11 composite with suitable composition can provide a balanced desulfurization, hydroisomerization and aromatization performance and benefit the upgrading of FCC gasoline. For this proposal, Fan et al. [2] synthesized HZSM-5/SAPO-34 composite catalyst for FCC gasoline upgrading. HZSM-5/SAPO-34 composite was synthesized by *in situ* overgrowth of SAPO-11 on HZSM-5. The as-synthesized HZSM-5/SAPO-11 composite existed in the form of a core-shell structure with HZSM-5 as the core and SAPO-11 as the shell. In the case of the physical mixture of HZSM-5 and SAPO-11, HZSM-5 was morphologically separate from SAPO-11. In addition, the composite had more mesopores and moderate acidity distribution than the mixture. Such properties accelerated the diffusion of substrates and enhanced the synergetic effect between Brönsted (B) and Lewis (L) acid sites. Due to these advantages of the composite, the composite-based Ni-Mo catalysts showed high liquid yield as well as good desulfurization activity and low coke amount in hydro-upgrading FCC gasoline with respect to the mixture.

Both SAPO-34 and HZSM-5 are the most well-known catalysts for methanol to olefins (MTO) reaction. SAPO-34 and HZSM-5 show different catalytic reactivity and stability in the reaction. Generally speaking, SAPO-34 has high hydrothermal stability and high ethylene yield, but it rapidly deactivates by coke. In case of HZSM-5, it has high propylene yield and long catalyst life; however, it has the low yield of total light olefins. Therefore, Chae et al. [3] thought that a combination of HZSM-5 and SAPO-34 might give rise to an optimal catalyst to meet the market demand for ethylene, propylene and butadiene. They developed HZSM-5/SAPO-34 composite catalysts to control the product distribution of light olefins in MTO reaction. HZSM-5/SAPO-34 composites were successfully synthesized by seed and series synthetic methods and they showed the characteristics distinguishing from physical mixture.

HZSM-5/SAPO-34 composite prepared by seed method was the configuration of HZSM-5 as core and SAPO-34 as shell; nevertheless, it did not show the good catalytic reactivity in MTO. HZSM-5/SAPO-34 composite synthesized by successive crystallization of SAPO-34 synthetic gel after HZSM-5 crystallization exhibited the relative high reactivity.

Fischer-Tropsch synthesis (FTS) is an important process for the conversion of syngas into high-quality fuel and chemicals. In order to transform part of FTS products into the desired gasoline fraction, HZSM-5/MnAPO-11 composite catalysts for the conversion of syngas to gasoline was prepared by direct crystallization of MnAPO-11 gel with the addition of HZSM-5 [4]. The as-prepared HZSM-5/MnAPO-11 composite showed the high gasoline yield and the high *iso*-paraffin selectivity than HZSM-5 and the physical mixture of HZSM-5 and MnAPO-11, due to the presence of more mesopores and moderate acid sites in the composite.

Up to date, various synthesis methods have been attempted for the preparation of aluminosilicate/silicoaluminophosphate zeolite composite. Normally, the zeolite composite with binary structure can be synthesized by one-step hydrothermal crystallization and two-step hydrothermal crystallization. In the one-step crystallization, aluminosilicate (or silicoaluminophosphate) zeolite gel directly crystallizes with the addition of silicoaluminophosphate (or aluminosilicate) zeolite. In case of the two-step synthesis, the successive crystallization of aluminosilicate (or silicoaluminophosphate) performs after the crystallization of silicoaluminophosphate (or aluminosilicate) zeolite gel. During the two-step crystallization produces, the rudimental template from the first crystallization may change the composite configuration in the second crystallization step. Thus, the one-step crystallization simplifies the synthesis procedure and avoids the reduction in the crystallinity as well as the configuration change of the composite.

Although these remarkable achievements have been made, it is still challenge to solve the important problems for providing new perspectives on the synthesis and application of zeolite composite: (1) the chemical synthesis of the composite with closely connected aluminosilicate and silicoaluminophosphate framework; (2) the adjustment of the weight proportion and framework Si/Al ratio of microphase in the composite for the requirement of different catalytic reaction; (3) the improvement of the catalytic reactivity and stability of the composite. To achieve these, it is necessary to optimize the synthesis conditions and to deeply understand the reactivity-properties relationship.

Propylene is one of basic raw chemicals for producing polypropylene, propylene oxides, acrylic acid and acrylonitrile, etc.. Nowadays, propylene is mainly produced from the steam cracking and catalytic cracking of naphtha. Since the need of propylene derivates grows fast, the production of propylene from these two units can not meet the market demand of propylene in recent years. Due to high oil price and the technology limitation, the increase of propylene production from these two units is restricted. Therefore, it is desirable to search an alternative route for increasing the production of propylene.

In this present work, HZSM-5/SAPO-34 and HBeta/SAPO-11 composites were prepared by one-step hydrothermal crystallization, respectively. The physicochemical properties of the composites were comparatively investigated by powder X-ray diffraction (XRD), infrared spectroscopy (IR), solid state nuclear magnetic resonances (NMR), N_2 isothermal adsorption-desorption, scanning electron microscope (SEM), temperature-programmed desorption of adsorbed ammonia (NH_3-TPD) and infrared spectroscopy of adsorbed pyridine (Py-IR), etc.. In addition, catalytic performance of HZSM-5/SAPO-34 composite for the conversion of ethanol to propylene and catalytic performance of HBeta/SAPO-11 composite for the cracking of 2-butylene to propylene were tested in detail. The composites showed the high catalytic reactivity and stability than the individual zeolite and the physical mixture of zeolites. The distinctive properties and synergetic catalysis were found in the composite, and the influences of the properties of these samples on their catalytic performance were discussed.

2. Synthesis, Characterization and Catalytic Performance of HZSM-5/SAPO-34 Composite for Conversion of Ethanol to Propylene

Recently, the conversion of ethanol to propylene, especially ethanol produced by the fermentation of biomass, has raised increasing attention. The conversion of ethanol to propylene is considered to be a carbon-neutral process for the increase of propylene production by the use of renewable biomass. A few works concerning ethanol to propylene have been reported so far, and these works mainly focused on the modification of the catalysts and the optimization of reaction conditions. Modified HZSM-5 zeolite, such as, Sr-HZSM-5 [5], Zr-HZSM-5 [6], P-HZSM-5 [7], Ga-HZSM-5 [8] and La-HZSM-5 [9], etc. were found to be promising catalysts for the reaction. The highest propylene yield of ca. 32% was obtained at 500-550 °C [6-9].

On the other hand, the nature of active site on the catalyst remains not fully understood. Furumoto et al. [8] considered that the strength of acid sites was curial factor for the conversion of ethanol to propylene. In order to obtain the high propylene yield, modified-HZSM-5 zeolite had stronger acidity than Ga-modified HZSM-5 [8]. Xia et al. [10] reported that the catalytic activity of the different MFI-type zeolite catalysts was proportional to the number of acid sites [10]. However, the catalytic reactivity and stability of the catalyst should be further improved for industrial application.

It is well known that HZSM-5 and SAPO-34 were good catalysts for ethanol to ethylene. In addition, HZSM-5 and SAPO-34 were found to exhibit the high activity for the direct conversion of ethylene to propylene [11, 12]. Oikawa et al. [11] reported that SAPO-34 showed highly selective conversion of ethylene to propylene due to the shape selectivity effect of the small-pore SAPO-34 and the modest acid strength of acidic protons. Lin et al [12] found that HZSM-5 exhibited the highest activity for the direct conversion of ethylene to propylene. The conversion of ethylene increased with the degree of H^+ exchange in the H-Na-ZSM-5 and the Al content in the HZSM-5 samples with different Si/Al ratios [12]. The strong B acid sites were proposed to account for the conversion of ethylene [12].

In this section, HZSM-5/SAPO-34 composite as catalyst for ethanol to propylene was prepared by hydrothermal synthesis. The catalytic properties and performance of HZSM-5/SAPO-34 composite, physical mixture of HZSM-5 and SAPO-34 (HZSM-5/SAPO-34 mixture), HZSM-5 and SAPO-34 were investigated by N_2 isothermal adsorption-desoprtion, XRD, IR, SEM, NH_3-TPD and catalytic test. Based on the obtained results, the influences of properties on the catalytic performance of these catalysts were discussed.

2.1. Experimental Methods

2.1.1. Preparation

Powder HZSM-5 (Si/Al_2=25, molar ratio) was obtained from Nankai Unv. SAPO-34 was prepared by hydrothermal synthesis described as following. During the preparation of SAPO-34, a calculated amount of pseudoboehmite (AlOOH, Xibei Chem.), tetraethylammonium hydroxide (TEAOH, 25 wt.%, Aladdin Chem.), orthophosphoric acid (85 wt.%, Ruijingte Chem.), silica solution (SiO_2=30w%, Xianshiji Chem.) and deionized water were mixed well to form gel under fast stirring. In this gel, Al_2O_3 : P_2O_5 : SiO_2 : TEAOH : H_2O=1 : 1 : 0.30 : 2.0 : 50 (molar ratio). The gel was charged into a stainless steel autoclave with Teflon lining. Crystallization was carried out at 200 °C for 72 h. The resultant solid was washed with distilled water and then dried at 110 °C for 6 h.

Hydrothermally synthesized HZSM-5/SAPO-34 composite was obtained by one step crystallization of the mixture of SAPO-34 synthetic gel with the same composition above and the powder HZSM-5 (Si/Al_2=25, molar ratio, Nankai Unv.) under the same conditions to the crystallization of SAPO-34. In the typical hydrothermal synthesis of HZSM-5/SAPO-34 composite, 10 g of pseudoboehmite (AlOOH, Xibei Chem.) was mixed with 98 g of tetraethylammonium hydroxide (TEAOH, 25 wt.%, Aladdin Chem.) as template to form mixture. Under the fast stirring, 16 g of orthophosphoric acid (85 wt.%, Ruijingte Chem.) and 6 g of silica solution (SiO_2=30w%, Xianshiji Chem.) were added dropwise into the above mixture to make a thick white paste. Subsequently, a certain amount of powder HZSM-5 (Si/Al_2=25, molar ratio, Nankai Unv.) and the white paste were mixed well. The mixture of HZSM-5 (Si/Al_2=25, molar ratio, Nankai Unv.) and the white paste was aged overnight to form gel. The final gel was charged into a stainless steel autoclave with Teflon lining. Crystallization was carried out at 200 °C for 72 h. The final product was washed with distilled water and dried at 110 °C for 6 h.

On the other hand, HZSM-5/SAPO-34 mixture was prepared by the physical mixture of powder HZSM-5 (Si/Al_2=25, molar ratio, Nankai Univ.) and the as-synthesized SAPO-34. The weight ratio of HZSM-5 to SAPO-34 was ca. 1 in HZSM-5/SAPO-34 mixture.

2.1.2. Characterization

N_2 isothermal adsorption-desoprtion of these samples was performed at the temperature of liquid nitrogen by Micromeritics ASAP400 adsorptionmeter. XRD patterns of these samples were collected by Rogaku Rotflex D/Max-C powder X-ray diffractometer with Cu $K\alpha$ radiation (λ=0.15046 nm) operated at 40 kV and 30 mA. SEM images of these samples were obtained by JEOL JSM-6360LV-EDS electron microscope meter. NH_3-TPD experiment was made in a continuous flow fixed-bed reactor system with thermal conduction detector (TCD, Beifen Ruili GC-3400, Beifen Co.). Before NH_3-TPD experiment, 0.1 g catalyst was placed in a quartz tube and heated at 200 °C in a N_2 flow (30 mL/min, 99.995%, Tianyuan Gas.) for 1 h. The catalyst was cooled to 50 °C and then NH_3 adsorption was performed by pulsing NH_3 (99.999%, Jining Tezhong Qiti Co.) to the sample until NH_3 adsorption became saturated. The catalyst was purged with N_2 flow (30 mL/min, 99.995%, Tianyuan Gas.) to remove NH_3 physically adsorbed on the catalyst surface. NH_3-TPD was carried out in 30 mL/min N_2 flow (99.995%, Tianyuan Gas.) and from 50 to 600 °C at heating rate of 10 °C/min.

2.1.3. Catalytic Test

Ethanol conversion was performed in a fixed-bed continuous-flow microreactor apparatus. A quartz tube reactor (650 mm length, 10 mm I.D.) mounted inside a tube furnace at atmospheric pressure. The space up and down the catalyst bed (0.5 g, 20~40 mesh) in the reactor was filled with quartz wool. The reaction temperature was controlled and measured by temperature controller (YUDIAN AI-518/518P) and K-type thermal couple. Ethanol (A.R., Beijing Chem. Co.) was injected into the catalyst bed by a syringe pump (LabAlliance series II). The gaseous mixture of ethanol and N_2 flow (99.995%, Tianyuan Gas.) passed through the catalyst bed. The off gas was kept at ca. 120 °C by heating belt. The feed and products were analyzed on-line by gas chromatography (Ruimin GC-2060, Ruimin Co.) equipped with flame ionization detector (FID) and thermal conductivity detector (TCD). FID with a KB-Al_2O_3/Na_2SO_4 capillary column (50 m × 0.32 mm × 15 μm) analyzed CH_4, C_2H_6, C_2H_4, C_3H_8, C_3H_6, C_4H_{10}, C_4H_8, C_5+ (aliphatic and aromatics hydrocarbon). TCD with a GDX-103 packed column detected ethanol. Mass balance more than 95% was used to calculate the conversion of ethanol and the yield of products. The conversion of ethanol to coke was neglected because the instantaneous formation of coke was dependent on reaction time and difficult to estimate.

2.2. Results and Discussion

2.2.1. Structure Properties

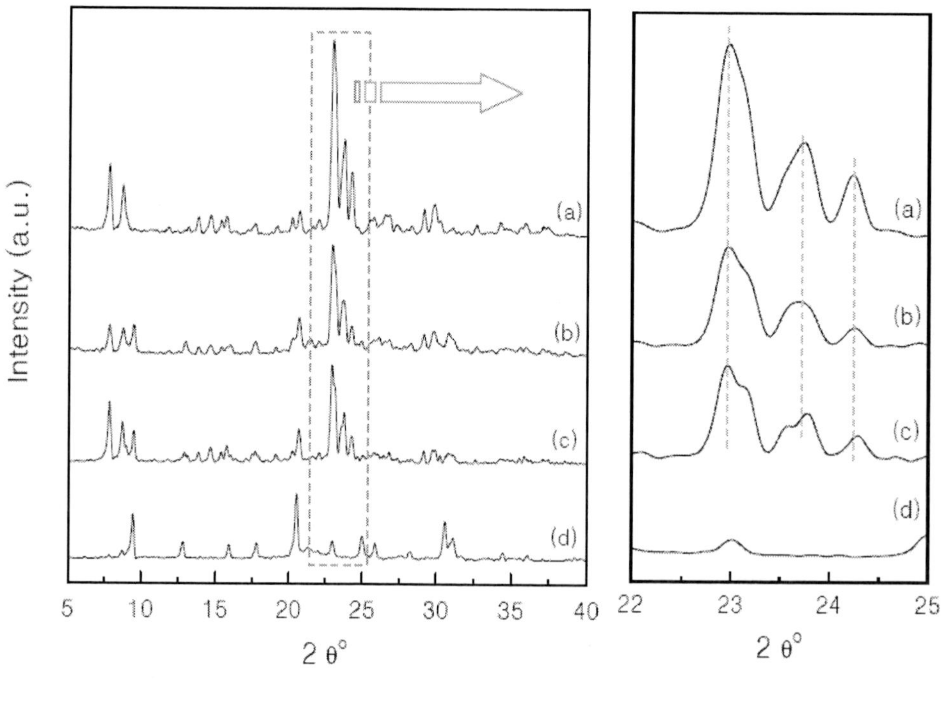

Figure 1. XRD patterns of these catalysts (a) HZSM-5, (b) HZSM-5/SAPO-34 mixture, (c) HZSM-5/SAPO-34 composite, (d) SAPO-34.

HZSM-5/SAPO-34 composite was characterized by XRD and IR to identify its structure. Figure 1 shows XRD patterns of HZSM-5/SAPO-34 composite, HZSM-5/SAPO-34 mixture, HZSM-5 and SAPO-34. Characteristic XRD peaks of HZSM-5 (2θ°=7.8, 8.7, 22.9, 23.8, 24.2) and SAPO-34 (2θ°=9.4, 20.6, 30.6, 31.2) were detected from HZSM-5/SAPO-34 composite and HZSM-5/SAPO-34 mixture. No diffraction peaks other than those of HZSM-5 and SAPO-34 zeolites were found. These results indicate that HZSM-5/SAPO-34 composite with the binary structure was successfully prepared by the hydrothermal synthesis and physical mixture.

It is observed that XRD peaks of HZSM-5 in HZSM-5/SAPO-34 composite slightly shifted toward high 2θ° relative to the corresponding peaks of individual HZSM-5. In contrast, XRD peaks of HZSM-5 in HZSM-5/SAPO-34 mixture were as similar as those of HZSM-5. These results suggest that the interaction of HZSM-5 and SAPO-34 occurred in HZSM-5/SAPO-34 composite and led to the change of HZSM-5 framework.

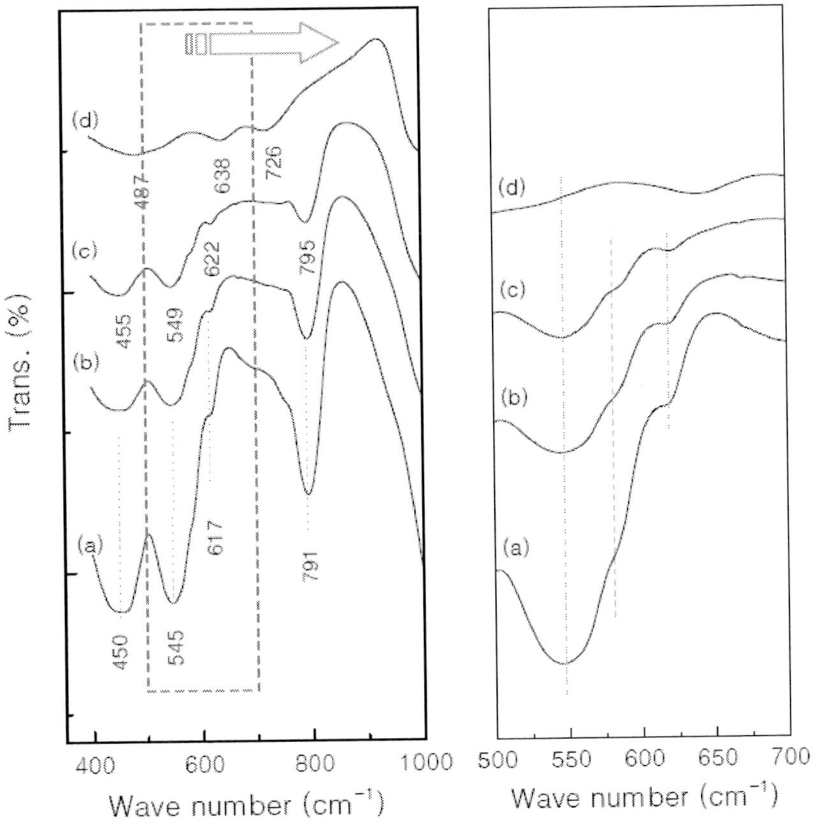

Figure 2. IR spectra of these catalysts (a) HZSM-5, (b) HZSM-5/SAPO-34 mixture, (c) HZSM-5/SAPO-34 composite, (d) SAPO-34.

Relative to HZSM-5, both HZSM-5/SAPO-34 composite and HZSM-5/SAPO-34 mixture showed the weak XRD peak intensity. The results suggest that both HZSM-5/SAPO-34 composite and HZSM-5/SAPO-34 mixture clearly appeared the reduction in the crystallinity. HZSM-5/SAPO-34 composite and HZSM-5/SAPO-34 mixture had the similar crystallinity, implying that they had the similar HZSM-5/SAPO-34 weight ratio. Namely, HZSM-5/SAPO-34 weight ratio was about 1 in HZSM-5/SAPO-34 composite.

Figure 2 shows IR spectra of these samples detected in wavenumber range of 400-1000 cm^{-1}. Useful information on the framework structure of these zeolites was gained. HZSM-5/SAPO-34 composite and HZSM-5/SAPO-34 mixture presented all characteristic vibration of TO_4 unit (795, 791 and 726 cm^{-1}) as well as the asymmetric and symmetric vibration of T-O binding/double rings (450, 455, 487, 545, 549, 617, 622 and 638 cm^{-1}) belonging to HZSM-5 and SAPO-34. It is noticed, IR bands of the T-O binding/double rings vibrations for HZSM-5/SAPO-34 composite (549, 585 and 622 cm^{-1}) slightly shifted toward the high wavenumber with respective to those for HZSM-5 and HZSM-5/SAPO-34 mixture. The results further suggest that the chemical interaction between HZSM-5 and SAPO-34 occurred in HZSM-5/SAPO-34 composite. Based on the results of XRD and IR characterization, the combination of HZSM-5 and SAPO-34 in HZSM-5/SAPO-34 composite was not physical mixture, which was different from that in HZSM-5/SAPO-34 mixture.

2.2.2. Texture Properties

Textural properties of HZSM-5/SAPO-34 composite, HZSM-5/SAPO-34 mixture as well as HZSM-5 and SAPO-34 were characterized by N_2 isothermal adsorption-desorption and the results are summarized in Table 1. Specific surface area (S_{BET}) of these samples was calculated by BET method, and the total volume of pore (V_{pore}) was calculated by *t*-plot analysis of adsorption isotherm. It can be seen in Table 1, S_{BET} of these catalysts decreased as the order of HZSM-5 > HZSM-5/SAPO-34 mixture > SAPO-34 > HZSM-5/SAPO-34 composite. The V_{pore} of these catalysts slightly fluctuated from 0.25 to 0.29 cm^3/g.

Table 1. Texture and acidity of these catalysts

Catalysts	S_{BET} (m^2/g)	V_{pore} (cm^3/g)	The distribution and concentration of acid sites (a.u./g)			
			Weak (50~150 °C)	Medium (150~350 °C)	Strong (> 350 °C)	Total
HZSM-5	353	0.25	75.6	110.0	67.9	253.5
ZS-MM	340	0.26	112.7	116.8	28.1	257.6
ZS-HS	276	0.27	134.0	146.6	22.6	303.2
SAPO-34	300	0.29	152.9	218.1	8.6	379.6

Figure 3 shows N_2 adsorption-desorption isotherms of HZSM-5/SAPO-34 composite, HZSM-5 and SAPO-3 catalysts. HZSM-5 appeared a large hysteresis loop at the relative pressure P/P_0 ranging from ca. 0.45 to 1.0, and SAPO-34 showed little hysteresis loop in N_2 isotherms. These results show that HZSM-5 had a plenty of mesopores whereas SAPO-34 hardly had mesopores. The N_2 adsorption-desorption isotherms of HZSM-5/SAPO-34 mixture (not shown) was similar to that of HZSM-5. Obviously, HZSM-5/SAPO-34 composite had a smaller hysteresis loop in N_2 isotherms than HZSM-5, suggesting that HZSM-5/SAPO-34 composite had less mesopores than HZSM-5. HZSM-5/SAPO-34 composite had the small S_{BET} and low mesopores comparing with HZSM-5 and HZSM-5/SAPO-34 mixture, possibly due to the formation of agglomerate crystal HZSM-5 and SAPO-34.

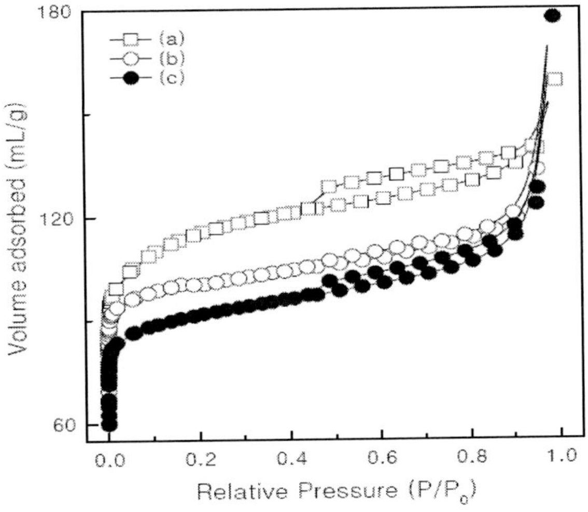

Figure 3. N_2 adsorption desorption isotherms of these catalysts (a) HZSM-5, (b) SAPO-34, (c) HZSM-5/SAPO-34 composite.

2.2.3. Morphology Properties

SEM was used to comparatively study the morphology and size of HZSM-5, SAPO-34, HZSM-5/SAPO-34 composite and HZSM-5/SAPO-34 mixture. Figure 4 shows SEM images of these samples. SAPO-34 appeared cubic particle and its particle size ranged from 2 to 3 μm. It is clear that HZSM-5 showed block-shaped particle with some agglomerates and the size of ca. 1 μm. The micrograph of HZSM-5/SAPO-34 mixture exhibited only discrete mixture of HZSM-5 and SAPO-34. Concerning HZSM-5/SAPO-34 composite, HZSM-5 crystals or its loose aggregates were evenly embedded and closely contacted with SAPO-34. These results indicate that the combination of HZSM-5 and SAPO-34 in HZSM-5/SAPO-34 composite was distinctively different from that in HZSM-5/SAPO-34 mixture.

Figure 4. SEM images of these catalysts (a) HZSM-5, (b) SAPO-34, (c) HZSM-5/SAPO-34 composite, (d) HZSM-5/SAPO-34 mixture.

2.2.4. Acidity Properties

Figure 5 shows NH_3-TPD plots of these samples. Both HZSM-5 and HZSM-5/SAPO-34 mixture showed three NH_3 desorption peaks at 118, 228 and 431~459 °C. HZSM-5/SAPO-34 composite displayed one broad peak at ca. 128 °C as well as a shoulder peak at 228 °C and a weak desoprtion at ca. 369 °C. The results indicate that three types of acid sites presented on HZSM-5, HZSM-5/SAPO-34 mixture and HZSM-5/SAPO-34 composites. SAPO-34 appeared two peaks with peak maxima temperature around 163 °C, suggesting that two types of acid sites mainly existed.

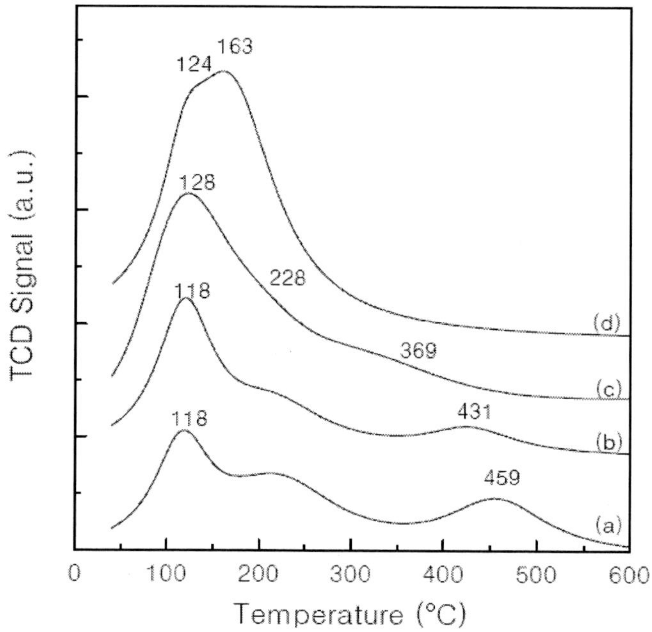

Figure 5. NH_3-TPD plots of these catalysts (a) HZSM-5, (b) HZSM-5/SAPO-34 mixture, (c) HZSM-5/SAPO-34 composite, (d) SAPO-34.

NH_3-TPD peak area was integrated at the temperature range of 50~150, 150~350 and > 350°C, respectively expressing the concentration of weak (50~150 °C), medium (150~350 °C) and strong (> 350 °C) acid sites. It can be seen in Table 1, there were weak, medium and strong acid sites on HZSM-5/SAPO-34 mixture, HZSM-5/SAPO-34 composite and HZSM-5. It is observed that on SAPO-34 more of NH_3 desorption occurred at low temperature (<350 °C), indicating that weak and medium acid sites were dominative on SAPO-34 surface. The total concentration of acid sites decreased as the order: SAPO-34 > HZSM-5/SAPO-34 composite > HZSM-5/SAPO-34 mixture > HZSM-5. In addition, the concentration of strong acid sites increased as the sequence: SAPO-34 < HZSM-5/SAPO-34 composite < HZSM-5/SAPO-34 mixture < HZSM-5. It is noticed that the NH_3-TPD plot and the acidic strength distribution of HZSM-5/SAPO-34 composite were alike those of SAPO-34. These results suggest that HZSM-5/SAPO-34 composite showed the acidity similar to SAPO-34. Hence, SAPO-34 might partially overlap on HZSM-5 in HZSM-5/SAPO-34 composite, due to the reason that HZSM-5 was used as seed and SAPO-34 crystallized on HZSM-5 during the hydrothermal synthesis of HZSM-5/SAPO-34 composite.

2.2.5. Catalytic Reactivity

Table 2. Catalytic reactivity of these samples in ethanol conversion

Catalyst	Conv. Ethanol (wt.%)	Yield (wt.%)							
		CH_4	C_2H_4	C_2H_6	C_3H_6	C_3H_8	C_4H_8	C_4H_{10}	C_5+
HZSM-5	100.0	0.9	10.4	4.3	16.4	35.5	8.0	21.6	2.9
HZSM-5/SAPO-34 composite	100.0	0.0	100.0	0.0	0.0	0.0	0.0	0.0	0.0
HZSM-5/SAPO-34 mixture	100.0	0.2	12.1	1.6	25.3	14.5	20.7	19.1	6.5
SAPO-34	100.0	trace	87.8	0.2	6.9	0.1	4.9	0.2	trace

Reaction conditions: 500 °C, 0.1 MPa, P_{C2H5OH}=20 kPa, total flow rate=80 mL.min^{-1}, 0.5 g catalyst, TOS=3 h.

The catalytic reactivity of HZSM-5/SAPO-34 composite was tested in ethanol conversion reaction. Table 2 shows the products distribution of ethanol conversion over these samples at 500 °C. The conversion of ethanol was almost 100% on all samples. It is found that only C_2H_4 formed on SAPO-34 under the reaction conditions. C_1-C_4 alkanes, C_2H_4, C_3H_6, C_4H_8 and C_5+ were produced during the ethanol conversion reaction on HZSM-5, HZSM-5/SAPO-34 composite and HZSM-5/SAPO-34 mixture. Hence, a series of reactions, such as, dehydration, cracking, hydrogen transfer and aromatization etc., occurred on these samples. 16.4% yield of C_3H_6 and 10.4% yield of C_2H_4 were gained on HZSM-5. HZSM-5/SAPO-34 mixture showed 87.8% yield of C_2H_4 and 6.9% yield of C_3H_6. It is worth noticing, HZSM-5/SAPO-34 composite showed the much higher C_3H_6 yield than the other investigated catalysts. Thus, the interaction of SAPO-34 and HZSM-5 occurred in HZSM-5/SAPO-34 composite, which favored the production of propylene in ethanol conversion.

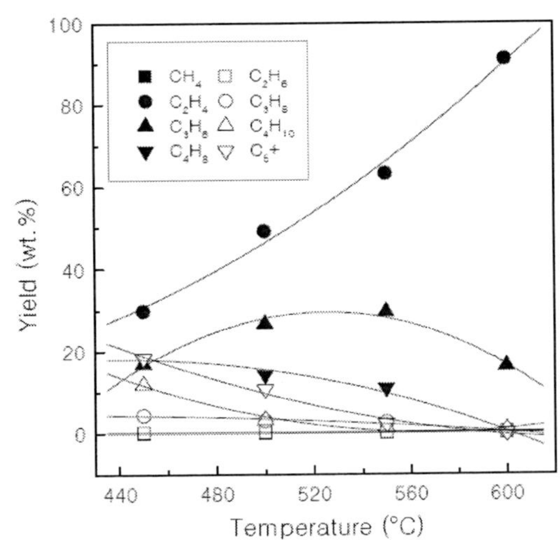

Figure 6. Effect of reaction temperature on ethanol conversion on HZSM-5/SAPO-34 composite. Reaction conditions: 0.1 MPa, P_{C2H5OH}=20 kPa, total flow rate=80 mL.min^{1}, 0.5 g catalyst, TOS=4 h.

Figure 6 shows effects of reaction temperature on ethanol conversion over HZSM-5/SAPO-34 composite. As the temperature rising, the yield of C_2H_4 gradually increased whereas the yield of C_1-C_4 alkanes and C_5+ decreased. The increase in the reaction temperature might enhance the cracking of hydrocarbon to C_2H_4; as a result, the yield of C_2H_4 increased with the rise of reaction temperature [5-9]. The yield of C_3H_6 initially went up with increasing the reaction temperature, reached a maximum value of 29.5% at 550 °C, and then decreased to ca. 16.0% at 600°C. Therefore, the moderate reaction temperature was required to obtain the highest propylene yield in ethanol conversion.

2.2.6. Catalytic Stability

Figure 7 shows the time course of ethanol conversion on HZSM-5/SAPO-34 composite and HZSM-5/SAPO-34 mixture at 500 °C. In case of HZSM-5/SAPO-34 mixture, C_3H_6 yield firstly increased to ca. 32.0% at reaction time of 0.5 h and then dropped to ca. 3.0% after reaction time up to 6 h; C_2H_4 yield momentously increased with increasing the reaction time while C_4H_8 yield gradually decreased. The results suggest that HZSM-5/SAPO-34 mixture showed poor catalytic stability under the reaction conditions. In contrast, HZSM-5/SAPO-34 composite exhibited no distinct reduction in the yield of C_3H_6, C_2H_4 and C_4H_8 during the reaction time of 6 h. Therefore, HZSM-5/SAPO-34 composite had better catalytic stability than HZSM-5/SAPO-34 mixture.

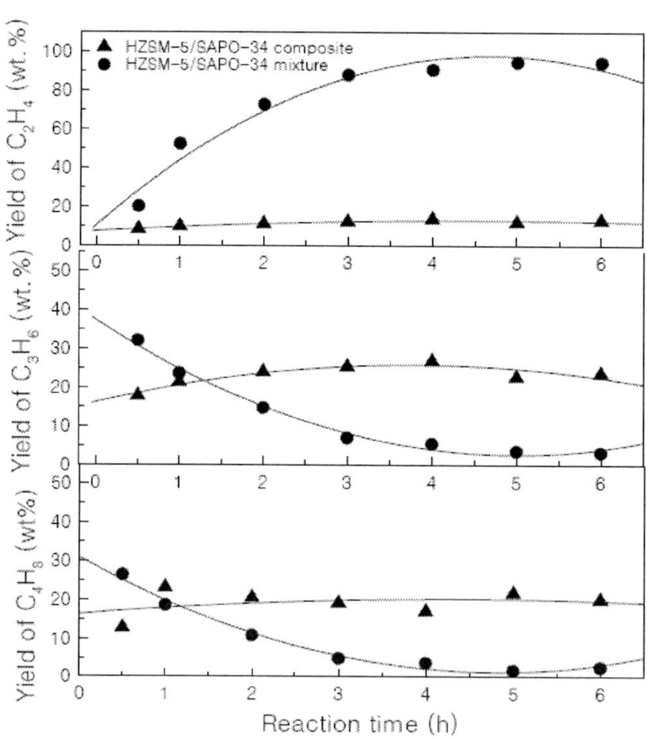

Figure 7. Time course of ethanol conversion on HZSM-5/SAPO-34 mixture and HZSM-5/SAPO-34 composite. Reaction conditions: 500 °C, 0.1 MPa, P_{C2H5OH}=20 kPa, total flow rate=80 mL.min^{-1}, 0.5 g catalyst.

On the basis of the above results, the chemical interaction of HZSM-5 with SAPO-34 occurred in HZSM-5/SAPO-34 composite. During the hydrothermal synthesis of HZSM-5/SAPO-34 composite, the growth of SAPO-34 occurred on HZSM-5 particles as seed; in addition, HZSM-5 could partially dissolve as silica source for the synthesis of SAPO-34. These phenomena led to the chemical interaction and the aggregation of monocrystals HZSM-5 and SAPO-34. Such interaction modified the properties of as-prepared HZSM-5/SAPO-34 composite and improved its catalytic reactivity in ethanol conversion to propylene.

The acidity of the HZSM-5/SAPO-34 composite had great roles on the catalytic reactivity of the composite in ethanol to propylene. Ethanol can transfer to light olefins and the latter can further convert to aromatics, alkane and coke, etc. [5-9]. For achieving the high yield of olefins, the catalyst should possess the moderate concentration and strength of acid sites so that it can enhance the formation of light olefins and hold back the secondary reactions of the light olefins [5-9]. In case of this work, HZSM-5/SAPO-34 composite exhibited higher yield of propylene and better catalytic stability than HZSM-5/SAPO-34 mixture, due to one of possible reasons that HZSM-5/SAPO-34 composite had the proper concentration and strength of acid sites.

2.3. Summary

Hydrothermally synthesized HZSM-5/SAPO-34 composite showed improved catalytic reactivity and stability in the conversion of ethanol to propylene compared with HZSM-5/SAPO-34 mixture, HZSM-5 and SAPO-34. For HZSM-5/SAPO-34 composite, SAPO-34 might partially overlap on HZSM-5 and the chemical interaction between HZSM-5 and SAPO-34 occurred. Such interaction modified the properties of the composite and improved its catalytic performance in ethanol to propylene.

3. Synthesis, Characterization and Catalytic Performance of HBETA/Sapo-11 Composite for Cracking of 2-Buylene to Propylene

Catalytic cracking of butylenes to propylene has attracted much attention because it represents an alternative route to produce propylene from low-cost petroleum resource [13]. A few of zeolite-based catalysts, such as, P modified HZSM-5 [14-18], P-W modified HZSM-5 [19], CC-16+5wt.% RPSA zeolite [20], ZSM-23 [21], ZSM-48 [22] and MCM-22 [23] have been developed for the reaction so far. The porous structure and acidity of these catalysts played important roles on their catalytic performance in butylenes cracking. In order to reach high yield of propylene, it is necessary to further optimize the structure and acidity of the zeolite-based catalysts for butylenes cracking.

Zeolite HBeta possesses an intergrowth of two or three polymorphs having a tridimensional system of interconnected 12-membered ring channels with pore diameters of 0.55×0.55 nm^2 and 0.76×0.64 nm^2 [24-26]. Owing to its high acidity and peculiar pore system, zeolite HBeta currently receives much attention as potential catalyst in catalytic

and HBeta/SAPO-11 mixture. The results indicate that binary structure HBeta/SAPO-11 composite was synthesized by hydrothermal method and the physical mixture.

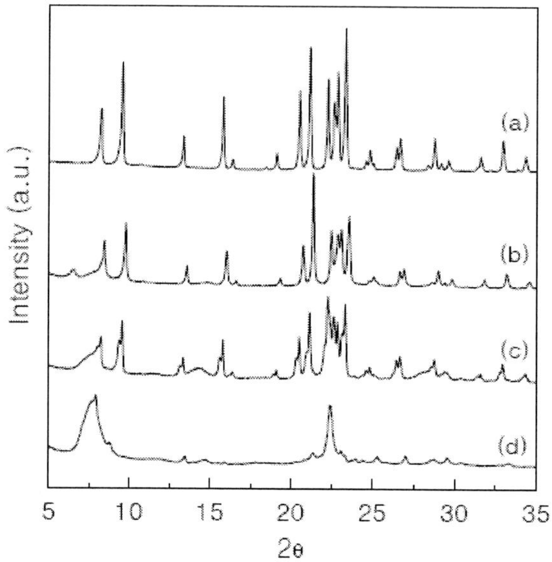

Figure 8. XRD patterns of these samples (a) SAPO-11, (b) HBeta/SAPO-11 mixture, (c) HBeta/SAPO-11 composite, (d) HBeta.

The composite and the mixture showed the lower diffraction peaks intensity than individual HBeta and SAPO-11, suggesting a reduction in the crystallinity of the composite and the mixture. On the other hand, XRD peaks of HBeta in the composite slightly shifted toward low $2\theta^{o}$ relative to the mixture and HBeta, possibly due to the formation of defect structure or lattice contraction in the framework of HBeta. The result indicates that the chemical interaction between SAPO-11 and HBeta occurred in the composite. During the synthesis process of the composite, HBeta severed as seed and at least partially solved as silica source for the crystallization of SAPO-11 over HBeta. Thus, microcrystal SAPO-11 interacted with HBeta strongly, which caused the defect structure and/or lattice contraction in the composite. The mixture had the HBeta/SAPO-11 weight ratio of ca. 0.82. It is found that XRD patterns of the mixture were similar to those of the composite. Thus, it is deduced that HBeta/SAPO-11 composite also possessed the HBeta/SAPO-11 weight ratio of ca. 0.82.

Figure 9 shows IR spectra of HBeta, SAPO-11, HBeta/SAPO-11 composite and HBeta/SAPO-11 mixture. The composite and the mixture showed all characteristic IR bands which could be attributed to HBeta (1218, 1171, 1085, 199,119, 600 566, 519, 466 cm^{-1}) and SAPO-11 (1224, 1125, 742, 712, 593, 546, 446 cm^{-1}) [31]. It is worth noticing, the composite showed new IR bands at ca. 895 and 942 cm^{-1} relative to the mixture as well as HBeta and SAPO-11. Bellussi et al. [32] found that titanium silicalite (TS-1) showed a new IR band near ca. 950 cm^{-1} compared with TiO_2 and SiO_2. They attributed the IR band to Si-O-Ti linkage in the Ti-substituted in TS-1 [32]. Wang et al. observed that Zr-MCM-41 presented new IR band at ca. 960 cm^{-1} with respect to MCM-41 [33]. The new IR band was due to the stretching vibration of Si-O-Zr formed from the interaction between Si and Zr, or an asymmetric Si-O-Si stretching perturbed by the presence of Zr in its environment [33]. In case of this work, the new IR bands at ca. 895 and 942 cm^{-1} were tentatively assigned to the framework combination of SAPO-11 and HBeta by chemical bonds and/or the defected TO_4 unit in the composite [32-37]. These results further suggest that the interaction of SAPO-11 and HBeta occurred in the composites.

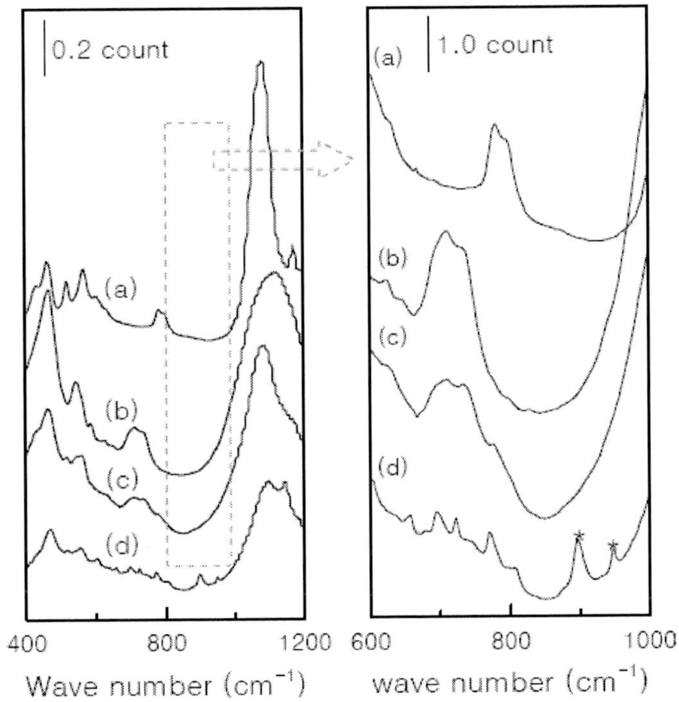

Figure 9. IR spectra of these samples (a) HBeta, (b) SAPO-11, (c) HBeta/SAPO-11 mixture, (d) HBeta/SAPO-11 composite.

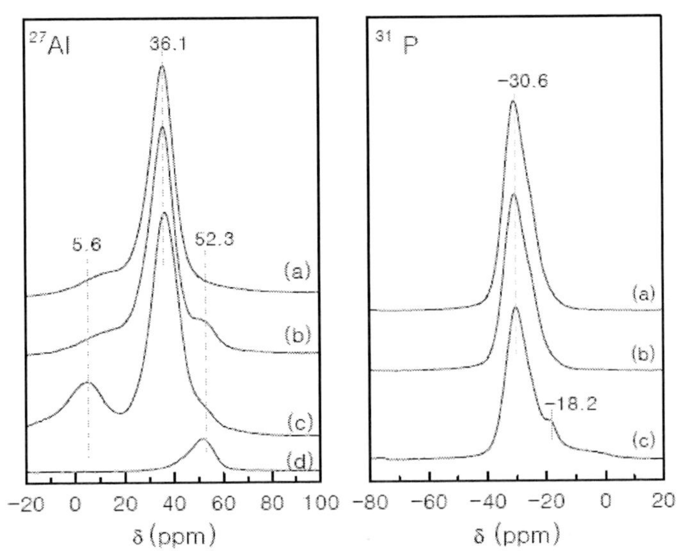

Figure 10. ^{27}Al MAS NMR and ^{31} P MAS NMR spectra of the sample (a) SAPO-11, (b) HBeta/SAPO-11 mixture, (c) HBeta/SAPO-11 composite, (d) HBeta.

Solid state MAS NMR spectroscopy was used to further characterize the chemical environment of framework and non-framework elements in these samples. Figure 10 shows ^{27}Al and ^{31}P MAS NMR spectra of HBeta, SAPO-11, HBeta/SAPO-11 composite and HBeta/SAPO-11 mixture. The mixture presented two signals at 36.1 and 52.3 ppm, due to the resonances of tetrahedral aluminum in the framework of SAPO-11 (36.1 ppm) and HBeta (52.3 ppm). It is worth noticing, the composite showed a new signal at 5.6 ppm relative to other samples. The weak signal at 5.6 ppm was possibly attributed to the presence of extra-framework aluminum and/or amorphous alumina in the composite [38, 39].

From ^{31}P MAS NMR spectra of SAPO-11, the composite and the mixture (Figure 10), a sharp signal at *c.a.* -30.6 ppm was observed, owing to the characteristic resonance of tetrahedral phosphorous in the framework of SAPO-11. It is worth noticing, a new signal at *c.a.* -18.2 ppm was detected for HBeta/SAPO-11 composite, tentatively due to the presence of $P(AlO_4)_n$ ($n<4$) or the phosphorous coordinating with Si originating from HBeta in the composite [40, 41]. Therefore, new phosphorous-oxygen species might form by the interaction of HBeta with SAPO-11 in the composite.

3.2.2. Texture Properties

S_{BET} and V_{pore} of HBeta, SAPO-11, HBeta/SAPO-11 composite and HBeta/SAPO-11 mixture were respectively measured by N_2 isothermal adsorption-desorption, and the results are shown in Table 3. As seen in Table 3, HBeta had the biggest S_{BET} and SAPO-11 was the smallest S_{BET} among these samples. The composite showed lower S_{BET} than the mixture. On the other hand, the V_{pore} of these samples gradually decreased as the sequence of HBeta > SAPO-11 > HBeta/SAPO-11 mixture > HBeta/SAPO-11 composite. These results imply that the surface area and pore volume of HBeta/SAPO-11 composite contracted.

Table 3. Texture and acidity of these samples

Sample	S_{BET} (m^2/g)	V_{pore} (cm^3/g)	Density of acid sites (a.u./g)						B/L ratio		
			200		350		total				
			L	B	L	B	L	B	Weak	Strong	Total
SAPO-11	188	0.17	19	12	13	1	32	13	0.63	0.07	0.40
HBeta/SAPO-11 composite	229	0.10	52	32	38	22	90	54	0.62	0.58	0.60
HBeta/SAPO-11 mechanical	244	0.13	65	28	48	15	113	43	0.43	0.31	0.38
HBeta	309	0.29	29	14	24	8	53	22	0.48	0.33	0.41

3.2.3. Morphology Properties

Figure 11 shows SEM images of HBeta, SAPO-11, HBeta/SAPO-11 composite and HBeta/SAPO-11 mixture. HBeta showed spherical particles and agglomerates with the size of ca. 0.4~0.5 μm. SAPO-11 clearly presented cubic particle and its particle size ranged from 4 to 5 μm. Concerning the composite, it is found that every HBeta crystal or its loose aggregates was evenly embedded in SAPO-11. The mixture exhibited only isolated mixture of spherical HBeta and cubic SAPO-11. Thus, HBeta and SAPO-11 closely combined in the composite with respect to the combination in the mixture.

Figure 11. SEM images of these samples. (a) HBeta, (b) SAPO-11, (c) HBeta/SAPO-11 composite, (d) HBeta/SAPO-11 mixture.

3.2.4. Acidity Properties

Acidity of HBeta, SAPO-11, HBeta/SAPO-11 composite and HBeta/SAPO-11 mixture was investigated by Py-IR. Figure 12 shows Py-IR spectra of these samples at specific desorption temperature of pyridine. IR bands at 1455 and 1545 cm^{-1} were observed on these samples, due to the characteristic IR bands of pyridine chemisorbed on L and B acid sites, respectively [42-45]. IR band at ca. 1490 cm^{-1} was attributed to pyridine co-adsorbed on both B and L acid sites [42-45]. These results indicate that both B and L acid sites existed on these samples.

When desorption temperature raised from 200 °C to 350 °C, IR bands belonging to B and L acid sites were still detected. Thus, there were weak/medium and strong acid sites on these samples. Nevertheless, the half-width and intensity of these bands reduced with the increase of temperature. The result suggests that the weak and medium acid sites decreased from the sample as increasing the temperature.

The concentration and distribution of acid sites (a.u./g) were estimated by the integration of the IR bands. It can be seen in Table 3, the concentration of B acid sites decreased as the order of HBeta/SAPO-11 composite > HBeta/SAPO-11 mixture > HBeta > SAPO-11, and the concentration of L acid sites reduced as the sequence of HBeta/SAPO-11 mixture > HBeta/SAPO-11 composite > HBeta > SAPO-11. HBeta/SAPO-11 mixture had the highest concentration of total acid sites among these samples. B/L ratio at 200 °C (B_{200}/L_{200}) and 350 °C (B_{350}/L_{350}) was calculated, respectively meaning weak/medium B/L ratio of weak/medium acid sites (B_{200}/L_{200}) and strong B/L ratio of strong acid sites (B_{350}/L_{350}). It is observed from Table 3 that B_{200}/L_{200} ratio increased as HBeta/SAPO-11 mixture < HBeta < HBeta/SAPO-11 composite < SAPO-11; and B_{350}/L_{350} ratio raised as SAPO-11 < HBeta/SAPO-11 mixture <

HBeta < HBeta/SAPO-11 composite. HBeta/SAPO-11 composite had the highest B_{total}/L_{total} (total acid sites) ratio among these investigated samples.

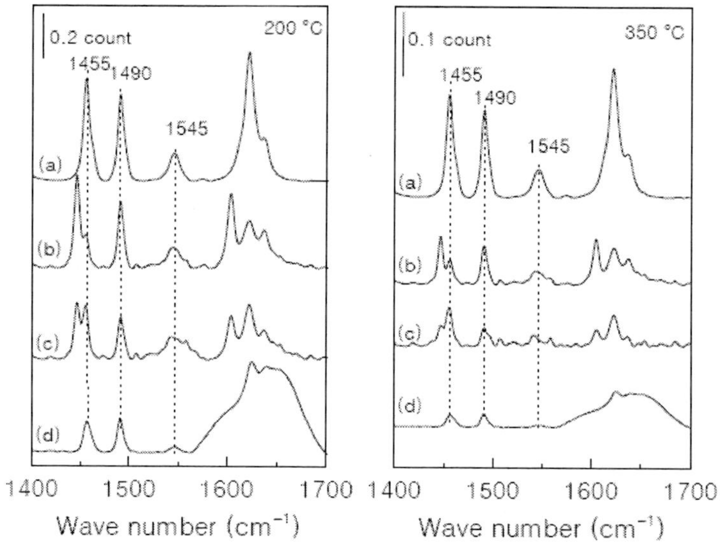

Figure 12. Py-IR spectra of these samples at 200 °C and 350 °C, (a) HBeta, (b) HBeta/SAP-11 composite, (c) HBeta/SAPO-11 mixture, (d) SAPO-11.

The concentration of weak/medium acid sites was more than that of strong acid sites on these samples (Table 3). Both the composite and the mixture had the more concentration of weak/medium acid sites than individual HBeta and SAPO-11. The mixture showed the most concentration of weak/medium acid sites among these samples.

The above results reveal that HBeta/SAPO-11 composite exhibited the different concentration and distribution of acid sites compared with HBeta/SAPO-11 mixture as well as individual HBeta and SAPO-11. HBeta/SAPO-11 composite showed the framework combination of SAPO-11 and HBeta by chemical bonds; in addition, the extra-framework aluminum and new phosphorus species formed in the composite. Hence, the strong interaction between SAPO-11 and HBeta occurred in HBeta/SAPO-11 composite and significantly modified the acidity of the composite.

3.2.5. Catalytic Reactivity

Catalytic reactivity of HBeta/SAPO-11 composite, HBeta/SAPO-11 mixture, HBeta and SAPO-11 in the cracking of 2-butylene to propylene was tested. It can be seen in Table 4, methane, ethane, ethylene, propane, propylene, n-butane, *iso*-butane, 1-butylene, *iso*-butylene, C_5+ (C_5-C_{12} hydrocarbons including aromatics) and coke, etc. were produced in 2-butylene conversion on these samples. The conversion of 2-butylene decreased as the order of HBeta > HBeta/SAPO-11 mixture > HBeta/SAPO-11 composite > SAPO-11. HBeta/SAPO-11 composite exhibited the higher selectivity and yield of propylene than any other samples. The yield of propylene on HBeta/SAPO-11 mixture was as similar as that on HBeta. These results further confirm that the strong interaction occurred in the composite.

Figure 13 shows effect of reaction temperature on the catalytic reactivity of HBeta/SAPO-11 composite and HBeta/SAPO-11 mixture in the reaction. With the increase of reaction temperature, the conversion of 2-butylene gradually increased on the two samples. When the reaction temperature rose, the yield of propylene and ethylene monotonously increased. The composite exhibited the higher propylene yield than the mixture at the investigated temperature.

Table 4. Catalytic reactivity of these samples in the cracking of 2-butylene

	HBeta/SAPO-11 composite	HBeta/SAPO-11 mixture	HBeta	SAPO-11
2-butylene conversion (wt. %)	68.8	76.7	80.5	42.7
Product selectivity (wt. %)				
Methane	0.1	0.2	2.6	trace
Ethane	0.3	3.0	7.2	1.0
Ethylene	6.5	2.6	3.2	1.1
Propane	1.8	3.0	11.5	4.3
Propylene	40.2	23.1	21.4	12.1
n-butane	3.0	5.1	4.8	2.7
iso-butane	2.5	5.6	5.9	13.5
1-butylene	2.3	10.6	9.7	5.2
iso-butylene	22.5	17.8	10.6	29.2
C_5~C_{12}	20.0	28.9	21.7	30.0
Coke	0.8	2.2	1.0	0.5
Product Yield (wt. %)				
Ethylene	3.1	2.0	2.6	0.5
Propylene	27.7	17.7	17.2	5.2
Propylene+Ethylene	30.8	19.7	19.8	5.7

Reaction conditions: 490 oC, 1 g. h-1, 0.1 MPa, TOS=240 min.

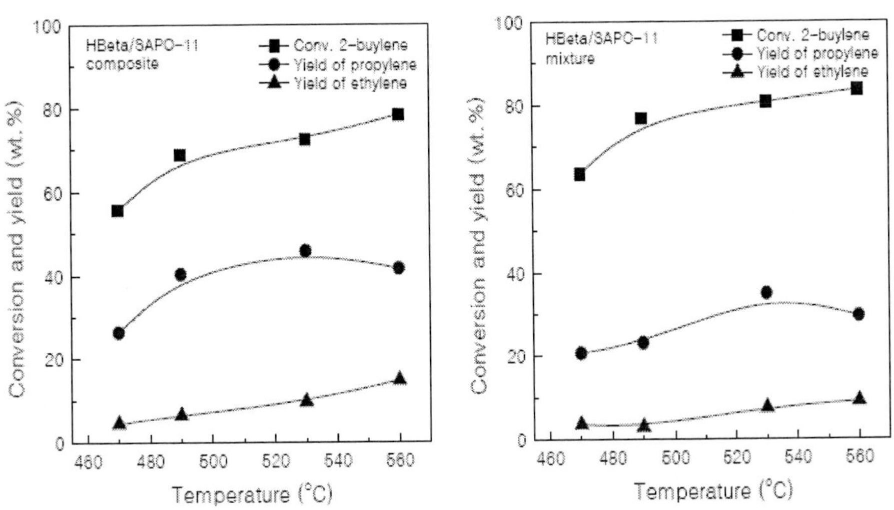

Figure 13. Effect of reaction temperature on the cracking of 2-butylene over HBeta/SAPO-11 composite and HBeta/SAPO-11 mixture. Reaction conditions: WHSV=1 h^{-1} and 0.1 MPa, TOS=120 min.

3.2.6. Catalytic Stability

Figure 14 shows the time course of 2-butylene cracking to propylene over HBeta/SAPO-11 composite and HBeta/SAPO-11 mixture. Under the reaction conditions, both the composite and the mixture almost kept the stable catalytic reactivity. During the initial reaction stage (~60 min), 2-butylene conversion on the composite and the mixture decreased. These phenomena might be due to the formation of coke on these samples surface. The deposition of coke covered the active sites on the samples, resulting in the reduction of 2-butylene conversion. In addition, the decrease in 2-butylene conversion on the mixture was faster than that on the composite, possibly owing to the reason that the mixture had the high total concentration of acid sites compared with the composite.

It is known that 2-butylene cracking is involved in complex reaction network. B acid sites are considered to be active sites for 2-buylene cracking to propylene; in addition, the concentration and strength distribution of B acid sites are close relative to 2-buylene cracking to propylene [30]. The solid acid catalyst with the high concentration of B acid sites and the high B/L ratio favor the production of propylene in the reaction [14-22, 30]. HBeta/SAPO-11 composite showed the high propylene yield, mainly due to its high concentration of B acid sites and the high B/L ratio.

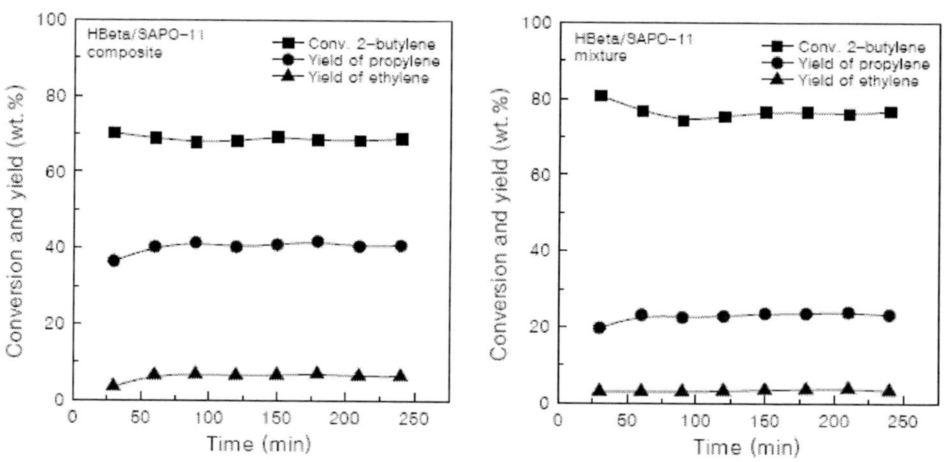

Figure 14. Time course of 2-butylene cracking to propylene over HBeta/SAPO-11 composite and HBeta/SAPO-11 mixture. Reaction conditions: 490 °C, WHSV=1 h^{-1} and 0.1 MPa.

3.3. Summary

HBeta/SAPO-11 composite was synthesized by the hydrothermal method. During the synthesis of HBeta/SAPO-11 composite, HBeta was seed and partially dissolved as silica source for the crystallization of SAPO-11. HBeta/SAPO-11 composite had the different physicochemical properties and catalytic performance of 2-butylene cracking to propylene, due to the occurrence of the strong interaction of SAPO-11 and HBeta in the composite.

CONCLUSION

Aluminosilicate/silicoaluminophosphate zeolite composite prepared by hydrothermal synthesis has distinctive physicochemical properties and catalytic performance. In this work, HZSM-5/SAPO-34 composite and HBeta/SAPO-11 composite could be successfully prepared by one-step hydrothermal crystallization. During the synthesis of these composites, aluminosilicate zeolite (HZSM-5 or HBeta) was as seed and partially dissolved as silica sources for the *in situ* crystallization of silicoaluminophosphate zeolite (SAPO-34 or SAPO-11) over aluminosilicate zeolite. XRD, IR, MAS NMR, SEM, NH_3-TPD and Py-IR were effective characterization technologies to reveal the chemical interaction of aluminosilicate with silicoaluminophosphate in these composites. Compared with the mixture, the composite presented the special structure, texture, morphology and acidity, mainly due to the chemical interaction of aluminosilicate and silicoaluminophosphate zeolites. HZSM-5/SAPO-34 composite was found to show the high catalytic reactivity for ethanol to propylene; and HBeta/SAPO-11 composite exhibited the high catalytic reactivity in 2-butylene to propylene. It is worth noticing that the composite showed the higher catalytic reactivity and better catalytic stability than the mixture. The strong interaction between individual aluminosilicate and silicoaluminophosphate zeoiltes in the composite was found and accounted for the special properties and the high catalytic reactivity of the composite.

ACKNOWLEDGMENT

We gratefully acknowledge the support of this work by Ministry of Education (No. NCET-10-878, 20096101120018, 2009-37th of SRFROCS), Shaanxi Province (No. 2009ZDKG-70, 09JK793), Northwest University (PR200905, 10YSY08) and State Key Lab of Chemical Resource Engineering (CRE-2011-C-304).

REFERENCES

[1] Fan, Y.; Bao, X.; Lei, D.; Shi, G.; Wei, W.; Xu, J.A. *Fuel.* 2005, 84, 435-442.
[2] Fan, Y.; Lei, D.; Shi, G.; Bao, X. *Catal. Today.* 2006, 114, 388-396.
[3] Chae, H.; Song, Y.; Jeong, K.; Kim, C.; Jeong, S. *J. Phys. Chem. Solid* 2010, 71, 600-603.
[4] Lee, Y.; Kim, Y.; Viswanadham, N.; Jun, K.; Wook B. *J. Appl. Catal. A* 2010, 374, 18-25.
[5] Goto, D.; Harada, Y.; Furumoto, Y.; Takahashi, A.; Fujitani, T.; Oumi, Y.; Sadakane, M.; Sano, T. *Appl. Catal. A* 2010, 383, 89-95.
[6] Song, Z.; Takahashi, A.; Mimura, N.; Fujitani, T. *Catal. Lett.* 2009, 131, 364-369.
[7] Inoue, K.; Inaba, M.; Takhara, I.; Murata, K. *Catal. Lett.* 2010, 136, 14-19.
[8] Furumoto, Y.; Harada, Y.; Tsunoji, N.; Takahashi, A.; Fujitani, T.; Ide, Y.; Sadakane, M.; Sano, T. *Appl. Catal A* 2011, 384, 262-267.
[9] Song, Z.; Takahashi, A.; Nakamura, I.; Fujitani, T. *Appl. Catal. A* 2010, 384, 201-205.

[10] Xia, W.; Takahashi, A.; Nakamura, I.; Shimada, H.; Fujitani, T. *J. Mole. Catal. A* 2010, 328, 114-118.
[11] Oikawa, H.; Shibata, Y.; Inazu, K.; Iwase, Y.; Murai, K.; Hyodo, S. Kobayashi, G. Baba, T. *Appl. Catal. A* 2006, 312, 181-185.
[12] Lin, B.; Zhang, Q.; Wang, Y. *Ind. Eng. Chem. Res.* 2009, 48, 10788-10795.
[13] Zhang, X.; Wang, J.; Zhong, J. *Petro. Chem. Tech.* 2004, 33, 781-787.
[14] Xue, N.; Liu, N.; Nie, L.; Gu, M.; Peng, L.; Guo, X.; Ding, W.; *J. Mol. Catal. A* 2010, 327, 12-19.
[15] Zhao, G.; Teng, J.; Xie, Z.; Jin, W.; Yang, W.; Chen, Q.; Tang, Y.; *J. Catal.* 2007, 248, 29-31.
[16] Jiang, G.; Zhang, L.; Zhao, Z.; Zhou, X.; Duan, A.; Xu, C.; Gao, J. *Appl. Catal. A* 2008, 340, 176-182.
[17] Wang, Z.; Jiang, G.; Zhao, Z.; Feng, X.; Duan, A.; Liu, J.; Xu, C.; Gao, J. *Energy & Fuel.* 2010, 24, 758-763.
[18] Xue, N.; Nie, L.; Chen, X.; Guo, X.; Ding, W.; Chen, Y.; Gu, M.; Xie, Z. *J. Catal.* 2007, 248, 20-28.
[19] Xue, N.; Nie, L.; Fang, D.; Guo, X.; Shen, J.; Ding, W.; Chen, Y. *Appl. Catal. A* 2009, 352, 87-94.
[20] Lu, Y.; He, M.; Shu, X.; Zong, B. *Appl. Catal. A* 2003, 255, 345-347.
[21] Wang, B.; Gao, Q.; Gao, J.; Ji, D.; Wang, X.; Suo, J. *Appl. Catal. A* 2004, 274, 167-172.
[22] Zhao, G.; Teng, J.; Zhang, Y.; Xie, Z.; Yue, Y.; Chen, Q.; Tang, Y. *Appl. Catal. A* 2006, 299, 167-174.
[23] Zhu, X.; Liu, S.; Song, Y.; Xie, S.; Xu, L. *Appl. Catal. A* 2005, 290, 191-199.
[24] Wadlinger, R. L.; Keller, G. T.; Rosinski, E.J. *US Patent.* 3 308 069, 1967.
[25] Treacy, M.M.J.; Newswam, J.M. *Nature.* 1988, 332, 249-251.
[26] Higgins, J.B.; Lapierre, R.B.; Schlenker, J.L.; Rohrman, A.C.; Wood, J.D.; Keller, G.T.; Rohrbaugh, W.J. *Zeolites.* 1988, 8, 446-452.
[27] Sun, M.; Sun, J.; Li, Q. *Chem. Lett.* 1998, 9, 519-520.
[28] Corma, A.; Martinez, A.; Arroyo, P.A.; Monteiro, J.L.F.; Sousa-Aguiar, E.F. *Appl. Catal. A* 1996, 142, 139.
[29] Bennett, J.M.; Richardson, J.W.; Pluth, J.J.; Smith, J.V. *Zeolites* 1987, 7, 160-162.
[30] Kissin, Y.V. *Catal. Rev.* 2001, 43(1/2): 85-146.
[31] Geidel, E.; Bohlig, H.; Peuker, C. *Stud. Surf. Sci. Catal.* 1991, 65, 511-516.
[32] Bellussi, G.; Carati, A.; Clerici, M.G.; Maddinelli, G.; Millini, R. *J. Catal.* 1992, 133, 220-230.
[33] Wang, X.; Lefebvre, F.; Patarin, J.; Basset, J.M. *Micro. Meso. Mater.* 2001, 42, 269-274.
[34] Wu, Z.; Zhao, Y.; Liu, D. *Micro. Meso. Mater.* 2004, 68, 127-136.
[35] Rakshe, B.; Ramashe, V.; Ramaswam, V.; Ramaswam, A.V. *J. Catal.* 1996, 163, 501-505.
[36] Dang, Z.; Anderson, A.G.; Amenomiya, Y.; Morrow, B.A. *J. Phys. Chem.* 1995, 99, 14437-14440.
[37] Liu, Z.; Davis, R.J. *J. Phys. Chem.* 1994, 98, 1253-1261.
[38] Briend, M.; Lamy, A.; Peltre, M.J.; Man, P.P.; Barthomeuf, D. *Zeolites* 1993, 13, 201-211.

[39] Prakash, A.M.; Satyanarayana, C.V.V.; Bagwe, R.P.; Ashtekar, S.; Chakrabarty, D.K. *Micro. Meso. Mater.* 1996, 6, 89-98.
[40] Blackwell, C.S.; Patton, R.L. *J. Phys. Chem.* 1984, 88, 6135-6139.
[41] Zhang, Z.; Zong, B.; *Chin. J. Catal.* 2003, 24, 856-861.
[42] Hughes; T. R.; White, H. M. *J. Phys. Chem.* 1967, 71, 2192-2201
[43] Eberly, P. E. *J. Phys. Chem.* 1968, 72, 1042-1049.
[44] Vedrine, J. C.; Aurox, A.; Bolis, V. *J. Catal.* 1979, 59, 248-262.
[45] Topsøe, N.Y.; Pdersen, K.; Derouane, E. *J. Catal.* 1981, 70, 41-52.

Reviewed by:

1. Prof. Dr. Hua-ming Li, School of Chemistry and Chemical Engineering, University of Jiangsu, Zhengjiang 212013, China. E-mail: lihm@ujs.edu.cn

2. Prof. Dr. Chuan-jin Huang, School of Chemistry and Chemical Engineering, Xiamen University, Xiamen 361005, China. E-mail address: huangcj@xmu.edu.cn

In: Zeolites: Synthesis, Chemistry and Applications
Editors: Moisey K. Andreyev and Olya L. Zubkov
ISBN: 978-1-61942-861-4

Chapter 5

STORAGE OF HYDROGEN IN ZEOLITES

Jing Li and Erdong Wu
Shenyang National Laboratory for Materials Science,
Institute of Metal Research
Chinese Academy of Sciences, Shenyang, China

1. INTRODUCTION

An efficient and practicable hydrogen storage and delivery system is a key for the success of hydrogen fuel for various ecological energy systems in the future, as the conventional hydrogen storage systems such as gas cylinders and liquid hydrogen have serious limitations in transportation. Consequently, the storages of hydrogen in solid materials have been extensively studied in the last a few decades because of their advantages as potential energy carriers [1-5]. Among these storage systems, the adsorption in porous materials has attracted increasing interest due to its simplicity, less drastic pressures and temperature requirements. Although attentions have been focused on light carbon microporous materials owing to their potential high gravimetric storage capacities, other microporous materials, such as organic frameworks and zeolites, have also been examined [6,7]. The consideration of hydrogen storage in zeolite materials is mainly due to their rich variety of porous structures, which compare favorably when adsorption capacities are measured on a volume basis. In addition, the high stability and low cost of zeolites are also important advantages for hydrogen storage media [8]. On the other hand, the selective adsorption properties of zeolites as molecular sieves have also been applied to the separation of the hydrogen isotopes in the potential nuclear fusion reactions for energy generation [9]. In the experimental and theoretical works devoted to the above applications, the physisorption and encapsulation of molecular hydrogen in tailored zeolites are the most available options.

A convenient classification of pore systems of solids according to the average width of the pores, *w*, is summarized in Table 1. The classification is originally proposed by Dubinin [10] and officially adopted by International Union Pure and Applied Chemistry (IUPAC). The adsorption in macropores is basically the same as that on open surfaces. The capillary condensation with hysteresis loop could happen in mesopores. Whereas, the amount adsorbed on micropores at a given relative pressure will be enhanced due to the significantly higher interaction potential between small molecular hydrogen and the walls of narrower pores. Therefore, the vast majority of zeolites as media for the hydrogen storage are in the class of micropores.

Up to now, the hydrogen storage properties of all common microporous zeolites and some mesoporous zeolites have been investigated at both room and cryogenic temperatures. Fig 1 shows the framework structures of some these zeolites. Among them, the microporous X zeolites of faujasite (FAU) frameworks are excellent ion-exchange agents and most extensively studied. The frameworks consist of sodalite cages which are connected through hexagonal prisms (Fig 1a). The cages and prisms form large cavities (supercages) with a diameter of ~ 12 Å and windows with 12-membered rings of width 7.4 Å. The microporous A zeolites of LTA frameworks are propitious to many experimental and theoretical investigations owing to their void volume of simple α cage formed by eight β cages through cubic prisms (Fig 1b). The cages allow hydrogen molecules to enter through 8-membered rings, which can vary its size from ~ 3 Å to ~ 5 Å by exchanged ions. Another studied microporous zeolite is ZSM-5, which is a pentasil-zeolite of MFI (mordenite framework inverted) type. The ZSM-5 is composed of pentasil units of [5^8] formed by eight 5-membered rings (Fig 1c). The units are linked together by oxygen bridges to form pentasil chains. The chains are further interconnected to form three dimensional frameworks containing corrugated sheets with 10-ring channels. The pore size of the channels parallel to the corrugations is 5.4 - 5.6 Å, which can accommodate hydrogen molecules. A typical mesoporous zeolite studied for hydrogen storage is MCM-41, which is an all-silica molecular sieve system consists of hexagonally arranged channels with diameters varying from 1.5 to 10 nm, and prepared using a molecule surfactant template method (Fig 1d). The hydrogen adsorption on MCM-41 with 3 nm-diameter pores suggest that the surfaces of the mesopores in MCM-41 adsorb hydrogen similar to the micropores in above zeolites, and the adsorbed hydrogen is more stable than on other oxidic mesoporous surfaces, such as SiO_2 [1].

Table 1. Clasification of pores according to their width, w

Type of pores	Width
micropores	less than ~ 2 nm
mesopores	between ~ 2 nm and ~ 50 nm
macropores	more than ~ 50 nm

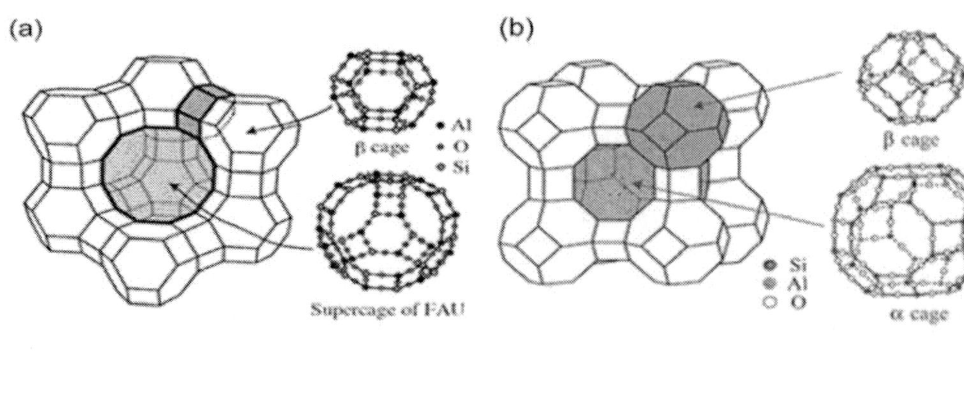

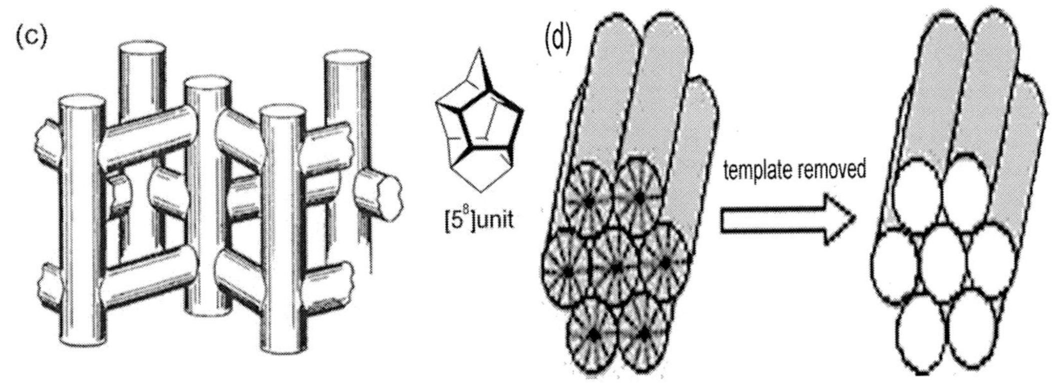

Figure 1. The framework structures of (a) X, (b) A, (c) ZSM-5 and (d) MCM-41 zeolites.

2. Hydrogen Adsorption on Zeolites

2.1. Adsorption on Porous Materials

Depending upon the strength of interaction, adsorption processes can be divided into two categories, i.e. the chemical and physical adsorptions. The former (or chemisorption) is characterized by large interaction potentials between adsorbents and adsorbates, leading to high heats of adsorption and approaching the value of chemical bonds. Whereas, the later (or physisorption) is characterized by van der Waals interactions between adsorbents and adsorbates, involving relatively low interaction potentials and heats of adsorption.

Hydrogen adsorption on porous materials involves the interaction of hydrogen molecules with internal surfaces of variety of pores. The governing stability of the adsorptive hydrogen, the optimal temperature of the adsorption, and the forces involved in the process are essentially physical (dispersive). The weak physisorption interaction usually leads to the requirement of cryogenic temperatures, typically 77 K of liquid nitrogen temperature, to achieve reasonable loadings. Upon adsorption, the entropy change of adsorbate, ΔS, is necessarily negative since the condensed state is more ordered than the gaseous state due to the loss of at least one degree of translational freedom. The spontaneity of the adsorption process required that the Gibbs free energy, ΔG, also be a negative quantity. Hence, the enthalpy, ΔH, will be negative, result in an exothermic process.

The physisorption involves subcritical and supercritical adsorptions. Various experiments clearly show that the adsorption behaviors of supercritical fluids on porous materials differ from that of the subcritical adsorption, i.e. the standard types of adsorption isotherms according to IUPAC classification [11]. Typical experimental isotherms under supercritical conditions go through distinct maxima with increasing pressure, which would not be found at the subcritical regions. The amount adsorbed under supercritical conditions is the excess amount of adsorbates present in the pores and on the surface of the adsorbent, which is defined as the different amount to that would be present under the same conditions in the absence of surface-gas interactions. If the adsorption density in the pore volume is greater than the bulk density, the surface mass excess is positive. However, it will be negative when the fraction of the pore volume where the adsorption density is less than the bulk density is greater than the remaining volume where the absorption density is greater, or when the adsorption density throughout the pore volume is less than the bulk density [12].

2.2. Measurements of Adsorption

The hydrogen adsorption results reported so far in the literatures sometimes do not coincide with one another. Therefore, the reproducible and precise measurement of hydrogen adsorption is very important for the development of hydrogen storage materials. There are generally three methods to measure the hydrogen storage capacities of adsorbents: (a) gravimetric, (b) volumetric and (c) temperature programmed desorption (TPD). Before measurements, the adsorbents are normally degassed at high temperature under vacuum. The equilibrium of adsorption can be estimated when the adsorbates are in contact with the surfaces of the adsorbents and the temperature and pressure or mass stabilized.

Figure 2 shows the typical diagrams of gravimetric and volumetric apparatuses. Since the buoyancy of adsorbents at gas phase has to be emended and the minute increment of samples could not be measured easily by the microbalance in the gravimetric method, most researchers favor the use of volumetric methods. The major component of the volumetric apparatus is an adsorption cell for adsorbents, which can be put in a thermostat to satisfy different temperatures. The capacity of adsorption can be calculated by knowing the pressure, temperature and equation of state of the adsorbed gas. On the other hand, the TPD procedure is mainly used to measure the encapsulated gases.

In the volumetric method involving high pressures and low temperatures, the compressibility factor, z, expressed as the equation of state of hydrogen, considerably affects the determination of the amount adsorbed. As a measure of deviation from ideal behavior, compressibility factor can be equal to or higher than 1, and the extent of its variation from unity indicates the degree of non-ideality for real gas. The relationship between z-value and pressure expressed by different types of equations at two typical temperatures are shown in Figure 3 [14]. The z-values all increase almost linearly with pressures at room temperature (291 K), and the standard deviation from different equations at the highest pressure of 5.7 MPa is only 0.69% (Figure 3a). However, at the near critical temperature of 35 K, the deviations of hydrogen from ideal gas at relatively high pressures become significant, and the z-values generated from different equations behave differently (Figure 3b). The z-values obtained from Viral and 8-terms Benedict-Webb-Rubin (BWR) equations become negative at high pressure, and are not valid for hydrogen in the near critical region. On the other hand, the z-values from Lee-Kesler (LK) and Soave-Redlich-Kwong (SRK) equations change the trends from decrease to increase at high pressure, hence are suitable for the region [14].

The measurement procedure to determine the adsorbed amount on zeolites could involve two possible routes: the step-by-step and the repeated measurements. The former changes the pressure step-by-step and sums up the adsorbed amount at each temperature. The method is most commonly used in the measurements of hydrogen adsorption, but could lead to accumulation of errors. On the other hand, the later evacuates the whole system for the measurement of each pressure point to reduce systematic errors, but the procedure is more time consuming due to the time required for reestablishment of pressure in the system [14].

Zeolites are hydrate, crystalline aluminosilicates, whose crystal size is commonly in the range of 1 - 10μm. Commercial and synthesized zeolite samples are usually powder and need to be shaped with binder material (clay, alumina, polymers, etc) into bodies (beads, pellets, extrudates, etc) prior to their applications as adsorbents. The presence of the binder changes the apparent density of zeolite adsorbents, inevitably affects the adsorptive properties. Zhou [15] concluded that the gravimetric density and weight percentage of stored hydrogen of activated carbon in powder and pellet at 77 K and 6 MPa were quite different. Whereas Li and Wu [14] indicated that the morphology of the zeolite particles affected the hydrogen storage capacity. Therefore, the macroscopic shapes and sizes of the zeolites should be specified in experimental investigations.

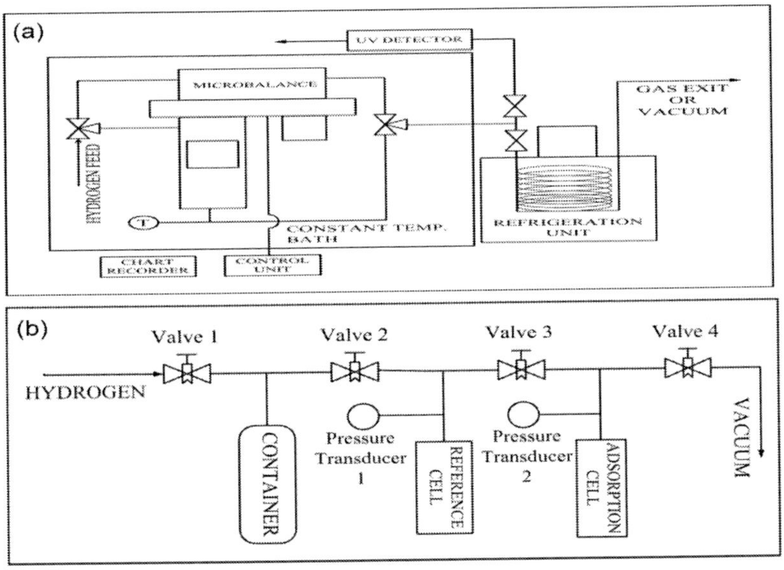

Figure 2. Schematic diagrams of (a) the high pressure flow gravimetric apparatus [13], and (b) the volumetric adsorption apparatus. The name or meaning of each unit in the apparatus is labeled on the symbol of the unit.

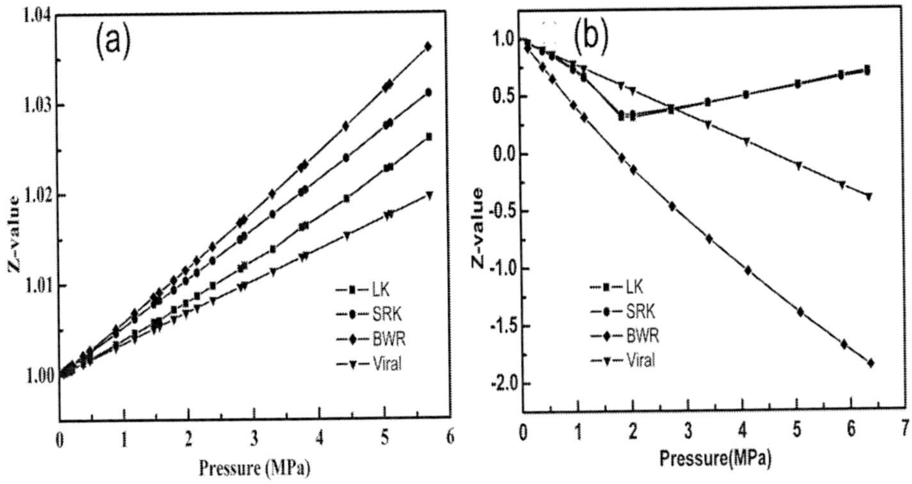

Figure 3. Compressibility factor, z, of hydrogen expressed by different equations of states at (a) 291K and (b) 35K.

2.3. Adsorptive Properties of Zeolites on Hydrogen

The challenge in the research on hydrogen storage materials is to pack the hydrogen atoms or molecules as close as possible. As expected from a physisorption process, the higher hydrogen adsorption is obtained at low temperature, and that hydrogen adsorption capacity increases with pressure. Owing to the low critical temperature (33.2 K) of hydrogen molecules, up to now the hydrogen adsorption studies on zeolites have been mostly limited to the temperature region of above liquid nitrogen temperature (77 K) for supercritical adsorption, where the accomplishment of experiments is relatively easy. In the case of subcritical conditions, the neutron diffraction technology has been used during the adsorption isotherm measurements to elucidate the detailed structural characterization of both the adsorbents and the confined phase [16, 17]. On the other hand, Li and Wu [14] have investigated the hydrogen adsorption on two adsorbents including NaX zeolite in the vicinity region of critical temperature at relatively high pressures, and observed a phenomenon of 'inverse capillary condensation' in the small pores of the adsorbents.

Basmadjian [18] was the first to measure the adsorption equilibrium of hydrogen and its isotope on NaX, CaA and NaA zeolites, and obtained the mass amounts of 0.84wt% and 0.97wt% for H_2 and D_2 respectively in NaX, 0.97wt% and 1.08wt% in CaA, and 0.79wt% and 0.89wt% in NaA, at the pressure of 1 bar and temperature of 75 K. Kotoh et al [19] recently carried out a similar experiment on CaA and derived consistent result, where the higher adsorption was observed for D_2. On the other hand, Levin et al [20] studied the equilibrium adsorptions of hydrogen isotopes on NaX zeolite in the range of pressures from 0.2 to 7 MPa and temperatures from 100 to 300 K, and derived the maximum adsorption capacities of 0.21 and 0.23wt% for H_2 and D_2, respectively, at 6 MPa and 100 K, thus confirmed that the adsorption capacities of D_2 on zeolites were always higher than that of H_2 under the same operation conditions.

Skazyvaey et al [21] investigated the hydrogen adsorption capacities of synthetic faujasites, mordenites, and types A, E, and I zeolites at the pressures of ~ 760 mmHg and 77 K. They found that the synthetic faujasites and type E zeolites adsorbed the highest amounts of hydrogen of ~ 0.9wt% with negligible difference between them. Langmi et al [22] investigated the hydrogen storage properties on X, Y, A and RHO zeolites containing various exchangeable cations at 77 K and a pressure range up to 15 bar. They showed that the adsorbed amounts increased with pressures, and a highest gravimetric storage capacity of 2.19wt% was observed for CaX among these zeolites.

Du and Wu [23] studied the hydrogen adsorption properties on A, X and ZSM-5 zeolites up to a high pressure of ~ 7 MPa at three supercritical temperatures of 77 K, 195 K and 293 K. As shown in Figure 4, the excess adsorption isotherms were basically type I, where the adsorbed amounts were low and increased linearly with pressure at 195 K and 293 K, whereas adsorption were high and saturated at 77 K. However, in the 77 K isotherms except for that of KA, more than one adsorbed H_2 per cations at their saturation pressures and descending adsorptions after the maximum at 2 - 5 MPa were observed. The typical supercritical phenomenon of adsorption descent at 77 K could be explained as a result of the higher growth rate of density of the gas phase than that of the adsorbed phase under increasing pressure. In general, the hydrogen adsorption capacities of these microporous zeolites appeared to be related to the sizes of their pores. X zeolite with the largest pore size had the highest adsorption capacity, the ZSM-5 was next, and the A zeolites had lower adsorption capacities.

In particular for the A zeolites, both the pore sizes and capacities followed the order of CaA > NaA > KA. On the other hand, the larger pore sizes inside of the microporous zeolites mean higher surface areas available for hydrogen adsorption. As shown in Figure 5, Du and Wu actually derived that the gravimetric storage capacities almost correlated linearly with the BET surface areas in the A, X and ZSM-5 zeolites.

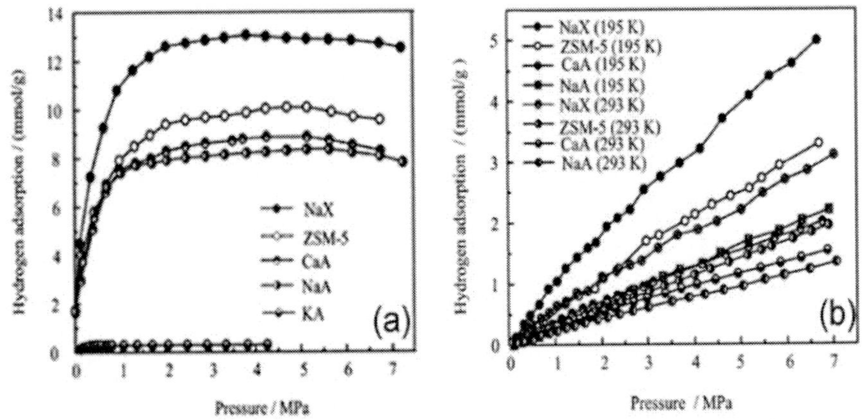

Figure 4. Hydrogen adsorption isotherms on zeolites at (a) 77 K and (b) 195 K, 293 K.

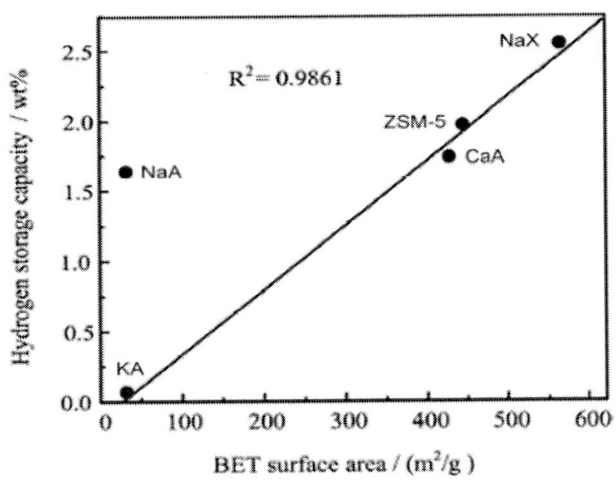

Figure 5. Correlation between hydrogen storage capacity and BET surface area, where the R2 is the correlation coefficient, and the name of each zeolite is labeled besides its symbol.

A large number of adsorption experiments on carbonaceous and silica-alumina-based sorbents suggested that the hydrogen uptake capacity of an adsorbent depended on its micropore volume or the BET surface area [12, 22, 24]. Jhung et al [25] determined the hydrogen adsorption capacities of several microporous molecular sieves including aliminosilicate zeolites of AlPO-5, SAPO-5, SAPO-34 at the pressure of 0 - 1atm. Among them, the SAPO-34 with the highest specific surface area achieved the maximum adsorption capacity of 160 ml/g at 77 K and 1 atm, which was likely attributed to the high micropore surface area (630 m^2/g), large micropore volume (0.29 cm^3/g) of this molecular sieve.

However, some studies indicated that the hydrogen adsorption capacity of a zeolite was related to the surface area of its pores with the sizes of 0.5 - 0.7 nm rather than the total microporous surface area [26, 27]. Li and Wu [7] studied the cryoadsorption of hydrogen on divalent cations (Mg^{2+}, Zn^{2+} and Ni^{2+}) exchanged X zeolites and their base form in Na^{+} at the low temperatures of 35 - 140 K and high pressures up to 7 MPa. They found that the hydrogen adsorption capacities of these zeolites at low temperatures were dependent on their cation's sizes, charges, and distributions in the exchanged zeolites rather than their BET surface areas. As shown in Figure 6, the hydrogen capacities of the zeolites followed an order of NaX > MgX > ZnX > NiX, which was the same as that of the stability of the cations, but opposite to the hydration tendency of the ions and the isosteric heat of the adsorption of the zeolites. The results indicated that the redox potential of the cation rather than specific surface area in the same type of zeolites was the dominant factor in hydrogen adsorption. On the other hand, the data measured at below 77 K complemented the inadequate information on the hydrogen adsorptive properties of zeolites at near critical temperatures. It was known that temperature was a controlling factor in physisorption of hydrogen, but the mass amounts adsorbed on various zeolites under supercritical conditions of around 77 K had been limited to less than 3wt%. The highest gravimetric capacity of 4.02 wt% or adsorbed density of 70.2 kg/m^3 in NaX achieved at 35 K and pressure of 960 KPa indicated that the hydrogen storage capacity could be significantly increased at near critical temperatures and reached the density close to that of the liquid hydrogen (70.8 kg/m^3).

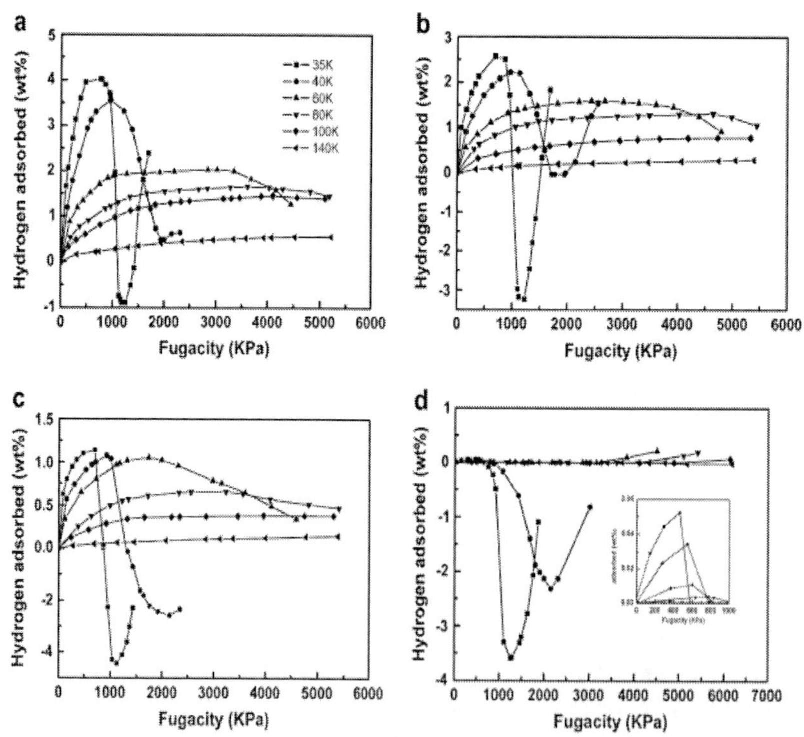

Figure 6. Hydrogen adsorption isotherms over (a) NaX, (b) MgX, (c) ZnX and (d) NiX zeolites on fugacity scale. The fugacity has a unit of pressure and contains all the information on the nonideality of the gas.

Li and Yang [28] studied the low silica type X zeolites (LSX, Si/Al=1) fully exchanged by alkali-metal cations (Li^+, Na^+, and K^+) for their hydrogen storage capacities. Hydrogen adsorption isotherms were measured separately at 77 K and 1 atm, and at 298 K and 10 MPa. It was observed that the hydrogen adsorption capacity depended strongly on the cationic radius and the density of the cations that were located on the exposed sites. The interaction energies between H_2 and the cations follow the order $Li^+ > Na^+ > K^+$, as predicted based on the ionic radii. The Li-LSX achieved an H_2 capacity of 1.5wt% at 77 K and 1 atm, and 0.6wt% at 298 K and 10 MPa, among the highest of known sorbents.

The metals of Pd, Ru, Rh and Ni have been known to be important catalysts for various hydrogenation reactions. Prasanth et al [29, 30] investigated the hydrogen capacities and adsorption properties of NaX and its Pd, Ru, Rh and Ni exchanged forms at 303 K and 333 K and pressures up to 5 bar in a gravimetric system and at 77 K in a static volumetric apparatus setup up to 1 bar. The chemisorptions were observed at 303 K and 333 K due to the chemical interaction between the transition metal cations and hydrogen molecules, and the maximum hydrogen capacity of about 0.77wt% was obtained for RhNaX-55 at 333 K. However, the hydrogen adsorption capacities at 77 K were found to decrease along with increase of the cation exchange levels, which was attributed to the lose of crystallinity during the ion-exchange process.

Jhung et al [31] measured the hydrogen adsorptions of several ion-exchanged Y zeolites by alkali metal cations and proton at 77 K, 87 K and 90 K and ~ 1 atm to study the effect of electrostatic field or basicity. They found that the adsorption capacity per surface area, isosteric heat of adsorption and micropore occupancy all increased with decreasing electronegativity of the metals in the zeolites, and the maximum adsorption capacities achieved at 77 K follows an order of HY (1.05wt%) < NaY (1.22wt%) < CsY (1.23wt%) < KY (1.41wt%).

Combined with diffuse reflectance Fourier transform (DRIFT) spectra, Kazansky et al [32] studied the hydrogen adsorption behaviors on X and Y-type zeolites with different Si/Al ratios and exchanged cations at a low pressure of ~ 60 KPa and 77 K. The number of H_2 adsorbed per Na^+ is the largest for the most basic NaX zeolite with Si/Al = 1.05, corresponding to 0.61wt% of hydrogen uptake or ~ 1 H_2 per Na^+ in NaX. In contrast, the hydrogen adsorption is the lowest in NaY with Si/Al = 2.4, corresponding to 0.17wt% of hydrogen uptake or ~ 0.5 H_2 per Na^+.

In addition to microporous zeolies, some mesoporous zeolites, such as MCM-41, had been investigated for hydrogen storage. Sheppard et al [33] measured the hydrogen adsorptions of MCM-41 at room temperature and 77 K using both the volumetric and gravimetric methods. As shown in Figure 7, the excess adsorption peaked to 1.6wt% at ~ 3.5 MPa upon increasing pressure at 77 K, but the gravimetric adsorption continued to increase and reached a high value of 2.7wt% by ~ 4.5 MPa. On the other hand, Park and Lee [34] studied the hydrogen storage capacity of MCM-41 containing nickel oxides at 298 K and 100 bar. The presence of Ni on mesoporous MCM-41 created hydrogen-favorable sites that leaded to a chemical reaction between Ni-n/MCM-41 (where the number (n) meant that the contained Ni in MCM-41 is n% in weight) and hydrogen molecules, hence enhanced the adsorption capacity by a spillover effect. The adsorption capacities increased linearly with pressure, and the order of the adsorption capacities of these zeolites at 298 K and 95 bar was: Ni-5/MCM-41 (0.68wt%) > Ni-2/MCM-41 (0.57wt)% > Ni-10/MCM-41 (0.45wt%) > MCM-41 (0.35wt%), consistent with the order of the pore volumes and sizes of the zeolites.

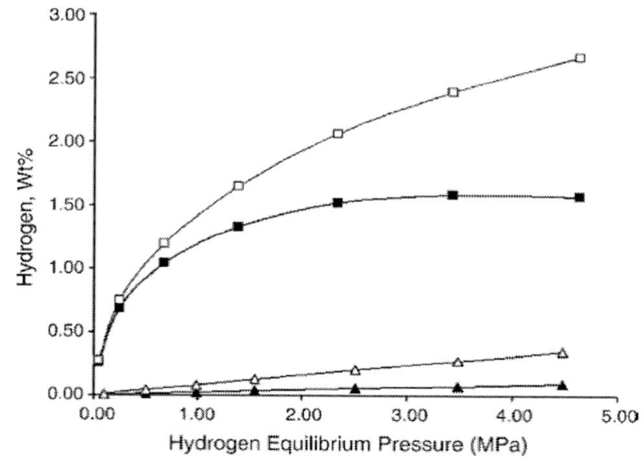

Figure 7. Hydrogen adsorption isotherms on MCM-41 [33], where (■) and (□) are excess and gravimetric adsorption, respectively, at 77 K, whereas (△) and (▲) are excess and gravimetric adsorption, respectively, at 296K.

3. Modification of Zeolites for Hydrogen Adsorption

Aluminosilicate zeolites are of interest for hydrogen adsorbents since they can be easily compacted and modified in their porous structures. The ion-exchange has been the most common modification method for improvement of hydrogen adsorption properties of zeolites. The effects of ion-exchange on hydrogen adsorption properties for some zeolites have been discussed in the section above. Moreover, there are two contradictory aspects for ion-exchanged zeolites on hydrogen adsorption. On one hand, as the size of the exchangeable metal cations increases, the electrostatic field or basicity of the zeolite increases due to the increased electronegativity differences between the cations, aluminum, silicon and oxygen. The electrostatic field may considerably increase the hydrogen adsorption capacity especially at low pressures. On the other hand, with the increasing size and number of the exchangeable cations, the available void volume per gram of zeolite would decrease, resulting in decreased hydrogen uptakes. Taking Ni/MCM-41 modification as an example [34], with Ni content of Ni-5/MCM-41, both total pore volume and pore size increased and hydrogen-storage capacity increased as well. However, with Ni content increasing to Ni-10/MCM-41, the hydrogen uptake was the lowest, because the excessive Ni could have an effect of pore blocking or narrowing on Ni/MCM-41. Therefore, it should take all aspects of effects of exchangeable cations into consideration on the modification.

A method of preparing defective structures as trapping centers for storage of hydrogen is mechanical ball milling, which has been effectively used in carbonic materials. Orimo et al. [35] investigated the hydrogen storage of nanostructured graphite ball milled under hydrogen pressure. The reactive high-power milling for 80 h leaded to a high hydrogen uptake of 7.4wt% determined by oxygen combustion. On the other hand, the effects of ball milling on the storage properties of hydrogen and its isotope on 5A-type zeolites were studied by Li et al. [36]. The zeolites were milled for different times under medium-power in air rather than in hydrogen environment. The measurements of hydrogen adsorption under a pressure of ~ 1

MPa at 77 K indicated that the amounts adsorbed were not enhanced upon milling, though the structure of the zeolite was not shown obvious damage. It appears that the creation of defect structures on the zeolites was not able to compensate the decrease of surface area caused by accumulation of the nanoparticles in or adjacent to the pores. However, as shown in Figure 8, a comparison study on adsorption of H_2 and D_2 revealed that appropriate ball mill modification of the zeolite could improve its separation efficiency on H_2 and D_2.

Another modification method applied to zeolites on hydrogen adsorption is transitional-metal loading. The metal oxides or inorganic salts can be dispersed onto the surfaces of supports during calcination to form spontaneous monolayer dispersion. The dispersion significant increases the entropy and formation of surface bonds, hence improves the activity of the surface for catalytic reactions [37]. Shi et al applied the method to modify NaX zeolite by MnO_2 loading and examined the hydrogen adsorption of the modified NaX at both 77 K and 295 K [38]. Monolayer dispersion of MnO_2 on the surface of NaX was obtained by calcination of NaX and MnO_2 together at both 583 K and 613 K for 24 h. As shown in Figure 9, the dispersion efficiency was affected by the calcinations conditions, and the increase of hydrogen adsorption capacity up to 30% was achieved for modified NaX at 295 K, though adsorption at 77 K did not show obvious improvement. The improved hydrogen adsorption properties on the MnO_2 modified NaX were attributed to the involvement of Kubas interaction, where Mn in MnO_2 acted as a *d*-electron donor for the enhanced binding of hydrogen molecules. On the other hand, Li et al [36] studied the effects of transition metal salts $NiCl_2$ and $CuCl_2$ loading on 5A zeolite on adsorption characteristics of hydrogen isotopes at 77 K and ~ 1 MPa. Although the adsorption amounts on the modified 5A were reduced, but the difference of adsorption amounts between H_2 and D_2 increased significantly for certain type of $NiCl_2$ loaded sample, suggesting that the transitional-metal loading could benefit the separation of H_2 and D_2.

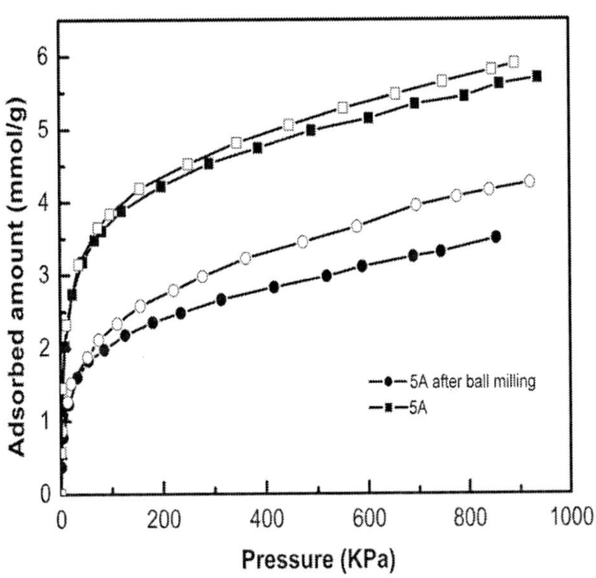

Figure 8. Hydrogen and deuterium adsorption isotherms on 5A and modified (30 min milling) molecular sieves at 77 K. The filled and open symbols are for H2 and D2, respectively.

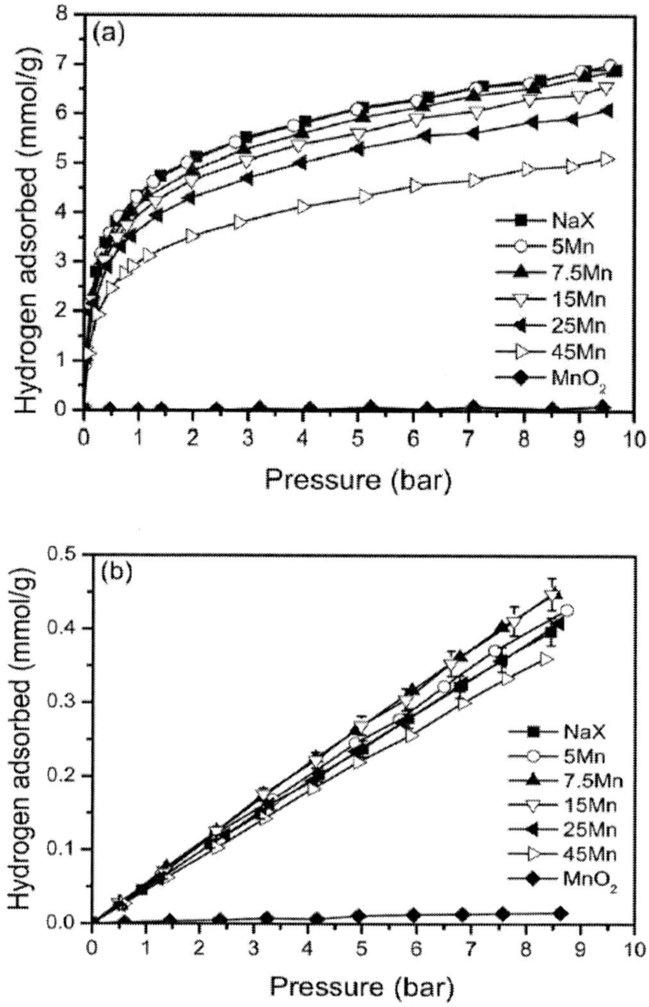

Figure 9. Hydrogen adsorption isotherms of different samples at (a) 77 K and (b) 295 K. The labels of 5Mn etc mean that the MnO2/NaX weight ratio is 0.05 etc. The error bars at interval positions are displayed, whereas those at the same level of the symbols are not displayed.

4. Computational Aspect of Hydrogen Adsorption on Zeolites

Molecular simulations have emerged as a rapid and cost effective method in evaluation of supercritical mechanisms of hydrogen adsorption and the states of hydrogen molecules in the micropores of zeolites. The simulations may be divided into several categories, e.g. empirical modeling, Monte Carlo simulations and ab-initio calculations.

The physical states and behaviors of hydrogen adsorption on microporous adsorbents at supercritical conditions are quite different from that of normal subcritical adsorption, and still not clearly known. For examples, the saturation vapor pressure of hydrogen does not exist above its critical temperature, and the specific amount of pseudo-saturated adsorbed molecules and the characteristic energy of adsorption system vary with the equilibrium temperature. Several equations, such as the modified van der Waals equition, modified Dubinin-Radushkevich (DR) equation, modified Dubinin-Astakhov's (DA) equation, Langmuir-Freundlich equation, have been proposed for the modeling of the supercritical adsorption of hydrogen on porous materials. On the other hand, Zhou et al. [39] has provided a simple isotherm equation derived from the distribution of energetic heterogeneity of adsorbent surface to model the experimental data involving monolayer adsorption at supercritical conditions.

Du and Wu [40] used the isotherm model based on Toth equation, Langmuir and Langmuir-Freudlich (LF) equation to analyze the supercritical isotherms based on the experimental results of hydrogen storage capacity on zeolite ZSM-5 at 77 K, 195 K and room temperature. As shown in Figure 10, the isotherm model based on Toth equation could preferably describe the experimental data over the entire experimental range, and a density of the adsorbed hydrogen of 55.6kg/m^3 at 77K was estimated by the Toth model.

On the other hand, Du and Wu [41] simulated the hydrogen adsorption isotherms on the A and X microporous zeolites under supercritical conditions using lattice density function theory (LDFT) based on the three-dimensional Ono-Kondo model. According to the sizes and shapes of zeolite pores, the local arrangements of adsorption sites within the pores in LDFT were modeled by the clusters of simple (S), face-centered (F), and body-centered (B) cubic lattice structures. As seen in Figure 11, the adsorption isotherms of hydrogen on the zeolites at 77 K predicted from the three LDFT models displayed multilayer adsorption behavior, and the adsorption amounts of the adsorption layers depended on the models for each zeolite. The adsorption amount of the first layer decreased according to the sequence of S, F, and B models, whereas that of the second layer increased according to the same sequence. As the adsorption sites on zeolite pore surfaces were definite in a restrict space, the adsorption amount of the first adsorption layer in S model was less than that in F model, and the derived adsorption isotherms were more consistent with the experimental isotherms, indicating that the LDFT simulation was effective in describing the multilayer or monolayer adsorption of hydrogen on A and X zeolites under supercritical conditions.

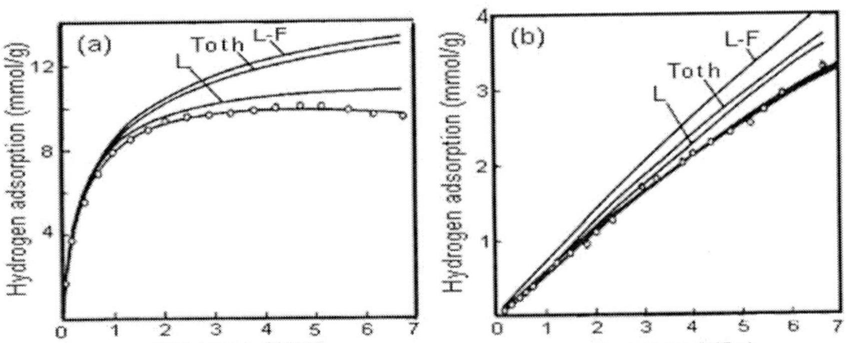

Figure 10. Surface excess and absolute adsorption isotherms at (a) 77K and (b) 195K and the fitting curves of different models. The symbol L is for Langmuir, L-F for Langmuir-Freundlich, and T for Toth equation. The open circles stand for experimental results of excess amount adsorbed and the least-square fitting curve through the opened symbols is based on Toth equation, and other three fitting curves originated from three equations stand for the absolute amount adsorbed.

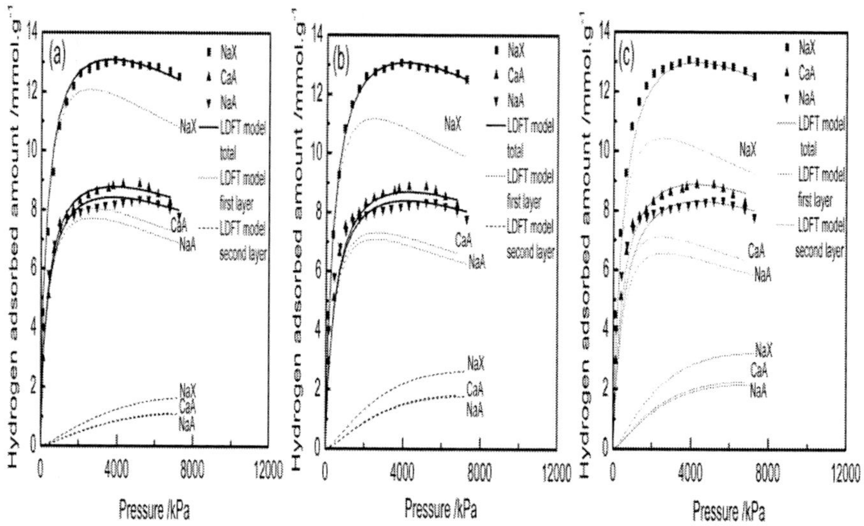

Figure 11. Different LDFT model fits of the adsorption isotherms to the experimental data for H2 adsorption on zeolites at 77K. (a) simple cubic lattice, (b) face-centered cubic lattice, and (c) body-centered cubic lattice.

Grand Canonical Monte Carlo (GCMC) simulation was a theoretical tool used in numerous works on adsorption in zeolites. The realization of GCMC simulation required knowledge of the atom positions in the zeolite crystal and a detailed description of interactions between the adsorbed molecules and adsorbent atoms. Darkrim et al [42] computed the amount of H_2 adsorbed in NaA zeolite at room temperature and high pressures between 10 MPa and 140 MPa by GCMC simulation. By selecting a set of parameters in Lennard-Jones potentials associated with van der Waals interactions between hydrogen and zeolite crystal from literature, good agreement with experimental results on H_2 adsorption at a temperature of 293 K in the pressure range 10 - 70 MPa was obtained in the simulation.

Vitillo and his co-workers [43] conducted evaluation of hydrogen storage capacities in twelve types of zeolites (CHA, DDR, FAU, FER, KFI, LTA, LTL, MEL, MFI, MOR, RHO, SOD) by means of classical atomistic simulations. The calculations were performed with the Discover module of Materials Studio package. The force field used was an enhanced version of the polymer consistent force field, in which parameters for hydrogen-hydrogen non-bond interactions were improved by fitting ab-initio data. The zeolite capacities are intrinsically limited by geometrical constraints to be at best 2.86wt%, well below the accepted target for mobile applications, but still in a range deserving interest. Also, the diffusion of hydrogen in zeolite cages could be modeled using classical molecular dynamics, allowing for flexibility of the host framework.

5. Hydrogen Storage by Encapsulation in Zeolites

Some gases, liquids and unstable compounds can be encapsulated in encapsulant analogues for isolation and protection to retain their properties. The size selectivity of zeolites offers a possible solution whereby if a zeolite is used, of which the largest pore aperture or window has a similar size as the passing molecule, the uptake and release can be controlled by the encapsulation behavior [44]. In different from the physisorption driven by van der Waals interaction between the adsorbents and gas molecules, the encapsulation is driven by thermal diffusion of gas molecules under a pressure difference between inside and outside of the zeolite pores, so that the gas molecules with high dynamic energies can be charged in and released from the pores through the expanded pore windows at high temperatures, and trapped in the pores by the shrunken windows at sharply dropped temperature.

The dihydrogen molecule is rather small, only 2.9 - 3.1 Å in diameter, whereas its kinetic diameter (2.98 Å) is more relevant to the encapsulation. The key in encapsulation of hydrogen is to control the window size of the zeolite pores to be slightly less than the kinetic size of the hydrogen molecules, so that the relatively immobile hydrogen would not penetrate the pore windows at ambient conditions and even in evacuated system, but be effectively transported at a moderately high temperature. Therefore, zeolite can be an efficient encapsulant for hydrogen encapsulation by controlling its pore window size and changing the system temperature and pressure.

Generally, the relationship between encapsulated amount of hydrogen in zeolite pores and the pressure of the system follows the Freundlich equation. In comparison with capsulation, the rapid release or decapsulation of trapped gas is more important for practical applications and difficult to achieve. The decapsulation methods include constant temperature diffusion (CTD) and temperature programmed diffusion (TPD). At present, the latter is more favorable and suitable to the conditions when the decapsulation temperature is much higher than the room temperature. Fraenkel [45] proposed that the best heating rate of TPD should be within the range of 0.08 - 0.35 C/s.

Subsquent to the report on inert gas encapsulation in zeolites [46], Barrer and Vaughan [47] predicted the possible encapsulation of hydrogen in several types of zeolites. The experimental work on encapsulation of hydrogen was conducted by Fraenkel and Shabtai [48] on KA, NaA, CaA and CsA zeolites at the temperatures of 200 - 400 C and pressure of 350 - 13300 psi. In the experiments, the zeolites were freshly activated and outgassed before encapsulation, quenched and evacuated after encapsulation, and the encapsulated hydrogen was released at the encapsulation temperature. The encapsulation efficiency of A-type zeolites increased with increase in ionic radius from Na^+ to K^+ but then decreased for larger (Rb^+, Cs^+) ions. Whereas, the effective critical pore size and available intra-crystalline void volume decreased, hence the encapsulated stability increased with increased ionic radius. Consequently, KA (H_2) lost 46% of its encapsulated hydrogen, and CsA only 7% after 5 days, indicating that in spite of their lower capacities, CsA and RbA are better hydrogen encapsulants. The encapsulation was restricted to β cages in NaA, since Na^+ did not effectively block the 8-ring apertures (4.2 Å) interconnecting the α cages. However, as Na^+ (r = 0.95 Å) was gradually replaced by Cs^+, the apertures became blocked, causing hydrogen encapsulation in α cages. Therefore, both α and β cages in $Cs_{5.4}Na_{6.6}A$ were occupied by hydrogen at 13300 psi and the amount encapsulated could reach as high as 0.6wt%. Weitkamp [49] carried out a similar experiment on these zeolites, but the encapsulated amounts of hydrogen they derived under the same experimental conditions were lower, who attributed the deviations to the preparation, pretreatment and exchanged degree of cations in the frameworks of the zeolites.

On the other hand, Krishnan et al [50] studied the capsulation of hydrogen on Cd^{2+} exchanged Cs-RHO zeolite using the TPD spectra as measured by mass spectrometry. An encapsulation capacity of 71 μmol/g was achieved in 2 h at 1 atm and 50 C. The capacities reached 161 and 620 μmol/g at 100 and 200 C, respectively. They attributed the high encapsulation capacities to the block effects of the Cd^{2+} and Cs^{+} located near the windows of the zeolites, and supported their arguments by the same experiment on H-RHO, where no encapsulation phenomenon was observed. As shown in Figure 12, the decapsulation of hydrogen as a function of temperature measured indicated that the encapsulation was initiated at 50C, and the increasing encapsulation time from 2 to 4 h increased the encapsulation uptake from 71 to 85 μmol/g. The increased time also resulted in the shift of the peak decapsulation temperature from 90 C to a high temperature of 102 C. On the other hand, the increase of the heating rate from 18 to 24 C/min, keeping the time of encapsulation fixed at 2 h, yielded 70 μmol/g of uptake, in agreement with the amount for 18 C/min, thus provided a check for the reproducibility of the encapsulation experiments.

Song and co-worker [51] used the Fermi-Dirac like statistics to study the properties of encapsulated molecules such as H_2, O_2, N_2, Ar and CH_4 in the Cs_3Na_9A zeolite. The density of state, distribution function, average binding energy and average activation energy of the encapsulated molecules were calculated. As the number of encapsulated molecules in zeolite cavities increased, the higher energy states in the cavities were gradually filled and, consequently, the activation energy for decapsulation was lowered. The activation energy for decapsulation depended on both the number of trapped gases and the decapsulated temperatures, and the increase of temperature was required for decapsulation of the trapped gases in the zeolites.

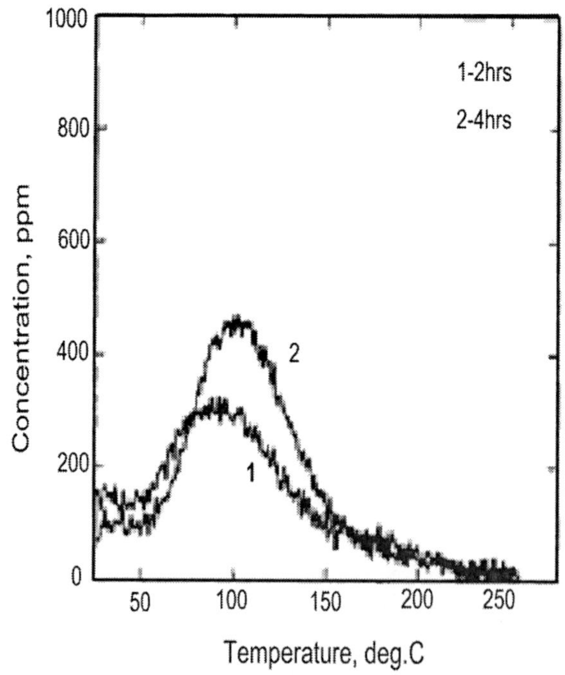

Figure 12. The TPD spectra of Cd2+ exchanged Cs-RHO, where the heating rate = 18°C/min.

Table 2. hydrogen encapsulation in zeolites

Zeolite	Framework type	Loading conditions	H_2 capacity (wt. %)
NaA	LTA	573 K,90 bar,15 min	0.032
$K_{8.6}Na_{3.4}A$	LTA	573 K,90 bar,15 min	0.141,0.124,0.120
$Rb_{5.5}Na_{6.5}A$	LTA	573 K,90 bar,15 min	0.118
$Cs_{5.4}Na_{6.6}A$	LTA	573 K,90 bar,15 min	0.086
$Cs_{5.4}Na_{6.6}A$	LTA	573K,91.7bar,15~90 min	0.581
NaA	LTA	573 K,100 bar,15 min	0.030
SAPO-42	LTA	573 K,100 bar,15 min	0.014
Sodalite	SOD	573 K,100 bar,15 min	0.083
NaX(NaY)	FUA	573 K,90 bar,15 min	0.023
RHO	RHO	573 K,90 bar,15 min	0.004
ZK-5	KFI	573 K,90 bar,15 min	0.004
Sigma-1	DDR	573 K,90 bar,15 min	0.018
Cd/Cs rho	RHO	473 K,1 bar, 2 h	0.125
H-rho	RHO	473 K,1 bar, 2 h	0
NaX	FAU	310 K,1 bar, 2 h	3.305×10^{-4}
Cs/NaX	FAU	310K,1bar, 2h	1.230×10^{-4}
Ni/NaX	FAU	310K,1bar, 2h	0.705×10^{-4}
Eu/NaX	FAU	310K,1bar, 2h	0
NaA	FAU	310K,1bar, 2h	2.278×10^{-4}
Eu/NaA	FAU	310K,1bar, 2h	5.120×10^{-4}

A survey of published results on hydrogen encapsulation in various types of zeolites is listed in Table 2. Generally, the stored amounts in thermally driven encapsulation process are still very low, as the high temperatures required to activate thermal diffusion disbenefit the filling of hydrogen into the pores. Therefore, researches on hydrogen storage in zeolites have emphasized more on the adsorption. However, hydrogen or its isotopes could be purified and separated by the 'sieve effect' of encapsulation at different temperatures [52]. On the other hand, hydrogen encapsulation in ion-exchanged zeolites may be used to identify the locations of the metal ions in different sites around the pores of the zeolites [50].

CONCLUSION

The zeolite materials are important candidate materials for hydrogen storage and isotope separation due to their rich variety of porous structures, high stability and low cost. Hydrogen molecules can be adsorbed or encapsulated in the tailored zeolites. The adsorption on the micropores is essentially physical, and can be enhanced due to the higher interaction potential between hydrogen and the walls of narrower pores, hence the vast majority of zeolites for hydrogen storage are microporous zeolites, and require cryogenic temperatures to achieve reasonable loadings. Owing to the low critical temperature of hydrogen, adsorption studies on zeolites are mostly limited to supercritical temperatures above that of liquid nitrogen. The main methods to measure hydrogen storage capacities of zeolites are gravimetric and volumetric, where the adsorbed amount is calculated by knowing pressure, temperature and equation of state of hydrogen. Up to now, the hydrogen storage properties of all common microporous zeolites, in particular, the X, A and ZSM-5 types, and some mesoporous zeolites, e.g. MCM-41, have been investigated at both room and cryogenic temperatures. The studies suggest that the hydrogen capacities depend on the micropore volums or surface area, the Si/Al ratio, and the size, charge, basicity and distribution of the cations of the zeolites. The highest gravimetric capacities of various zeolites determined at room and nitrogen temperature are in the levels of close to 1 and 3wt%, respectively. However, the hydrogen capacity of ~ 4wt% can be achieved at near critical temperature and reaches a density close to that of liquid hydrogen. Ion-exchange has been a common and useful modification method to improve the hydrogen storage performance, though it could damage the crystallinity of the zeolites. Besides, the ball milling and transition-metal loading have also been applied to the modification of zeolites, and shown beneficial effects on isotope separation. The molecular simulations, such as empirical adsorption modeling, lattice density function theory and Monte Carlo method, have been successfully used in evaluation of supercritical mechanisms of hydrogen adsorption and the states of hydrogen molecules in some microporous zeolites. On the other hand, hydrogen encapsulation in zeolites has been achieved at high temperatures and pressures, though the up-takes are relatively low with a maximum of ~ 0.6wt%. The encapsulation efficiency and stability are affected by the ionic radius, effective critical pore size and available intra-crystalline void volume of the zeolites, and the A-type zeolites are most commonly used for encapsulation.

However, despite of many progresses in hydrogen storage researches on zeolite materials, there still exit many questions to be answered and problems to be solved. For examples, whether the big cages or small cages contribute to the hydrogen storage in some zeolites are still not clearly known; the reports on storage capacities are often not coincident; the mechanisms of adsorption on the micorpores are poorly understood; and no zeolites possess satisfactorily combined qualities required for the storage media: high adsorption capacity, high bulk density and low thermal effects. Further researches therefore are essential to meet these challenges.

REFERENCES

[1] M.G. Nijkamp, J.E.M.J. Raaymakers, A.J. van Dillen, K.P. de Jong, Hydrogen storage using physisorption – mateials demands, *Appl. Phys*. A 2001 (72): 619-623.

[2] L. Zhou, Progress and problems in hydrogen storage methods, *Renew. Sustain. Energy Rev*. 2005 (9): 395-408.

[3] S.K. Bhatia, A.L. Myers, Optimum conditions for adsorptive storage, *Langmuir* 2006 (22): 1688-1700.

[4] M.J. Beneyto, F.S. García, D.L. Castelló, D.C. Amorós, A.L. Solano, Hydrogen storage on chemically activated carbons and carbon nanomaterials at high pressues, *Carbon* 2007 (45): 293-303.

[5] K.A. Tuan, M.A. David, Exploiting the Kubas interaction in the design of hydrogen storage materials, *Adv. Mater.* 2009 (21): 1787-1800.

[6] N.L. Rosi, J. Eckert, M. Eddaoudi, D.T. Vodak, J. Kim, M. O'Keeffe, O.M. Yaghi Hydrogen storage in microporous metal-organic frameworks, *Science* 2003, 300 (16): 1127-1129.

[7] J. Li, E.D. Wu, J.W. Song, F.S. Xiao, C.J. Geng, Cryadsorption of hydrogen on divalent cation-exchanged X-zeolites, *Inter. J. Hydrogen Energ.* 2009 (34): 5458- 5465.

[8] S.M. Auerbach (University of Massachusetts, Amherst, U.S.A.), K.A. Carrado (Argonne National Laboratory, Argonne, Illinois), P.K. Dutta, Handbook of Zeolite Science and Technology, 1 ed. *Marcel Dekker Inc.* 2003.

[9] X.Z. Chu, Y.P. Zhou, Y.Z. Zhang, W. Su, Y. Sun, L. Zhou, Adsorption of hydrogen isotopes on micro- and mesoprous adsorbents with orderly structure, *J. Phys. Chem.* B 2006 (110): 22596-22600.

[10] M.M. Dubinin, The potential theory of adsorption of gases and vapors for adsorbents with energetically nonuniform surfaces, *Chem. Rev.* 1960 (60): 235-241.

[11] IUPAC Manual of Symbols and Terminology, Appendix 2, Pt. 1, Colloid and Surface Chemistry. *Pure Appl. Chem.* 1972 (31): 577-638.

[12] D.D. Do, H.D. Do, Adsorption of supercritical fluids in non-porous and porous carbons: analysis of adsorbed phase volume and density, *Carbon* 2003 (41): 1777-1791.

[13] A.M. Jwayyed, R. Humayun, D.L. Tomasko, High pressure flow gravimetric apparatus for supercritical fluid extraction studies, *Rev. Sci. Instru.* 1997 (68): 4542-4548.

[14] J. Li, E.D. Wu, Adsorption of hydrogen on porous materials of activated carbon and zeolite NaX crossover critical temperature, *J. Superc. Fluids* 2009 (49): 196-202.

[15] L. Zhou, Y.P. Zhou, Y. Sun, Studies on the mechanism and capacity of hydrogen uptake by physisorption-based materials, Intern. *J. Hydrogen Energ.* 2006 (31): 259-264.

[16] N. Floquet, J.P. Coulomb, G. André, Hydrogen sorption in MCM-41 by neutron diffraction study. Characterization of the porous structure of MCM-41 and the growth mode of the hydrogen confined phases, *Microp. Mesop. Mater.* 2004 (72): 143-152.

[17] J.P. Coulomb, N. Floquet, Y. Grillet, P.L. Llewellyn, R. Kahn, G. André, Dynamic and structural properties of confined phases (hydrogen, methane and water) in MCM-41 samples (19Å, 25Å, and 40Å), *Stud. Surf. Sci. Catal.* 2000 (128): 235-242.

[18] D. Basmadjian, Adsorption equilibria for hydrogen, deuterium and their mixtures. Part I, Cand. *J. Chem.* 1960 (38): 141-148.

[19] K. Kotoh, T. Nishikawa, Y. Kashio. Multi-component adsorption characteristics of hydrogen isotopes on synthetic zeolite 5A type at 77.4K. *J. Nucl. Sci. Tech.* 2002 (39): 435-441.

[20] M.A. Levin, M.B. Gorbunov, T.A. Moskovskaya, V.V. Serpinskii, T.S. Yakubov, Adsorption of hydrogen isotopes at high pressures. *Russ. Chem. Bull.* 1986 (35): 2406-2408.

[21] V.E. Skazyvaev, S.S. Khovshchev, S.P. Zhdanov, Low-temperature adsorption of hydrogen on synthetic zeolites, *Russi. Chem. Bull.* 1975(24): 1143-1147.

[22] H.W. Langmi. D. Book, A. Walton, S.R. Johnson, M.M. Al-Mamouri, J.D. Speight, P.P. Edwards, I.R. Harris, P.A. Anderson, Hydrogen storage in ion-exchanged zeolites. *J. Alloy Comp.* 2005 (404-406): 637-642.

[23] X.M. Du, E.D. Wu, Physisorption of hydrogen in A, X and ZSM-5 type of zeolites at moderately high pressure, *Chin. J. Chem. Phys.* 2006 (19): 457-463.

[24] M.J. Beneyto, F.S. García, D.L. Castelló, D.C. Amorós, A.L. Solano, Hydrogen storage on chemically activated carbons and carbon nanomaterials at high pressures, *Carbon* 2007 (45): 293-303.

[25] S.H. Jhung, J.W. Yoon, H-K Kim, J-S Chang, Low temperature adsorption of hydrogen on nanoporous materials, *Bull. Korean Chem. Soc.* 2005 (26): 1075-1078.

[26] M.A. Casa- Lillo, F.L. Darkrim, D.C. Amoros, A.L. Solano, Hydrogen storage in activated carbons and activated carbon fibers, *J. Phys. Chem. B* 2002 (106): 10930-10934.

[27] Q. Wang, K. Johnson, Molecular simulation of hydrogen adsorption in single-walled carbon nanotubes and idealized carbon slit pores, *J. Chem. Phys.* 1999 (110): 577-586.

[28] Y.W. Li, R.T. Yang, Hydrogen storage in low silica type X zeolites, *J. Phys. Chem. B* 2006 (110): 17175-17181.

[29] K.P. Prasanth, Renjith S. Pillai, Sunil A. Peter, H.C. Bajaj, R.V. Jasra, Hydrogen uptake in palladium and ruthenium exchanged zeolite X, *J. Alloy Comp.* 2008 (466): 439-446.

[30] K.P. Prasanth, Renjith S. Pillai, H.C. Bajaj, R.V. Jasra, Adsorption of hydrogen in nickel and rhodium exchanged zeolites X, *Inter. J. Hydrogen Energ.* 2008 (33): 735-745.

[31] S.H. Jhung, J.S. Lee, J.W. Yoon, D.P. Kim, J-S. Chang, Low temperature adsorption of hydrogen on ion-exchanged Y zeolites, Inter. *J. Hydrogen Energ.* 2007 (32): 4233-4237.

[32] V. B. Kazansky, V.Y. Borovkov, A. Serich, H.G. Karge, Low temperature hydrogen adsorption on sodium forms of faujasites: barometric measurements and drift spectra, *Microp. Mesop. Mater.* 1998 (22): 251-259.

[33] D.A. Sheppard, C.F. Maitland, C.E. Buckley, Preliminary results of hydrogen adsorption and SAXS modeling of mesoporous silica: MCM-41. *J. Alloy Comp.* 2005 (404-406): 405-408.

[34] S.J. Park, S.Y. Lee, A study on hydrogen-storage behaviors of nickel-loaded mesoporous MCM-41, *J. Collo. Interf. Sci.* 2010 (346): 194-198.

[35] S. Orimo, G. Majer, T. Fukunaga, A. Zűttel, L. Schlapbach, H. Fujii, Hydrogen in the mechanically prepared nanostructured graphite, *Appl. Phys. Lett.* 1999 (75): 3093-3095.

[36] J. Li, J.S. Shi, E.D. Wu, S. Liu, Study of hydrogen and deuterium adsorption on modified molecular sieves, *Atom. Energ. Sci. Tech.*, 2012, In press (in Chinese).

[37] S.M. Rivera-Jiménez, A.J. Hernández-Maldonado, Nickel(II) grafted MCM-41: a novel sorbent for the removal of naproxen from water, *Microp. Mesop. Mater.* 2008 (116): 246-252.

[38] J.S. Shi, J. Li, E.D. Wu, Adsorption of hydrogen and deuterium in MnO_2 modified NaX zeolites, *Microp. Mesop. Mater.* 2012, In press.

[39] L. Zhou, J.S. Zhang, Y.P. Zhou, A simple isotherm equation for modeling the adsorption equilibria on porous solids over wide temperature ranges, *Langmuir* 2001 (17): 5503-5507.

[40] X.M. Du, E.D. Wu, Hydrogen adsorption property of zeolite molecular sieve ZSM-5 under high-pressure, *Chin. J. Mater. Res.* 2006 (20): 591-596. (in Chinese)

[41] X.M. Du, E.D. Wu, Modeling the supercritical adsorption of hydrogen on A- and X-type zeolites with a lattice density function, *Acta Phys. – Chim. Sin.* 2009 (25): 1823-1828.

[42] F. Darkrim, A. Aoufi, P. Malbruno, Hydrogen adsorption in the NaA zeolite: a comparison between numerical simulations and experiments, *J. Chem. Phys.* 2000 (112): 5991-5999.

[43] J.G. Vitillo, G. Ricchiardi, G. Spoto, A. Zecchina, Theoretical maximal storage of hydrogen in zeolitic frameworks, *Phys. Chem. Chem. Phys.,* 2005 (7): 3948-3954.

[44] J. Li, E.D. Wu, C.J. Geng, X.M. Du, Hydrogen storage by encapsulation on porous materials, *Prog. Chem.,* 2010 (22): 2238-2247 (in Chinese).

[45] D. Fraenkel, A. Levy, Analysis of temperature-programmed diffusion chromatograms obtained with zeolite-gas systems. *J. Chem. Soc. Faraday Trans.* 1988 (84): 1817-1834.

[46] W.J. Sesny, L.H. Shaffer, *US Patent* 3 316 69, 1967.

[47] R.M. Barrer, D.E.W. Vaughan, Trapping of inert gases in sodalite and cancrinite crystals. *J. Phys. Chem. Solids* 1971 (32): 731-743.

[48] D. Fraenkel, J. Shabtai, Encapsulation of hydrogen in molecular sieve zeolites. *J. Am. Chem. Soc.* 1977 (14): 7074-7076.

[49] J. Weitkamp, S. Ernst, F. Cubero, W. Frank, S. Wolfgang, Nitrido-sodalite Zn6 [P12N24] as a material for reversible hydrogen encapsulation. *Adv. Mater.* 1997 (9): 247-248.

[50] V.V. Krishnan, S.L. Suib, D.R. Corbin, S. Schwarz, G.E. Jones, Encapsulation studies of hydrogen on cadmium exchanged zeolite rho at atmospheric pressure, *Catal. Today* 1996 (31): 199-205.

[51] M.K. Song, J.S. Kim, K.T. No, A new theoretical approach to the encapsulation of small molecules in zeolites, *Appl. Surf. Sci.*, 2007 (253):5696-5700.

[52] G. Eklund, O. Von Krusenstierna, Storage and transportation of merchant hydrogen, Inter. *J. Hydrogen Energ.* 1983 (8): 463-470.

In: Zeolites: Synthesis, Chemistry and Applications
Editors: Moisey K. Andreyev and Olya L. Zubkov
ISBN: 978-1-61942-861-4

Chapter 6

SYNTHESIS OF NANOCRYSTALLINE ZEOLITES AND THEIR APPLICATIONS

Teruoki Tago**, Yuta Nakasaka and Takao Masuda
Division of Chemical Process Engineering, Faculty of Engineering, Hokkaido University, Hokkaido, Japan

1. INTRODUCTION

1.1. Nano-Sized Zeolite

Zeolites, crystalline aluminosilicate materials, possess 3-dimensionaly connected framework structures constructed from corner-sharing TO_4 tetrahedra, where T is any tetrahedrally-coordinated cation such as Si and Al. These framework structures are composed of n-rings, where n is the number of T-atoms in the ring (e.g. 4-, 5-, and 6-rings), and large pore openings of 8-, 10-, and 12-rings are framed by these small rings. The sizes of the intracrystalline pores and nanospaces, depending on the type of zeolite providing the framework, are close to the molecular diameters of lighter hydrocarbons. Moreover, strong acid sites exist on the nanopore surfaces within the zeolite crystal and on the external surface of the crystal, enabling the zeolites to be used in reaction processes (shape-selective catalysts such as fluid catalytic cracking of heavy oil, isomerization of xylene and synthesis of ethyl-benzene) and separation process (adsorptive- and membrane-separation).

However, compared to the sizes of micropores exhibiting a molecular-sieving effect, the crystal sizes of zeolites are very large, approximately 1–3 μm. When the zeolite is used as a shape-selective catalyst, the diffusion rates of reactant molecules within the zeolite crystals are lower than the intrinsic reaction rates. This resistance to mass transfer limits the reaction rates and low product selectivities. Moreover, since effective active sites (acid sites) for catalytic reactions are distributed on the internal surfaces of the main channels and the external surfaces of the crystal, the pore mouths are easily plugged due to coke deposition, leading to short lifetimes for the catalysts. Faster mass transfer is required to avoid these

* Corresponding author: T. Tago, tago@eng.hokudai.ac.jp.

serious problems, and two primary strategies have been proposed; one is the formation of meso-pores within zeolite crystals [1-3], and the other is the preparation of nano-crystalline zeolites [4-16]. In nano-crystalline zeolites, the diffusion length for reactant hydrocarbons, assignable to the crystal size, decreases, and the external surface area of the crystal increases as the crystal size decreases. The increase in external surface area and the decrease in diffusion resistance are effective for improving catalytic activity in gas-solid and liquid-solid heterogeneous catalytic reactions. Because of these favorable properties of nano-sized zeolites for catalytic reactions, the preparation methods for several types of zeolite nanocrystals have been reported and reviewed [17, 18].

In this chapter, a method is introduced for preparing nano-sized zeolites in a solution consisting of a surfactant, an organic solvent, and water (denoted as the emulsion method [19-23] hereafter). As mentioned above, nano-sized zeolite is expected to be a promising material for increasing the external surface area as well as decreasing the diffusion resistance of the organic reactant within the micropores, thereby improving the catalytic activity and lifetime. Moreover, mono-dispersed zeolite nanocrystals are expected to be promising candidates as seed crystals for forming structured materials such as hierarchical materials [24-26] and uniform zeolite membranes [27-29]. In the application of nano-sized zeolites to the heterogeneous catalysts and structured materials, acid sites and pore size modification of the zeolite using a organic silane compound is effective in improving the zeolitic properties. We also introduce a method based on the catalytic cracking of silane method [30-32] (denoted as the CCS method hereafter). In this method, regioselective deactivation of acid sites can be achieved using silane compounds with a variety of organic substituents.

2. Synthesis of Nano-Sized Zeolite Crystals

2.1. Synthesis of Nano-Crystalline Zeolites in Water/Surfactant/Organic Solvent

Usually, zeolites are prepared in an aqueous alkaline solution containing Si and Al sources and alkaline metal ions (sodium or potassium). In the synthesis of some types of zeolite, an organic structure directing agent (OSDA), such as an ammonium alkyl cation, is also necessary to form the zeolite framework (tetra-n-propyl ammonium hydroxide (TPA-OH) for MFI zeolite, and tetra-n-ethyl ammonium hydroxide (TEA-OH) for BEA and MOR zeolites). To prepare the zeolite, solutions containing these inorganics and the OSDA are simply poured into a Teflon-sealed stainless bottle and heated to a desirable temperature.

One method to prepare nano-sized zeolites is to quench the hydrothermal treatment upon nucleation. However, because unreacted Si and Al species still remain in the solution, an amorphous phase tends to form on the surface of the zeolite crystals. Another method involves increasing the nucleation rate by increasing the concentrations of the Si and Al sources. By increasing the nucleation rate, the crystal size of the zeolite decreases. However, zeolites obtained by this method have a broad size distribution due to simultaneous nucleation and crystal growth during the hydrothermal treatment. Accordingly, in order to obtain nano-sized zeolite, it is important to separate the nucleation and growth stages. Recently, there has been growing interest in the synthesis of nano-crystalline zeolites obtained with the addition of surfactants, including those prepared in a water/surfactant/organic mixture [19-23, 33-35] (emulsion method).

In the emulsion method, two solutions are prepared; an aqueous solution containing the Si and Al sources previously diluted with an OSDA aqueous solution, and a solution of the surfactant in an organic solvent. By varying the concentrations of the Si and Al sources in the aqueous solution along with the molar ratio of Si to OSDA and by changing the types of OSDA used, zeolite crystals with different properties can be prepared. The aqueous solution is added to the surfactant organic solution, and the mixture is magnetically stirred to obtain a homogeneous solution. The resulting mixture of water/surfactant/organic solvent is then placed in a Teflon-sealed stainless steel bottle, heated to 373 to 423 K, and held at the desired temperature for 12 to 120 h with stirring to yield the nano-sized zeolites (Figure 1).

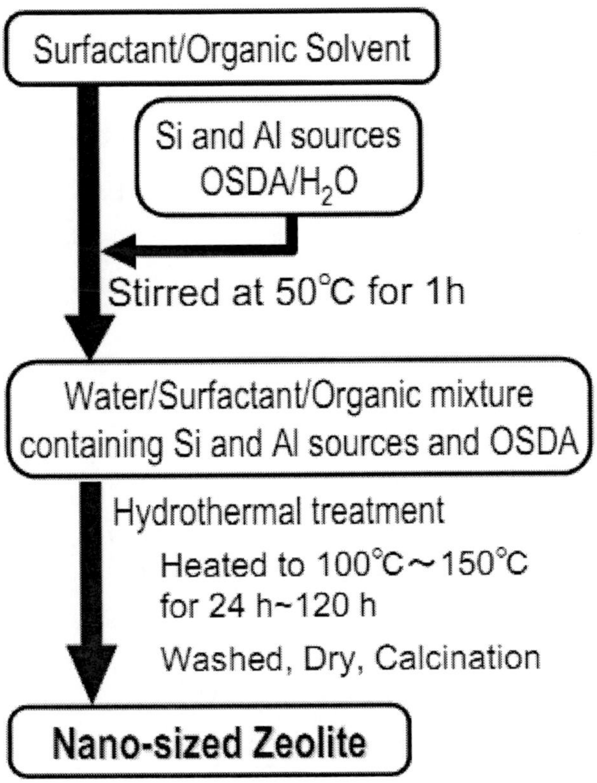

Figure 1. Preparation procedure for nano-sized zeolite by emulsion method.

Because surfactant molecules have hydrophobic and hydrophilic organic groups in the molecules, they adsorb to the solid surface, thus reducing the interface energy difference between the solid surface and the solvent and consequently enhancing nucleation of the metal and/or metal oxide nano-particles. Here, two additional parameters, the concentration of surfactant in the organic solvent [22, 23] and the molar ratio of water to surfactant [21], are important in the emulsion method, because these parameters affect the morphology and crystal sizes.

2.2. Effect of Ionicity of Surfactant on Preparation of Nano-Sized Zeolites

Next we examined the effect of the surfactant on morphologies for the silicalite-1 (MFI zeolite) prepared using various surfactants at a hydrothermal temperature of 140 °C and a surfactant concentration in the organic solvent of 0.50 mol/L. In general, there are four types of surfactants; anionic, cationic, non-ionic, and bipolar surfactants. In the present work, we use as nonionic surfactants, polyoxyethylene (15)oleylether (O-15), polyoxyethylene (15) nonylphenylether (N-15), and poly(oxyethylene) (15) cethylether (C-15), and as ionic surfactants, sodium bis (2-ethylhexyl) sulfosuccinate (AOT) and cetyltrimethyl ammonium bromide (CTAB) (Figure 2). Cyclohexane or 1-hexanol is used as an organic solvent.

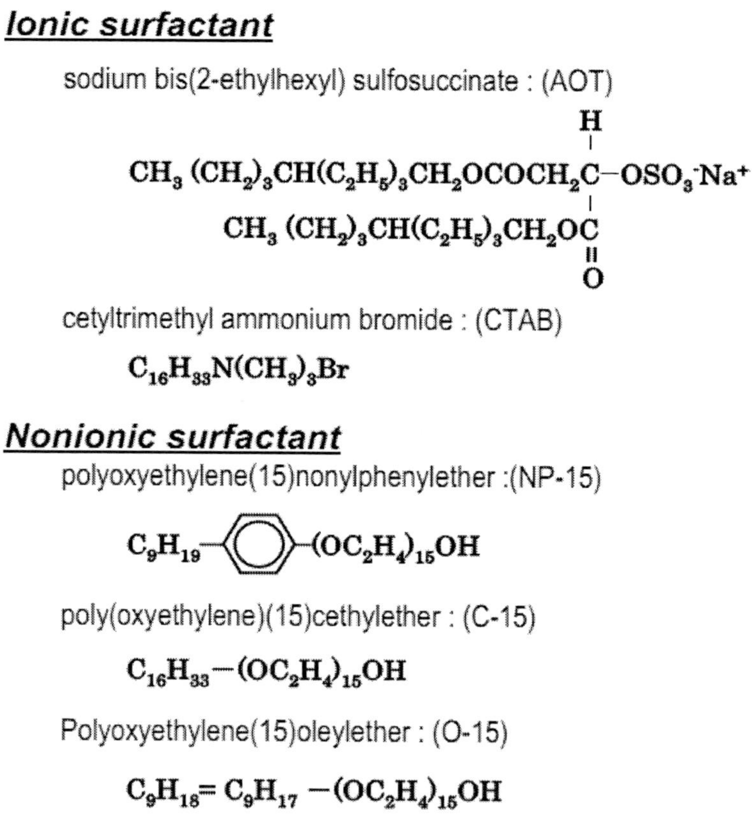

Figure 2. Molecular structures of surfactants used in emulsion method.

Figure 3 and 4 show X-ray diffraction (XRD) patterns and FE-SEM images of the obtained samples, respectively. In the XRD pattern for the sample prepared in the AOT/cyclohexane solution, peaks corresponding to sodium sulfate rather than silicalite-1 are seen. Moreover, the sample showed an irregular morphology. In this case, it is apparent that the TPA-OH molecules do not act as an OSDA in the synthetic solution. This result is ascribed to the fact that the AOT surfactant and the TPA-OH used as the OSDA have opposite ionic charges.

When using CTAB/1-hexanol, silicalite-1 crystals about 1.0 μm in size were found to be embedded in the amorphous SiO_2 as seen by SEM observation. The coexistence of silicalite-1 crystals and amorphous SiO_2 was also revealed by X-ray diffraction analysis. Since the pH of the synthetic solution is alkaline in this case, the surface of the SiO_2 has a negative charge. Accordingly, it is considered that CTA^+ and TPA^+ independently adsorb on the surface of SiO_2 because of their cationic charge. Correspondingly, silicalite-1 crystals and amorphous SiO_2 are formed separately from the independent adsorption of TPA^+ and CTA^+, respectively.

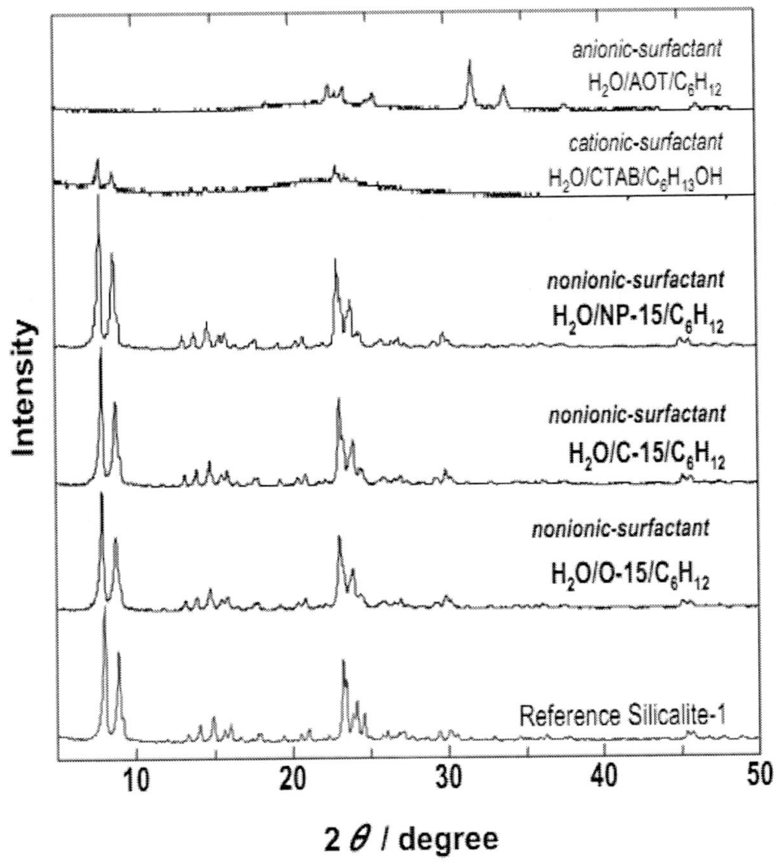

Figure 3. X-ray diffraction patterns of samples prepared in water/surfactant/organic solvent. Effects of ionicity of the surfactant on crystallinity of the silicalite-1 (MFI zeolite).

In the case of the nonionic surfactants, C-15, NP-15, and O-15 (the nonionic surfactant/cyclohexane system), the XRD patterns of the samples showed peaks corresponding to an MFI zeolite (the reference silicalite-1), and mono-dispersed silicalite-1 nanocrystals were obtained (as seen by SEM observation). In contrast, the silicalite-1 crystals prepared in water (without surfactant) exhibit a heterogeneous structure and exist as smaller crystals (diameter of approximately 30 nm) formed upon larger crystals (approximately 120 nm), indicating that nucleation, crystallization and crystal growth occurred simultaneously. However, in the emulsion method, aggregation of the silicalite-1 nuclei is inhibited by the surfactants adsorbed on the surface during hydrothermal treatment, and mono-dispersed nanocrystals can be obtained. These results indicate that the ionicity of the hydrophilic groups in the surfactant molecules plays an important role in the formation and crystallization processes of the nano-sized zeolites [19, 20].

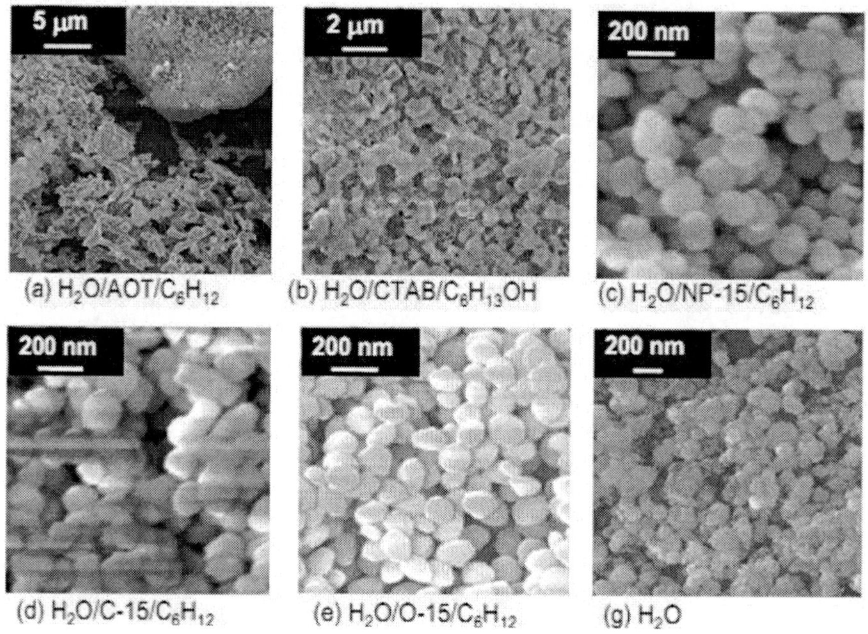

Figure 4. FE-SEM micrographs of samples prepared in water/surfactant/organic solvent. Effects of ionicity of the surfactant on morphology of the silicalite-1 (MFI zeolite). Si concentrations in water are (a)–(e) 0.63 mol/l and (f) 2.7 mol/l.

2.3. Crystal Size Control

In a catalytic reaction using a zeolite, the reaction occurs on acid sites located on the outer surface of the crystal as well as the inside pore surface. The crystal size of zeolite affects the outer surface area and the diffusion length, assignable to the crystal size, for reactant hydrocarbons within the crystal. Moreover, in preparation of zeolite membrane, zeolite seed crystals are typically immobilized on porous supports before secondary growth of the zeolite. Accordingly, it is very important to develop preparation methods for zeolite nanocrystals with different crystal sizes.

In the emulsion method, it is considered that adsorption of non-ionic surfactants on the surfaces of the zeolite precursors suppresses re-dissolution of the precursor in solution, inducing and enhancing nucleation of the zeolite. Moreover, the surfactant concentration in the solution can be easily changed. Accordingly, the effects of varying surfactant concentration on the morphology of MFI and MOR zeolites were examined for the O-15 surfactant. Figure 5 shows FE-SEM images of the obtained samples. The surfactant concentrations were varied in the range from 0.5 to 0.75 mol/L, and preparation of zeolites from water solutions without any surfactant/organic solvent (the conventional method) was also carried out for comparison. Interestingly, the crystal size and morphology depended on the surfactant concentration, despite the same concentrations of the Si, Al and OSDA being used throughout. Nano-sized zeolites with an average size of below 100 nm were obtained at a surfactant concentration of 0.5 mol/L. As the concentration increased, growth of MFI and MOR zeolites was clearly observed, with the crystal size reaching ~1.0 μm at 0.75 mol/L [21-23].

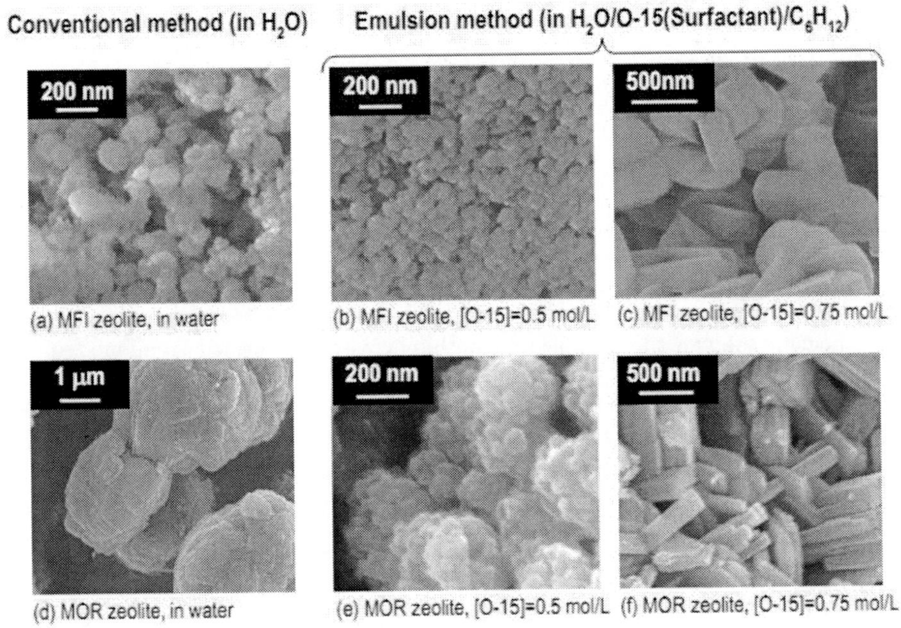

Figure 5. FE-SEM micrographs of MFI (silicalite-1) and MOR (Si/Al=12.5) zeolites with different crystal sizes.

2.4. Mechanism

In order to consider the mechanism of zeolite nanocrystals formation in water/surfactant/organic mixture, it is necessary to investigate the relationship between the ionicity of the surfactant and the OSDA molecule, and the ionic charges of the surface of SiO_2 and/or zeolite precursor in detail. In the method, the cyclohexane used as the organic solvent contributes to stabilization of the hydrophobic group of the surfactant. The effects of the ionicity of the hydrophilic group of the surfactant on the morphology and crystallinity of the obtained silicalite-1 zeolite nanocrystals are discussed above (Figures 3 and 4). A schematic figure showing the possible relationship between the surfactant and the zeolite surface is shown in Figure 6. The hydrophilic group of the surfactant O-15 is composed of poly-oxyethylene chains, making it a non-ionic surfactant. The surface of the zeolite precursor and crystal is composed of a hydrophobic surface (-Si-O-Si-) and hydrophilic silanol groups (-O-Si-OH). The hydrophilic groups of the surfactant can be adsorbed on the hydrophilic silanol groups, leading to stabilization of the silanol groups. On the hydrophobic surface, hydrophobic hydration occurs in the conventional preparation method without a surfactant [36-39]. It is considered that the hydrophilic groups of the surfactant are adsorbed onto the water molecules hydrophobically-hydrated on the zeolite surface. Moreover, because the hydrophilic property of the poly-oxyethylene chains in the surfactant is much smaller than that in water, it is considered that the hydrophilic poly-oxyethylene chains can be directly adsorbed onto the zeolite surface due to their higher affinity to the zeolite surface than water molecules. These adsorbed surfactants can contribute to the stabilization of the surface of zeolite precursor and crystals, which will influence the nucleation rate of zeolite, so that the MFI and MOR zeolite nanocrystals can be obtained.

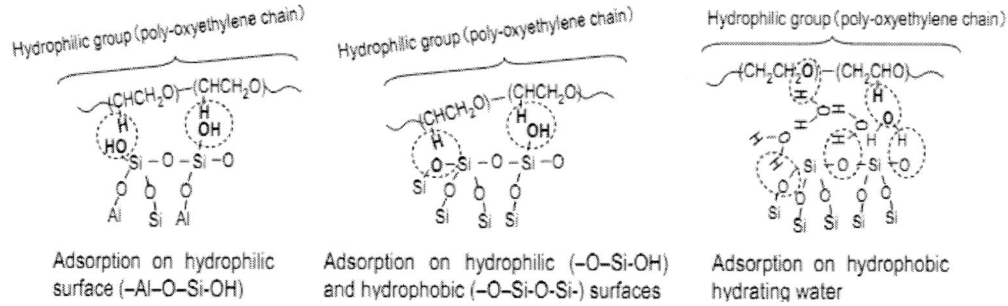

Figure 6. A schematic showing the possible relationship between the surfactant and the zeolite surface.

Three-dimensional cylindrical networks of the micellar structure composed of water and surfactant molecules likely exist in water/O-15/cyclohexane solvent, wherein silica-based cylindrical [40] and spherical [41-43] materials including metal and/or metal-oxide nano-particles are obtained at 323 K. Because these network structure lead to an increase in the viscosity of the solvent, it is considered that the viscosity of the solvent affected the nucleation rate. Moreover, as the surfactant concentration is increased, the excessive stabilization of the precursors and the high viscosity of the solution lead to a decrease in the nucleation rate of zeolite. Since the growth rate of the crystal is much higher than the nucleation rate, the growth of zeolite crystal is dominant at high surfactant concentrations, so that the crystal size and morphology can be controlled by the surfactant concentration.

3. Deactivation of Acid Site and Modification of Pore Size

3.1. Catalytic Cracking of Silane Method

Because zeolites contain micropores of a specific diameter almost equal to the molecular sizes of lighter hydrocarbons, they exhibit a remarkable molecular sieving effect for these hydrocarbons in reaction and adsorption processes, in which the spatial limitations within the zeolitic pores are expected to provide effective reaction/separation fields. Unfortunately, in some cases zeolites exhibit a poor spatial limitation effect. For example, reactions over acid sites located on the external surface of the zeolite crystals can give rise to coke deposition, leading to a short catalyst lifetime. To overcome these problems, selective deactivation of acid sites located near the external surface of zeolite crystal is required [31, 32]. Another problem occurs with systems in which all of the component molecules have sizes smaller than the pore diameter. In this case, methods for modifying the pore wall and/or decreasing the pore size are needed for effective separation [30], and therefore a strong demand exists for methods that allow regioselective deactivation and/or pore-size modification to improve the molecular sieving effect of zeolites.

Niwa *et al.* [44–46] have developed a method utilizing a chemical vapor deposition (CVD) technique with tetraethoxysilane and/or tetramethoxysilane precursors to form a silica (SiO_2) thin layer on the external surface of zeolite crystals. We have also proposed a method for SiO_2 formation based on the catalytic cracking of silane (the CCS method) using organic silane compounds, in which SiO_2 formation selectively occurs on the acid sites of the zeolite [30, 32].

In the CCS method, modification of the zeolite acidity and pore size can be achieved by SiO_2 unit formation, which occurs via a silane compound chemisorbed on the acid site of the zeolite. Accordingly, the procedure for the CCS method is outlined as follows [30, 32]. First, after air calcination, the zeolite is exposed to a vapor of the silane compound at 373 to 393 K in a N_2 stream, and the feed of the silane compound is then stopped to remove the physically adsorbed silane compounds on the zeolite surface. Here, the silane compounds are adsorbed on the acid sites of the zeolite by chemisorption. Next, the sample is heated to 823 K in a N_2 stream to decompose the silane compounds adsorbed on acid sites, resulting in the formation of silicon-coke composites on the acid sites. Finally, the gas feed flowing around the zeolite is switched from N_2 to air, and the temperature of the reactor is held at 823 K in an air stream for 60 min, allowing the silicon–coke compounds formed on the acid sites to be oxidized to SiO_2 units. Thus, the formation of SiO_2 units occurs on the acid sites at which the silane compounds initially chemisorb. By repeating the sequence described above, the formation of SiO_2 units occurs again on the acid sites, leading to a decrease in the zeolite acidity and pore size.

3.2. Regioselective Deactivation of Acid Sites by the CCS Method

In the CCS method, several types of silane compounds can be employed to deactivate the acid sites of zeolite, including those with methyl, ethoxy, and phenyl substituents (diethoxymethylsilane (DEM-silane), phenylsilane (P-silane), diphenylsilane (DP-silane), diphenylmethylsilane (DPM-silane), and triphenylsilane (TP-silane)). Because the molecular size of the silane compounds depends on the organic groups bonded to the Si atom, the order of the molecular sizes of the silane compounds is as follows (with the pore size of the MFI zeolite included for comparison); TP-silane > DPM-silane ≈ DP-silane ≈ pore diameter of MFI (10-membered rings) > P-silane > DEM-silane. Therefore, it is expected that selective deactivation of the acid sites located on the external surface of the zeolite crystals can be achieved by utilizing the molecular sieving effect of the zeolites for the silane compounds depending on their size [32]. Moreover, when the silane compounds smaller in size than the zeolite pores are used, SiO_2 formation may occur on the acid sites located on the pore surfaces within the crystals, leading to modification of the pore size [30].

For organic silane compounds having almost the same molecular size as the pore size of the zeolite (e.g. DP-, DPM- and TP-silanes), selective deactivation occurs at acid sites located near the external surface of the ZSM-5 crystal. In contrast, for silane compounds with the molecular sizes smaller than the pore size of the zeolite (e.g. P-silane and DEM-silane), acid sites are accessible on the pore surfaces within the crystal as well as near the external surface, enabling modification of the pore size. As a result, different distributions of acid sites on the MFI zeolite are accomplished by regioselective deactivation using silane compounds with different molecular sizes.

4. Applications of Nano-Sized Zeolite

4.1. Heterogeneous Catalytic Reaction

4.1.1. Application of Nano-Sized Zeolite to Catalytic Reactions

The acid sites in zeolites are located on the external and internal surfaces. In nano-crystalline zeolite, the external surface area of the crystal increases and the diffusion length for reactant hydrocarbons, assignable to the crystal size, decreases as the crystal size decreases. Due to these favorable properties, the nano-crystalline zeolites have been used as heterogeneous catalysts.

Zeolites exhibit a molecular sieving effect for lighter hydrocarbons, hence, the hydrocarbon molecules with sizes larger than the pore mouth size are mainly adsorbed on the external surface. Accordingly, an increase in the external surface area is effective for increasing catalytic activity. Song et al. [10, 11] reported the preparation of nano-sized MFI zeolites and they evaluated the effects of crystal size on external surface area and adsorption properties of toluene on the zeolite. Since the molecular size of toluene is almost the same as the pore mouth size in MFI zeolite, the external surface area, which is a function of the crystal size, affects the adsorption capacity for toluene. Zeolites with crystal sizes less than 100 nm have a higher adsorption capacity for toluene. Botella et al. [47] has reported the Beckmann rearrangement reaction over nanosized BEA zeolite where the reactions mainly proceed on the external surface of zeolite. Serrano et al. [48] have reported catalytic cracking of polyolefins over nano-crystalline MFI zeolite. Because the molecular size of polyolefins is larger than the pore size, cracking of polyolefins mainly occurs over the acid sites on the external surface, while the formation of lighter olefins proceeds on the internal pore surface. Jia et al. [49] have reported dehydration of glycerol to produce acrolein over nano-sized MFI zeolite. Since the hydrophilic glycerol molecules are strongly adsorbed on the external surface of the zeolite, nano-crystalline MFI zeolite is effective for this reaction due to the high surface area. In these reports listed above, the increase in the external surface area of the zeolite nanocrystal is important to improve the catalytic activity and lifetime.

Nano-sized zeolites exhibit a low diffusion resistance for lighter hydrocarbon molecules, hence, a decrease in the crystal size is effective for improving catalytic activity and stability. Sakthivel et al. [50] have reported the isomerization/cracking of hexane over nano-sized BEA zeolite. Since the molecular size of hexane is smaller than the pore size of MFI zeolite, the cracking reaction proceeds mainly on the acid sites located inside pores. Moreover, the small crystal size of BEA zeolite leads to low diffusion resistance of the reactant hexane and the products. Serrano et al. [51] have reported epoxide rearrangement reactions over MFI zeolite, where the high external surface area and low diffusion resistance within the micropores due to the small zeolite crystal size enhance the rearrangement reaction without diffusion restriction of the epoxide molecules.

4.1.2. Light Olefins Synthesis over Nano-Sized Zeolites

Light olefins such as ethylene, propylene and isobutylene are some of the most important basic raw materials for the petrochemical industry, and demands for these olefins are increasing every year. Ethylene has been mainly produced by steam cracking of naphtha, and propylene is a by-product of the cracking process. Recently, alternative production methods and/or processes for these light olefins have been required due to wild ups and downs in oil prices. Promising candidate routes for producing light olefins are catalytic cracking of naphtha and oxygen-containing chemicals to olefins, such as methanol-to-olefins [52, 53], ethanol-to-olefins [54-57], and acetone to olefins [58-64]. Because the molecular size of light olefins is smaller than that of aromatics, the spatial limitations within zeolite channels are expected not only to prevent the formation of aromatics, but also to provide effective reaction fields for increasing olefin yields.

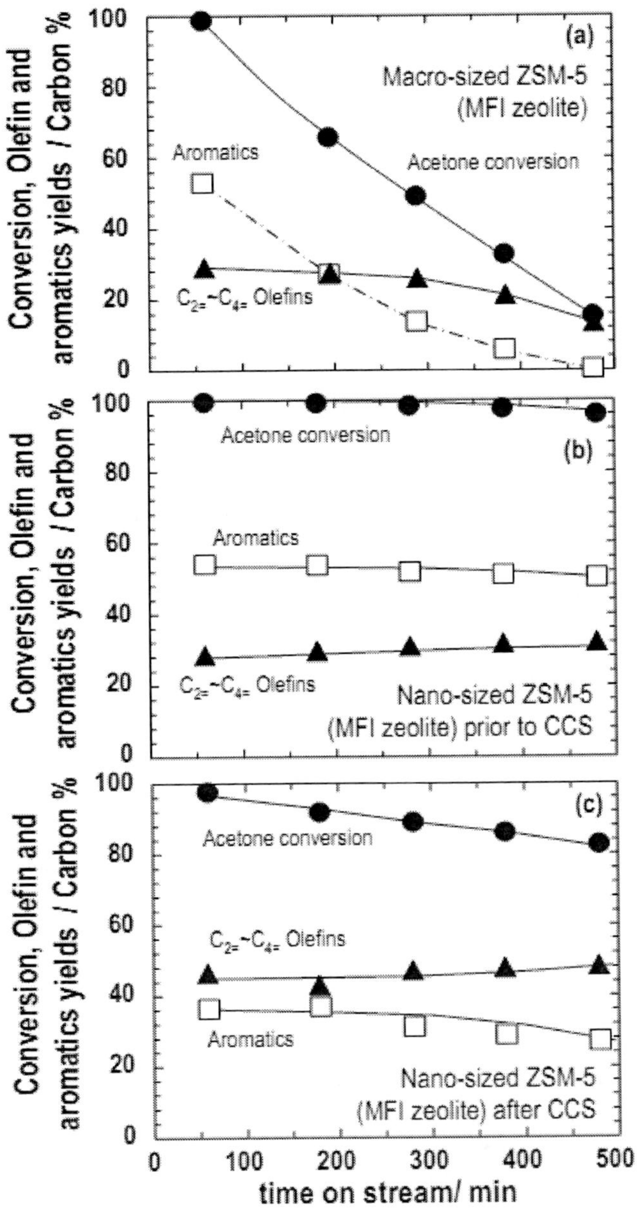

Figure 7. Change in acetone conversion and product selectivities with time over (a) macro-, (b,c) nano-sized zeolites prior to and after CCS treatment. The reactions were run at 673 K with a *W*/*F* of 0.5 kg-cat/(kg-acetone/h).

As discussed above, the light olefins are mainly produced within the zeolite crystal. Accordingly, the diffusion resistances of the light olefins within the micropore affect the product yields and catalytic stability, and application of nano-sized zeolites is effective in achieving faster mass transfer of these light olefins within the micropore during naphtha cracking [65], methanol-to-olefin [66, 67] and acetone-to-olefin [32] reactions.

Here, we introduce the effect of the crystal sizes of MFI zeolite on olefins yield and catalyst lifetime during acetone to olefins reactions. In the acetone conversion using a zeolite catalyst, isobutylene is first produced observed by decomposition of aldol condensation products of acetone (acetone dimer and trimer), followed by olefins such as ethylene and propylene productions by dimerization and cracking of the isobutylene [32].

Olefin synthesis from acetone was carried out over ZSM-5 zeolites with different crystal sizes. The crystal sizes of the ZSM-5 zeolite used as catalysts were 2000 nm and 40 nm, denoted as macro- and nano-sized zeolites, respectively. The changes in acetone conversion, product selectivity and olefin yield with reaction time over the macro- and nano-sized zeolites can be seen in Figure 7. Although the zeolites used as catalysts possessed nearly the same crystallinity and acidity, the change in acetone conversion with time was very different for the two zeolites. While acetone conversion in the macro-sized zeolite decreased to approximately 30% after 500 min, at that time the conversion for the nano-sized zeolites was above 80%. As the crystal size decreased, it was expected that the diffusion length for the intermediates decreased and the external surface area increased, leading to the stable and high activity of the nano-sized zeolite. However, the selectivity for aromatics was much higher than for the other products in the nano-sized zeolite. Because the nano-sized zeolite possesses a large external surface area compared to the macro-sized zeolite, it was concluded that the produced olefins easily reacted with each other on the acid sites located on the external surface of the crystal, where there is no spatial limitation for formation of aromatics. In order to increase the olefin yield, it was therefore necessary to inhibit aromatic formation on the acid sites located near the external surface. The nano-sized zeolite after CCS treatment using DP-silane was then applied to this reaction. The nano-sized zeolite after CCS treatment exhibited higher olefins and lower aromatics yields than the zeolite prior to CCS treatment, indicating that the selective deactivation of the acid sites near the external surface improving product yields of olefin.

4.2. Structured Materials Composed of Zeolite

4.2.1. Hierarchical Zeolite and Zeolite Membrane

As discussed above, the diffusion resistances of the reactant acetone as well as the light olefins within the micropore affect the product yields and catalytic activity. Since the nano-sized zeolites have large external surface areas and low diffusion resistance to hydrocarbons, they are potential candidates for seed crystals from which a structured material can be prepared.

Hybrid zeolitic-mesoporous materials [68-70] have been reported with the properties of both MCM-41 and a zeolite using zeolite seeds by assembling around cetyltrimetyl ammonium bromide micelles. In the hybrid zeolites, zeolite crystals are coated with and/or embedded in meso-porous silica, and hydrocarbon molecules can contact with the acid sites of zeolite through the meso-pore. Recently, hierarchical zeolites [71, 72] have been attracted attention of many researchers, where the micro- and meso-pores formed among nano-sized zeolites are characteristic of the structured materials. The hierarchical materials composed of nano-sized zeolites possess mainly two advantages. One is large meso-pores allowing large molecules to reach active sites, and the other is large external surface area. Accordingly, the hierarchical zeolites composed of nano-sized ZSM-5 [73] and TS-1 [74] exhibited high activity for polyolefin cracking and olefin epoxidation, respectively. Moreover, because design of the micro- and meso-pores is important, Serrano *et al.* [75, 76] have developed a method to fabricate hierarchical materials using organosilane bonded zeolite nanocrystals, where the degree of aggregation of the zeolite can be controlled.

Other structured materials composed of zeolite crystals are zeolite membranes and films. Since the preparation and application of ZSM-5 zeolite membranes by Sano *et al.* [77, 78], a great deal of research has been focussed on the use of zeolite membranes for a variety of potential applications, such as selective-reaction membranes [79, 80] and separation membranes [81–84]. To prepare a zeolite membrane by hydrothermal synthesis, zeolite seed crystals are deposited on a porous support, upon which secondary-growth of the zeolite occurs to form a zeolite membrane. Because the uniformity of the membrane affects its reaction/separation properties, seeding of the zeolites and the secondary-growth process must occur uniformly [85–88]. Therefore, nano-sized zeolite crystals are expected to be potential candidates for seed crystals, as their smaller size allows for greater control of the seeding and growth processes.

4.2.2. Membrane Composed of Nano-Sized Zeolite

Zeolite membranes combine the properties of zeolites with those of inorganic membranes. The zeolitic properties such as strong acidity and molecular sieving effect allow the zeolite membranes to be used as simultaneous reaction and separation systems [80, 89–91]. The zeolite membrane reactors reported to date falls into two configurations: arrangements in which a catalyst is placed on top of a zeolite layer [92–94], and homogeneous membranes containing zeolite crystals [95–100]. Few reports have described zeolite membrane reactors based on a catalytic zeolite layer [101-104]. In such a reactor, as molecules are transported to the external surface of the zeolite layer and penetrate into the layer, they undergo reactions over the active sites of the zeolite. These reactions then cease when the molecules have passed throughout the layer.

Other application of zeolite membrane is a pervaporation membrane. In the pervaporation method, the feed liquid mixtures are in contact with one side of the membrane (feed side), and the target components are removed from the mixture as a vapor from the other side of the membrane (permeate side) after permeation through the membrane. The relative affinities (*e.g.* hydrophilic or hydrophobic interaction) between the membrane and the various components in the liquid solution allows for selective permeation that is superior to that obatined by distillation, which relies on a vapor-liquid equilibrium to achieve separation. Therefore, for dehydration of organic solvents such as ethanol, iso-propanol, and acetone, several types of zeolite membranes have been prepared and studied as potential pervaporation membranes [105–110] due to their high stability at high temperatures and their strong resistance to organic solvents.

In the preparation of zeolite membranes, after the deposition and/or seeding of the nano-crystalline zeolites on the base materials, secondary growth of zeolite was carried out by a hydrothermal treatment. Wang et al. [111] have reported nano-structured zeolite 4A membranes, wherein nano-crystalline zeolites 4A were seeded on an alumina filter by a dip-coating method. Hasegawa et al. [87] and Tago et al. [112, 113] have reported Silicalite-1 nanocrystal-layered membranes, where the Silicalite-1 nanocrystals are piled up on an alumina filter, followed by hydrothermal synthesis to form a Silicalite-1 layer on the nanocrystal layer (Figure 8). The size of the nano-sized silicalite-1 crystals was 60 nm. As can be seen from the figure, a nanocrystal layer with a thickness of approximately 3.0 μm could be clearly observed on the alumina filter, along with a protective layer having a columnar crystal orientation of silicalite-1 with a thickness of approximately 1.0 μm on top of the nanocrystal layer. In these membranes, the thickness of the nanocrystal layer (amount of nano-crystalline zeolite used as seed crystals) can be easily changed, so that the separation properties of these membranes can be controlled [86, 114]. Moreover, because the zeolite crystals are seed crystals for the membranes, it is expected that the crystal sizes affect the membrane performances such as separation ability and flux of permeate molecules.

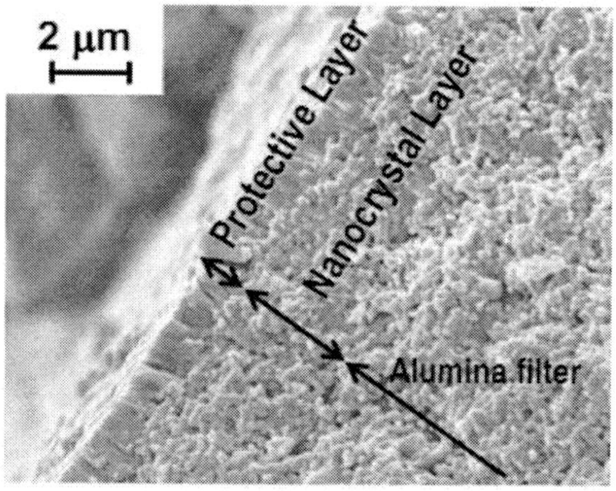

Figure 8. Cross-sectional SEM image of the silicalite-1 nanocrystal-layered membrane.

Conclusion

In this chapter, we introduce the preparation method for nano-sized zeolite and their applications to heterogeneous catalysts and structured materials, and it was revealed that a reduction in crystal size led to improved catalytic and separation properties. Accordingly, control of the crystal size and the tailoring the mesopores and macropores among the crystals are important factors for these applications. In order to obtain nano-crystalline zeolites, it is important to control the nucleation rate during hydrothermal synthesis. The additional of a surfactant into synthetic solutions of zeolites is a promising method to control the crystal size and improve the crystallinity due to adsorption of the surfactant on the zeolite surface. In petrochemical processes, reaction and separation process are very important operations for which zeolites have been used widely as catalysts and adsorbents. Therefore, downsizing of zeolite crystal size is one of the most promising approaches not only to improve existing chemical processes but also to develop a new application.

ACKNOWLEDGMENT

This work was partly supported by the Research Grant Program from the New Energy and Industrial Technology Development Organization (NEDO) of Japan.

REFERENCES

[1] Ogura, M., Shinomiya, S., Tateno, J., Nara, Y., Nomura, M., Kikuchi, E., Matsukata, M., *Appl. Catal. A. Gen.*, 219, 33-43 (2001).
[2] Groen, J. C., Peffer, L. A. A., Moulijn, J. A., Perez-Ramirez, J., *Micropore. Mesopor. Mat.*, 69, 29, (2004).
[3] Groen, J. C., Jansen, J. C., Moulijn, J. A., Perez-Ramirez, J., *J. Phys. Chem. B*, 108, 13062 (2004).
[4] Tsapatsis, M., Lovallo, M., Davis, M. E., *Micropor. Mat.*, 5, 381 (1996).
[5] Ravishankar, R., Kirschhock, C., Schoeman, B. J., Vanoppen, P., Grobet, P. J., Storck, S., Maier, W. F., Martens, J. A., DeSchryver, F. C., Jacobs, P. A., *J. Phys. Chem. B*, 102, 2633 (1998).
[6] Mintova, S., Olson, N. H., Valtchev, V., Bein, T. *Science*, 283, 958 (1999).
[7] Grieken, R. V., Sotelo, J. L., Menendez, J. M. and Melero, J. A., *Micropor. Mesopor. Mat.*, 39, 135 (2000).
[8] Mintva, S., Valtchev, V., *Micropor. Mesopor. Mat.*, 55, 171 (2002).
[9] Hincapie, B. O., Garces, L. J., Zhang, Q., Sacco, A., Suib, S. L., *Micropor. Mesopor. Mat.*, 67, 1, 1387 (2004).
[10] Song, W., Justice, R. E., Jones, C. A., Grassian, V. H., Larsen, S. C., *Langmuir*, 20, 4696 (2004).
[11] Song, W., Justice, R. E., Jones, C. A., Grassian, V. H., Larsen, S. C., *Langmuir*, 20, 8301 (2004).
[12] Song, W., Grassian, V. H., Larsen, S. C., *Chem. Comm.*, 23, 2951 (2005).
[13] Larlus, O., Mintova, S., Bein, T., *Micropor. Mesopor. Mat.*, 96, 405 (2006).
[14] Mintova, S., Valtchev, V., Onfroy, T., Marichal, C., Knozinger, H., Bein, T., *Micropor. Mesopor. Mat.*, 90, 237 (2006).
[15] Morales-Pacheco, P., Alvarez, F., Del Angel, P., Bucio, L., Dominguez, J. M., *J. Phys. Chem. C*, 111, 2368 (2007).
[16] Kumar, S., Davis, T. M., Ramanan, H., Penn, R. L., Tsapatsis, M. *J. Phys. Chem. B*, 111, 3393 (2007).
[17] Tosheva, L., Valtchev, V. P., *Chem. Mater.*, 7, 2494 (2005).
[18] Larsen, S. C. *J. Phys. Chem. C*, 111, 18464 (2007).
[19] Tago, T., Nishi, M., Kouno, Y., Masuda, T., *Chem. Lett.*, 33, 1040 (2004).
[20] Tago, T., Iwakai, K., Nishi, M., Masuda, T., *Stud. Surf. Sci. Catal.*, 159, 185 (2006).
[21] Tago,T., Iwakai, K., Nishi, M., Masuda, T., *J. Nanosci. Nanotechnol.*, 9, 612 (2009).
[22] Tago, T., Aoki, D., Iwakai, K., Masuda, T., *Top. Catal.*, 52, 865 (2009).
[23] Iwakai, K., Tago, T., Konno, H., Nakasaka, Y., Masuda, T., *Micropor. Mesopor. Mat.*, 141, 167 (2011).

[24] Rhodes, K. H. , Davis, S. A., Caruso, F., Zhang, B. J., Mann, S., *Chem. Mater.*, 12, 2832 (2000).
[25] Perez-Ramirez, J, Christensen, C. H., Egeblad, K., Christensen, C. H., Groen, J. C., *Chem. Soc. Rev.*, 37, 2530 (2008)
[26] Hartmann, M, *Angewandte Chemie*, 43, 5880 (2004)
[27] Bein, T., *Chem. Mater.*, 8, 1636 (1996).
[28] Tavolaro, A., Drioli, E., *Adv. Mater.*, 11, (12), 975 (1999).
[29] Caro, J., Noack, M., Kolsch, P., Schafer, R., *Micropor., Mesopor., Mater.*, 38, 3 (2000).
[30] Masuda, T., Fukumoto, N., Kitamura, M., Mukai, S. R., Hashimoto, K., Tanaka, T., Funabiki, T., *Micropor. Mesopor. Mat.*, 48, 239 (2001).
[31] Tago, T., Sakamoto, M., Iwakai, K., Nishihara, H., Mukai, S. R., Tanaka, T., Masuda, T., *J. Chem. Eng. Jpn*, 42, 162 (2009)
[32] Tago, T., Konno, H., Sakamoto, M., Nakasaka, Y., Masuda, T., *Appl. Catal. A*, 403, 183 (2011).
[33] Kuechl, D. E., Benin, A. I., Knight, L. M., Abrevaya, H., Wilson, S. T., Sinkler, W., Mezza, T. M., Willis, R. R., *Micropor. Mesopor. Mat.*, 127, 104, (2010).
[34] Naik, S. P., Chen, J. C., Chiang, A. S. T., *Micropor. Mesopor. Mat.*, 54, 293, (2002).
[35] Lee, S., Carr, C. S., Shantz, D. F., *Langmuir*, 21, 12031 (2005).
[36] Burkett, S. L., Davis, M. E., *J. Phys. Chem.*, 98, 4647 (1994).
[37] Burkett, S. L., Davis, M. E., *Chem. Matter.*, 7, 920 (1995).
[38] DeMoor, P. P. E. A., Beelen, T. P. M., Komanschek, B. U., Beck, L. W., Wagner, P., Davis, M. E., Santen, R. A., *Chem. Eur. J.*, 5, 2083 (1999).
[39] DeMoor, P. P. E. A., Beelen, T. P. M., Van Santen, R. A., *J. Phys. Chem. C.*, 103, 1639 (1999).
[40] Matsune, H., Tago, T., Shibata, K., Wakabayashi, K., Kishida, M., *J. Nanoparticle Research*, 8, 1083 (2006).
[41] Tago, T., Hatsuta, T., Miyajima, K., Kishida, M., Tashiro, S., Wakabayashi, K. *J. Am. Ceram. Soc.*, 85, 2188 (2002).
[42] Tago, T., Tashiro, S., Hashimoto, Y., Wakabayashi, K., Kishida, M., *J. Nanoparticle Research*, 5, 55 (2003).
[43] Takenaka, S., Orita, Y., Matsune, H., Tanabe, E., Kishida, M., *J. Phys. Chem. C*, 111, 7748 (2007).
[44] Niwa, M., Murakami, Y., *J. Phys. Chem. Sol.*, 50, 487 (1989).
[45] Niwa, M., Yamazaki, K., Murakami, Y., *Ind. Eng. Chem. Res.*, 30, 38 (1991)
[46] Katada, N., Niwa, M., *Chem. Vap. Dep.*, 2, 125 (1996).
[47] Botella, P., Corma, A., Iborra, S., Montón, R., Rodríguez, I., Costa, V., *J. Catal.*, 250, 161 (2007).
[48] Serrano, D. P., Aguado, J., Escola, J. M., Rodriguez, J. M., *J. Anal. Appl. Pyrol.*, 74, 353 (2005).
[49] Jia, C. J., Liu, Y., Schmidt, W., Lu, A. H., Schuth, F. *J. Catal.*, 269, 71 (2010).
[50] Sakthivel, A., Iida, A., Komura, K., Sugi, Y., Chary K. V. R., *Micropor. Macropor. Mat.*, 119, 322 (2009).
[51] Serrano, D. P., Van Grieken, R., Melero, J. A., Garcia, A., Vargas, C., *J. Mol. Catal. A: Chem.*, 318, 68 (2010).
[52] Stocker, M., *Micropor. Mesopor. Mater.*, 29, 3 (1999).
[53] Keil, F. J., *Micropor. Mesopor. Mater.*, 29, 49 (1999).

[54] Dahl, I. M., Wendelbo, R., Andersen, A., Akporiaye, D., Mostad, H., Fuglerud, T., *Micropor. Mesopor. Mater.*, 29, 159 (1999).
[55] Oikawa, H., Shibata, Y., Inazu, K., Iwase, Y., Murai, K., Hyodo, S., Kobayashi, G., Baba, T., *Appl. Catal. A*, 312, 181 (2006).
[56] Mentzel, U. V., Holm, M. S., *Appl. Catal. A*, 396, 59 (2011).
[57] Gayubo, A. G., Valle, B., Aguayo, A. T., Olazar, M., Bilbao, J., *Ind. Eng. Chem. Res.*, 49, 123 (2010).
[58] Novakova, J., Kubelkova, L., Dolejsek, Z., *J. Molec. Catal.*, 39, 195 (1987).
[59] Kubelkova, L., Cejka, J., Novakova, J., *Zeolite*, 11, 48 (1991).
[60] Kubelkova, L., Novakova, J., *Zeolite*, 11, 822 (1991).
[61] Biaglow, A. I., Sepa, J., Gorte, R. J., White, D., *J. Catal.*, 151, 373 (1995).
[62] Panov, A. G., Fripiat, J. T., *J. Catal.*, 178, 188 (1998).
[63] Panov, A., Fripiat, J. J., *Langmuir*, 14, 3788 (1998).
[64] Song, W. G., Nicholas, J. B., Haw, J. F., *J. Phys. Chem. B*, 105, 4317 (2001).
[65] Mochizuki, H., Yokoi, T., Imai, H., Watanabe, R., Namba, S., Kondo, J. N., Tatsumi, T., *Micropor. Mesopor. Mater.*, 145, 165 (2011).
[66] Firoozi, M., Baghalha, M., Asadi, M., *Catal. Commun.*, 10, 1582 (2009).
[67] Nishiyama, N., Kawaguchi, M., Hirota, Y., Van Vu, D., Egashira, Y., Ueyama, K., *Appl. Catal. A*, 362, 193 (2009).
[68] Karlsson, A., Stocker, M., Schmidt, R., *Micropor. Mesopor. Mater.*, 27, 181 (1999).
[69] Huang, L. M., Guo, W. P., Deng, P., Xue, Z. Y., Li, Q. Z., *J. Phys. Chem. B*, 104, 2817 (2000).
[70] Tao, Y. S., Kanoh, H., Abrams, L., Kaneko, K., *Chem. Rev.*, 106, 896 (2006)
[71] Rhodes, K. H., Davis, S. A., Caruso, F., Zhang, B. J., Mann, S., *Chem. Mater.*, 12 2832 (2000).
[72] Perez-Ramirez, J., Christensen, C. H., Egeblad, K., Christensen, C. H., Groen, J. C., *Chem. Soc. Rev.*, 37, 2530 (2008).
[73] Serrano, D. P., Aguado, J., Escola, J. M., Rodriguez, J. M., Peral, A., *J. Catal.*, 276, 152 (2010).
[74] Serrano, D., Sanz, R., Pizarro, P., Moreno, I., *Chem. Commun.*, 11, 1407 (2009).
[75] Serrano, D. P., Aguado, J., Escola, J. M., Rodriguez, J. M., Peral, A., *Chem. Mat.*, 18, 2462 (2006).
[76] Serrano, D. P., Aguado, J., Escola, J. M., Rodriguez, J. M., Peral, A., *J. Mater. Chem.*, 18, 4210 (2008).
[77] Sano, T., Kiyozumi, Y., Kawamura, M., Mizukami, F., Takaya, H., Mouri, T., Inaoka, W., Toida, Y., Watanabe, M., Toyoda, K., *Zeolite*, 11, 842 (1991).
[78] Sano, T., Yanagishita, H., Kiyozumi, Y., Mizukami, F., Haraya, K., *J. Membr. Sci.*, 95, 221 (1994).
[79] Igarashi, H., Uchida, H., Suzuki, M., Sasaki, Y., Watanabe, M., *Appl., Catal., A*, 159, 159 (1997)
[80] Hasegawa, Y., Kusakabe, K., Morooka, S., *J. Membr. Sci.*, 190, 1 (2001).
[81] Kita, H., Horii, K., Ohtoshi, Y., Tanaka, K., Okamoto, K., *J. Mat. Sci. Lett.*, 14, 206 (1995).
[82] Bakker, W. J. W., Kapteijn, F., Poppe, J., Moulijn, J. A., *J. Membr. Sci.*, 117, 57 (1996).
[83] Kondo, M., Komori, M., Kita, H., Okamoto, K., *J. Membr. Sci.*, 133, 133 (1997).

[84] Kusakabe, K., Kuroda, T., Murata, A., Morooka, S., *Ind., Eng. Chem. Res.*, 36, 649 (1997).
[85] Masuda, T., Hara, H., Kouno, M., Kinoshita, H., Hashimoto, K., *Micropor. Mesopor. Mat.*, 3, 565 (1995)
[86] Pera-Titus, M., Llorens, J., Gunill, F., Mallada, R., Santamaria, J., *Catal. Today*, 104, 281 (2005).
[87] Hasegawa, Y., Ikeda, T., Nagase, T., Kiyozumi, Y., Hanaoka, T., Mizukami, F., *J. Membr. Sci.*, 280, 397 (2006).
[88] Hasegawa, Y., Nagase, T., Kiyozumi, Y., Mizukami, F., *J. Membr. Sci.*, 294, 186 (2007).
[89] Armor, J., N., *Catal. Today*, 25, 199 (1995).
[90] Hasegawa, Y., Sotowa, K., Kusakabe, K., Morooka, S., *Micropor. Mesopor. Mat.*, 53, 37 (2002)
[91] McLeary, E.E., Jansen, J.C., Kapteijn, F., *Micropor. Mesopor. Mat.*, 90, 198 (2006).
[92] Van der Puil, N., Rodenburg, E. C., van Bekkum, H., Jansen, J. C., *Stud. Surf. Sci. Catal.*, 97, 377 (1995).
[93] Tanaka, K., Yoshikawa, R., Ying, C., Kita, H., Okamoto, K., *Catal. Today*, 67, 121 (2001).
[94] Liu, B.S., Gao, L.Z., Au, C.T., *Appl. Catal. A.*, 235, 193 (2002).
[95] Parton, R. F., Vankelecom, I. F. J., Casselman, M. J. A., Bezoukhanova, C. P., Uytterhoeven, J. B., Jacobs, P. A., *Nature*, 370, 541 (1994).
[96] Cot, L., Ayral, A., Durand, J., Guizard, C., Novnanian, N., Julbe, A., Larbot, A., *Solid State Sci.*, 2, 313 (2000).
[97] Lin, Y. S., *Separ. Purif. Technol.*, 25, 39 (2001).
[98] Wu, S.Q., Gallot, J.E., Bousmina, M., Bouchard, C., Kaliaguine, S., *Catal. Today*, 56, 113 (2000).
[99] Langhendries, G., Baron, G.V., *J. Membr. Sci.*, 141, 265 (1998)
[100] Vital, J, Ramos, A.M., Silva, I.F., Castanheiro, J.E., *Catal. Today*, 67, 217 (2001)
[101] Antia, J. E., Govind, R. *Ind. Eng. Chem. Res.*, 34, 140 (1995).
[102] Tuan, V.A., Li, S.G., Falconer, J.L., Noble, R.D., *Chem. Mater.*, 14, 489 (2002).
[103] Masuda, T., Asanuma, T., Shouji, M., Mukai, S. R., Kawase, M., Hashimoto, K., *Chem. Eng. Sci.,* 58, 649 (2003).
[104] Tago, T., Iwakai, K., Morita, K., Tanaka, T., Masuda, T., *Catal. Today,* 105, 662 (2005).
[105] Matsukata, M., Kikuchi, E., *Bull. Chem. Soc. JPN*, 70, 2341 (1997)
[106] Mizukami, F., *Stud. Surf. Sci. Catal.*, 125, 1 (1999).
[107] Noack, M., Kolsch, P., Caro, J., Schneider, M., Toussaint, P., Sieber, I., *Micropor. Mat.*, 35, 253 (2000).
[108] Lin, X., Kikuchi, E., Matsukata, M., *Chem. Comm.*, 11, 957 (2000).
[109] Morigami, Y., Kondo, M., Abe, J., Kita, H., Okamoto, K., *Separ. Purif. Technol.*, 25, 251 (2001).
[110] Bowen, T. C., Noble, R. D., Falconer, J. L., *J. Membr. Sci.*, 245, 1 (2004).
[111] Wang, H., Huang, L., Holmberg B. A., Yan, Y., *Chem. Comm.*, 16, 1708 (2002).
[112] Tago, T., Nakasaka, Y., Kayouda, A., Masuda, T., *Separ. Purif. Tecnol.*, 58, 7 (2007).
[113] Tago, T., Nakasaka, Y., Kayouda, A., Masuda, T., *Micropor. Mesopor. Mat.*, 115, 176 (2008).

[114] Hedlund, J., Noack, M., Kolsch, P., Creaser, D., Caro, J., Sterte, J. *J. Membr. Sci.*, 159, 263 (1999).

In: Zeolites: Synthesis, Chemistry and Applications
Editors: Moisey K. Andreyev and Olya L. Zubkov
ISBN: 978-1-61942-861-4

Chapter 7

TAILORING ACID SITES IN ZEOLITE FOR CATALYTIC REACTIONS: SYNTHESIS, CHARACTERIZATION AND APPLICATIONS

Mojgan Zendehdel*
Department of Chemistry, College of Science, Arak University, Arak; Iran

1. ABSTRACT

Zeolites are widely used as a solid acid catalyst in many industrial processes. The acidity of zeolite could be due to the Bronsted and Lewis acid sites. Exchange of monovalent cations with polyvalent ion creates strong Bronsted centers by the hydrolysis phenomena, which is useful for some of acid catalyst reactions such as alcohol dehydration, acylation, estrification, and Diels-Alder reaction. The number of proton sites can be decreased and increase the number of Lewis sites by dehydration reaction, which is carried out above 875 K°. The time of the cation exchange process can be reduced under microwave irradiation. The Lewis sites can be increased when the metal / zeolite is prepared in the solid state phase ion-exchange under microwave irradiation. Interestingly, in some reactions, including intramolecular acylation, isomerization and conversion of acids to benzimidazoles, the Lewis sites are the active centers. Despite the many reports a good catalytic activities for the solid ion exchanged zeolites, the interaction mechanism of microwave radiation with a mixture of metal ion salt/zeolite has not been yet completely understood. In particular it has been found that, as a result of microwave treatment, the metal ions may occur in at least three different forms over or within zeolites: i) external coating of zeolite crystals by metal clusters , ii) residual metal salts dispersed over the surface of zeolites, and iii) metal ions incorporated in zeolite channels and cages by ion-exchange . FT-IR spectroscopies with various probe molecules, such as ammonia, quinoline, pyridine and carbon monoxide have been used for the characterization of acid sites. Recently, Rietveld refinement method can be used for characterization of these sites. In this chapter, we introduce a selective method for tailoring Bronsted and Lewis sites in zeolite .Also; explain about some methods for characterization these sites. Finally, mechanism of some organic reactions with Bronsted and Lewis sites will be considered.

* Department of Chemistry, College of Science, Arak University, Arak; Iran; Email: m-zendehdel@araku.ac.ir.

2. Introduction

2.1. Zeolite Acidic Catalyst

Many research shows that the carbonium ion intermediate catalyzed by acidic zeolite such as Freidel-Craft, cracking, isomerization and esterification. The acidity could be due to Bronsted sites (proton donor) arising from hydroxyls in zeolite pore structre, or Lewis sites (electron acceptor), such as tricoordinated aluminum atoms of zeolite surface [1].The nature of acid sites related to following subjects:

I. The Al content of a zeolite may be expressed as the ratio of Al/Al+Si (molar) fraction of AlO_4 and SiO_4 tetrahedral or as the ratio Si/Al. Since the acidity arises from the replacement of tetravalent Si^{4+} atom by trivalent Al^{3+} atom in the aluminosilicate framework, the number of acid sites generated should be parallel to the Al content .Furthermore, any change in Si/Al ratio should be reflected to catalytic properties. In fact, an increase in the Si/Al ratio improves several zeolite properties.

II. In zeolite dehydration reaction occurs above 873K, which decreases the number of protons and increases the Lewis sites.

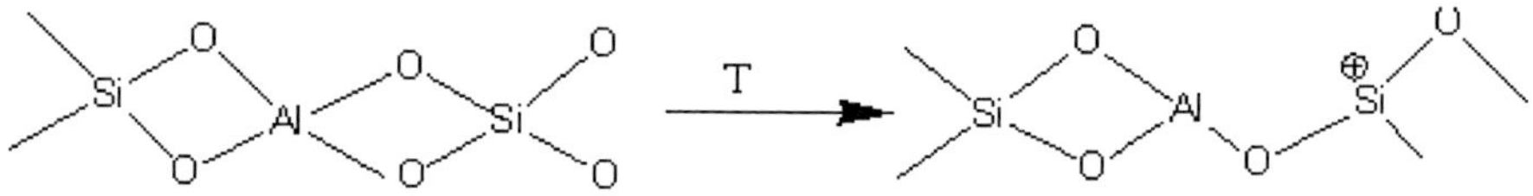

III. The thermal decomposition of ammonium exchanged zeolites yield the hydrogen form or Bronsted acid site. This zeolite is dehydrated and the number of Lewis sites increase at temperatures higher than 500K[2,3]

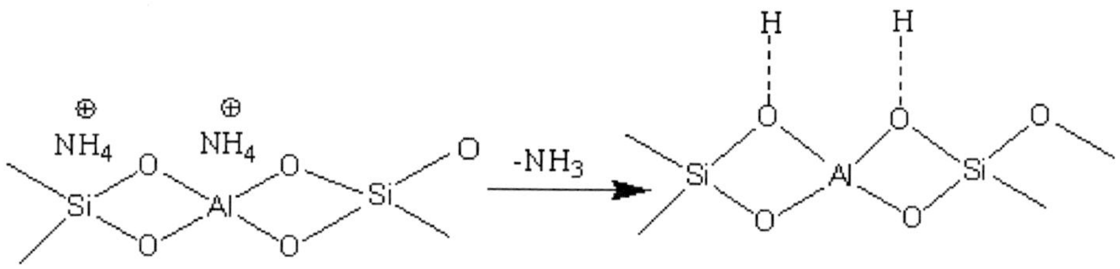

IV. The acid catalysis of zeolite is strongly depending on proton content and acidity strength .Both parameters vary with remaining cations. The exchange of monovalent ions by polyvalent cations improves the catalytic properties. Those highly charged cations create very strong Bronsted acid centers by hydrolysis phenomenon. The ion exchange properties of zeolite are important in acid catalyzed reactions. Most of studies refer to ion exchange that performed in aqueous solution on the other hand, it is well established that the ion exchange can be also achieved in the solid state phase by keeping in contact a zeolite and a metal containing salt at elevated temperatures with prolonged heating. Such a kind of ion exchange process can be very time consuming. However, the processing time can be reduced significantly by microwave heating [4-6]. Despite of the many reports a good catalytic activity for the solid ion exchanged zeolites [7-9], the interaction mechanism of microwave radiation with a mixture of metal ion salt/zeolite has not been yet completely understood [10]. In particular, it has been found that, as a result of microwave treatment, the metal ions may occur in at least three different forms over or within zeolites: i) external coating of zeolite crystals by metal clusters [11], ii) residual metal salts dispersed over the surface of zeolites, and iii) metal ions incorporated in zeolite channels and cages by ion-exchange [12, 13].

3. Method for Increasing of Acid Sites on Zeolite

Majority of studies show that introducing metal into zeolite by aqueous solution through an ion exchange process increases Bronsted sites .Generally for increasing Bronsted sites to a solution of metal salts zeolite was added. The mixture was stirred for 24 h and then filtered. The obtained solid was washed with water until a colorless filtrate observed. The final zeolite products were dried at room temperature. Although, introducing transition metal cations into zeolite pores import Lewis acidity to these catalysts and improve their catalytic activity but there aren't many report about this site on modified zeolite. In this section we introduce a simple method for increasing Lewis sites a mixture of zeolite and metal salts was mechanically mixed, ground and heated in microwave oven for 10-20 min at 900 W powers [6,14].

4. Characterization Acid Sites

For determination of zeolite acidity, especially Bronsted sites, various methods such as titration with Hammett indicators [15], microcalorimetry [16], and 1H, ^{13}C, ^{15}N and ^{31}P-NMR spectroscopy have been developed [17]. FT-IR spectroscopy with different probe molecules, such as ammonia [18], quinoline, pyridine [19] and carbon monoxide have been used for the characterization of acid sites. But in this chapter, we try to focus on the nature of acid sites in NaY zeolite that ion-exchanged in solid state by microwave irradiation.

4.1. Pyridine Adsorption

Pyridine adsorption is one of the best methods for characterization of zeolite. For this characterization the ion-exchanged zeolite with microwave irradiation was activated under vacuum at 140 °C for 1h. The samples were cooled down to room temperature under vacuum, and then immediately exposed to pyridine vapors for 2h. Desorption of physically adsorbed pyridine was carried out by evacuation for 1h at 140 °C. Finally samples were removed and analyzed by FT-IR and DSC. FT-IR spectrum of microwave some metals ion-exchanged NaY zeolite is shown in Fig.1-5. For all Metal/zeolite a new band appeared about 830cm^{-1}, and since such a signal is absent in the parent NaY zeolite and physically mixture of NaY + Metal chloride , it is reasonable to associate it with a T-O-M (II) (T=Al, Si) vibration [6,12]. After adsorption of pyridine the infrared spectra of M(II) /zeolite shows the ring variation modes of pyridine. The most sensitive vibrations with regard to nature of intermolecular interactions via nitrogen are lone pair electrons. In the IR spectra of pyridine adsorbed zeolites, 1200-1800 cm^{-1} regain, pyridinium ion and H-bonded pyridine can easily be distinguished [20]. These modes are observed at 1630 and 1547 cm^{-1} for NH_4Y zeolite, and at 1625, 1494 and 1454 cm^{-1} for AlHY zeolite [21, 22]. The band observed at 1646 cm^{-1} can be attributed to deformation vibrations of water which covered two bands at 1622 and 1630 cm^{-1} that related to interaction of pyridine to Bronsted and Lewis sites, respectively. The band about 1500 cm^{-1} can be associated to the interaction of pyridine to Bronsted site which due to the heavier atomic mass of metals in comparison with aluminum shifted to lower frequency. The bands about 1426 cm^{-1} can be assigned to Al (III) Lewis acid sites that shifted to lower frequencies to some extent. Also, a band related to metal and pyridine interaction at 1100cm^{-1} observed. The results obtained from literature review demonstrate that in the microwave solid state ion exchange method the Lewis acid sites increased. The increase of Lewis sites can be related to metal bonded to zeolites. The thermal behaviors of metal salts zeolite have been investigated by thermo gravimetric analysis. The percentages of weight lose before and after absorption of pyridine show that the higher amount of pyridine adsorbed related to the new acid sites in zeolite [14].

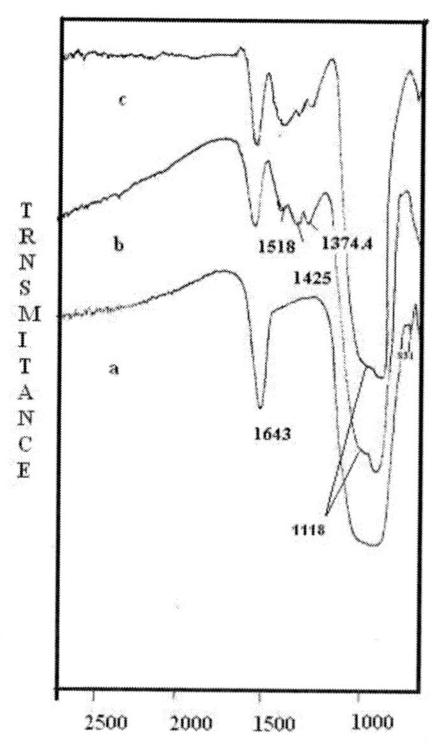

Figure 1. Infrared spectra **a.** activated V (Ш) /zeolite **b.** pyridine absorbed on V (Ш) /zeolite **c.** pyridine adsorbed on V (Ш) /zeolite.

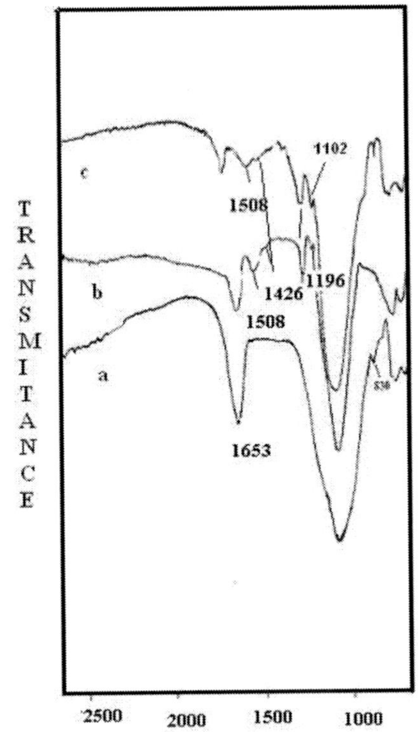

Figure 2. Infrared spectra **a.** activated Cr (Ш) /zeolite **b.** pyridine absorbed on Cr (Ш) /zeolite **c.** pyridine adsorbed on Cr (Ш) /zeolite.

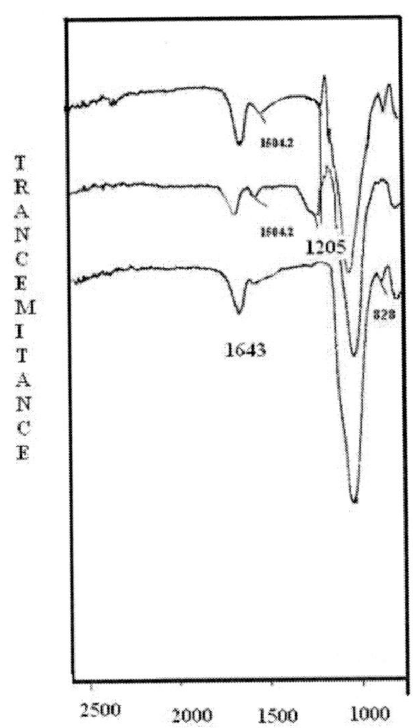

Figure 3. Infrared spectra a. activated Cu (II) /zeolite b. pyridine absorbed on Cu (II) /zeolite c. pyridine adsorbed on Cu (II) /zeolite.

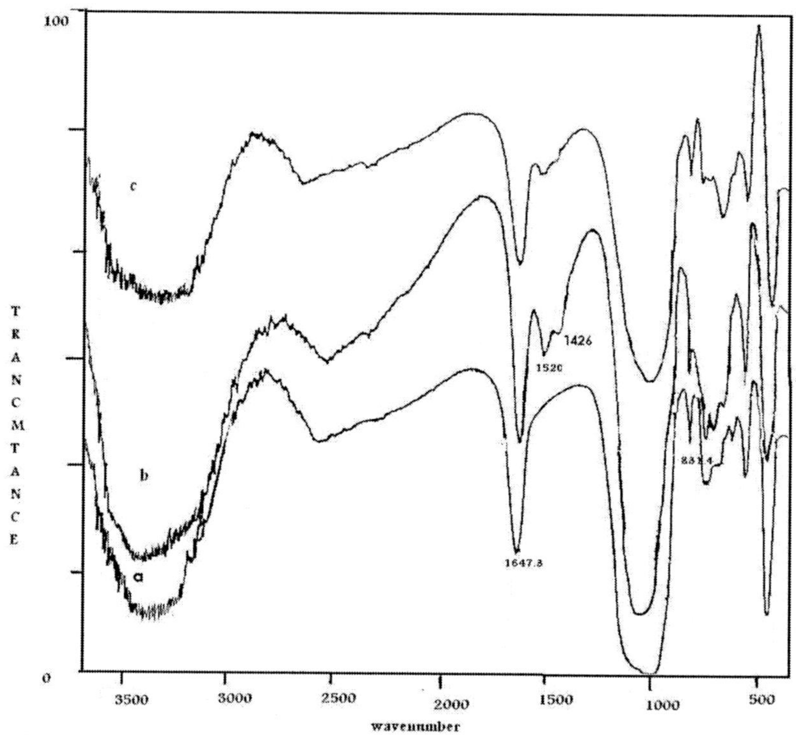

Figure 4. Infrared spectra **a.** activated Co (II) /zeolite **b.** pyridine absorbed on Co (II)) /zeolite **c.** pyridine adsorbed on Co (II) /zeolite

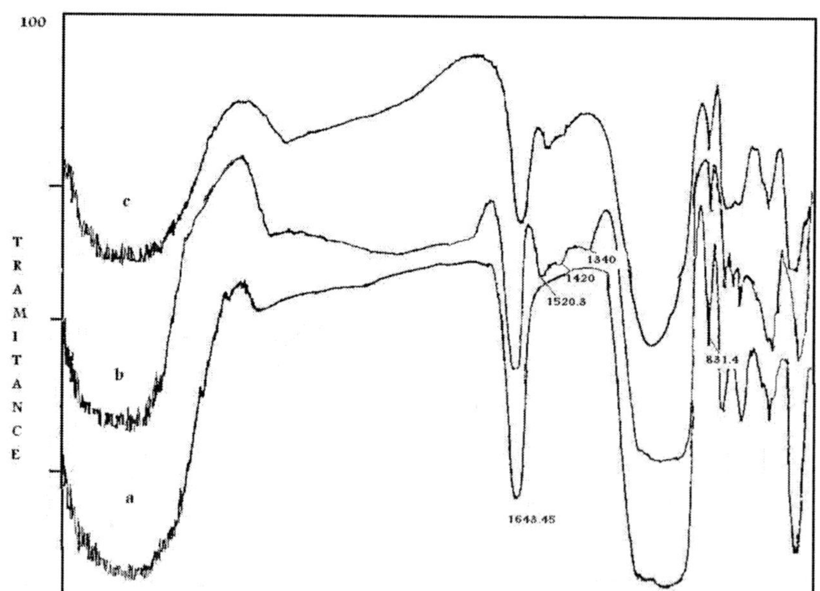

Figure 5. Infrared spectra **a.** activated Ni (II) /zeolite **b.** pyridine absorbed on Ni (II)/zeolite **c.** pyridine adsorbed on Ni (II) /zeolite.

4.2. X-Ray Powder Diffraction and Rietveld Refinements

The simple comparison of zeolite powder patterns in the synthesized and metal-exchanged forms show that, as the overall, no remarkable modifications occurred in the zeolite framework (Figure 6). However, when the peak intensities in selected 2θ regions are compared, it becomes clear that a detectable structural variation occurred most likely in the extra-framework zeolite content. In particular, the intensity of the (220) diffraction line shows the largest difference before and after the cation exchange/microwave treatment. The larger structure factor of the (220) Bragg peak in the ion-exchanged zeolite can be interpreted as due to the scattered intensity from metal ions (Ni, Co, and Cu) within the zeolite unit cell which interfere constructively on the long range. Such a feature would not be observed if the metal ions were just randomly dispersed on the outer surfaces of zeolite crystals. Rietveld refined lattice parameters of the metal-containing zeolite were compared with those of Na-mordenite with similar ratio [23]. In the case of metal- zeolite unit cell volume decrease with incorporated metal to zeolite (Table 1).The decrease in lattice parameters is attributed to the weak effective charge of metal ion than that of Na ion. The results of Rietveld refinements indicate that in the all three cases (Ni, Co, and Cu) the distribution and coordination of transition metal ions are very similar: three new sites have been found, although with slightly different position and population. The first site M^{2+} (X) is in a almost perfect boat-shaped coordination with framework oxygen's (O2, O3 and O8) that can be increased Lewis sites. The second site M^{2+}(C) is poorly coordinated, the third site M^{2+} (F) is in the 12-ring tetrahedral and metal cation coordinate with water molecules in the D, E, G, H and J sites (Fig.7), that can be caused Lewis and Bronsted sites respectively, which similarly to before reported in other ion-exchanged zeolite [24,25]. The former is the site which contributes to the largest extent to the (220) structure factor. Despite its low refined occupancy, neglecting this site would cause disappearing of the calculated intensity of (220) peak.

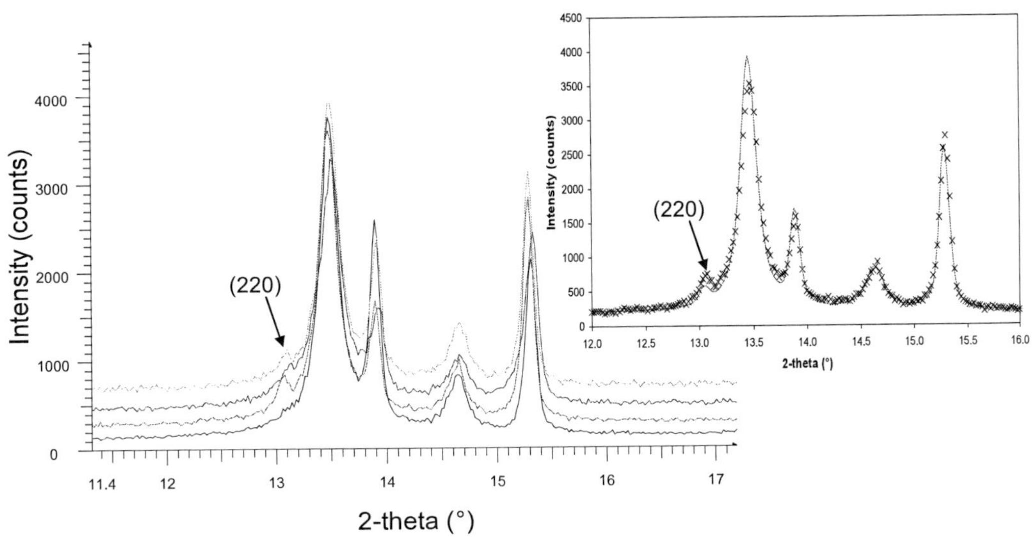

Figure 6. Left, Comparison of powder diffraction patterns for mordenite in the as synthesized and Co-, Ni-, and Cu-exchanged forms. ; right: selected region of Rietveld fit for microwave Ni-exchanged mordenite.

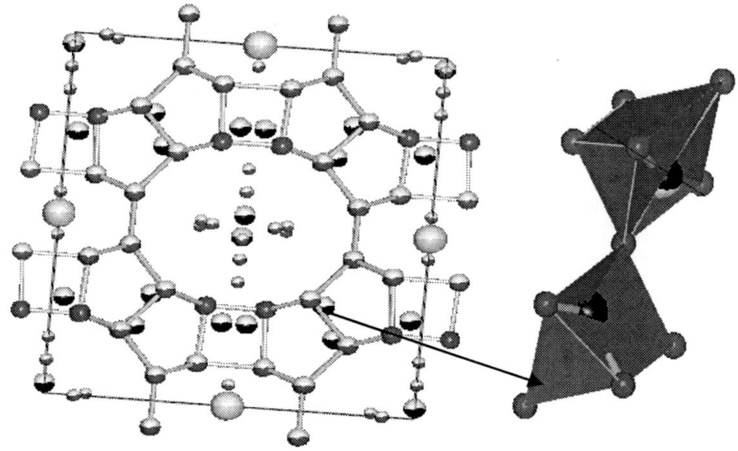

Figure 7. Three new sites for metal ions in Ni-, Co-, and Cu-exchanged mordenite.

Table 1. Interatomic distance for metal –oxygen in M-mordenite obtained by the Rietveld refinements

	Interatomic distance		Interatomic distance		Interatomic distance
CoX-O2	2.48(6)	NiX-O2	2.16(8)	CuX-O2	2.06(7)
CoX-O2	2.48(6)	NiX-O2	2.16(8)	CuX-O2	2.06(7)
CoX-O3	2.47(7)	NiX-O3	2.51(8)	CuX-O3	2.41(6)
CoX-O3	2.47(7)	NiX-O3	2.51(8)	CuX-O3	2.41(6)
CoX-O8	2.18(7)	NiX-O8	1.90(5)	CuX-O8	1.82(4)
CoX-O8	2.18(7)	NiX-O8	1.90(5)	CuX-O8	1.82(4)
CoF-WD	2.46(6)	NiF-WD	2.71(6)	CuF-WD	2.26(7)
CoF-WE	2.00(27)	NiF-WE	1.93(33)	CuF-WE	1.86(4)
CoF-WG	2.06(31)	NiF-WG	2.03(28)	CuF-WG	1.86(4)
CoF-WH	2.53(30)	NiF-WH	2.33(34)	CuF-WH	2.17(26)
CoF-WJ	2.40(4)	NiF-WJ	2.21(6)	CuF-WJ	1.92(5)
CoF-WJ	2.03(4)	NiF-WJ	1.70(6)	CuF-WJ	1.93(6)
CoC-O5	2.73(33)	NiC-O5	2.30(4)	CuC-O5	2.50(32)
CoC-WD	2.35(4)	NiC-WD	2.30(4)	CuC-WD	2.12(5)

4.3. Diffuse Reflectance UV-Vis Spectra

The UV-Vis DR spectra of Metal/zeolite obtained in many works show that there are a good interaction between metal and oxygen in zeolite frameworks. For example Fig. 8 are shown UV-Vis DR spectra for metal /zeolite that ion exchange by microwave irradiation . The two bands observed in the diffuse reflectance UV-Vis spectra of the Co-mordenite at ~ 14400 and ~16000 cm^{-1} can be respectively assigned to $^4T_1g \rightarrow ^4T_1g$ (p) and $^4T_1g \rightarrow ^4A_2g$ (F) d-d electronic transitions, and are in agreement with an octahedral coordination of the metal ion (Fig. 8a). The peak present at ~ 8000 cm^{-1} can be also related to $^4T_1g \rightarrow ^1T_2g$ (F) transition [26]. The electronic spectrum of Ni-mordenite shows two bands at ~13000 and 21000 cm^{-1}, which may be assigned to to $^3A_2g \rightarrow ^3$ T_1g (F) and $^3A_2g \rightarrow ^3T_1g$ (P) transitions that are consistent with octahedral nickel, and a tail of the $^{3}T_{2}g$ is contributed below 9000 cm^{-1} [27,28] (Fig. 8b). In the case of Cu-mordenite, Fig. 2c shows one $^{2}T_{2}g \rightarrow {}^{2}Eg$ spin-allowed d-d transition band at 12800 cm^{-1} which can be related to octahedral coordination on Cu^{2+} [24,29,30]. With note to these methods for characterization of acid sites, we can increase Lewis sites with ion exchange of zeolite by microwave irradiation.

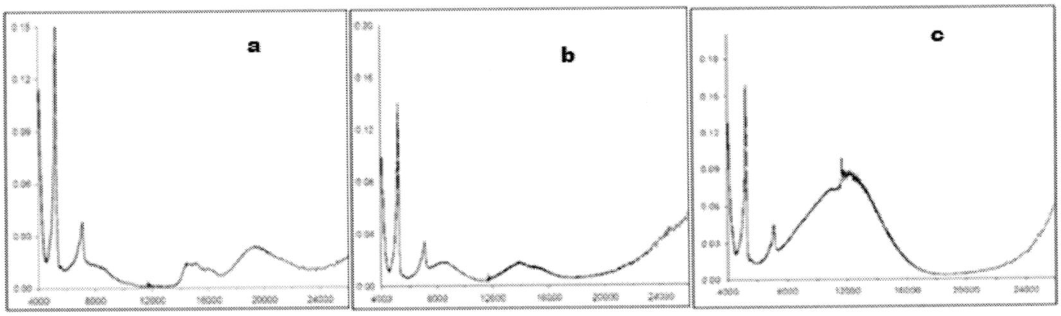

Figure 8. DRS for a. Co^{2+}, b.Ni^{2+}, c.Cu^{2+}mordenite after microwave irradiation

5. Organic Reactions

Recently, there has been interested in the use of inorganic solid acids in the organic synthesis [31, 32]. Among the various solid acid catalysts, zeolites have received an increasing attention because of their suitable acidity, thermal stability and low cost. A literature survey on organic reactions catalyzed by zeolite indicates that in the majority of them, Bronsted and Lewis acid centers are simultaneously important in the catalytic activity of zeolites. Interestingly, it has been demonstrated that in some reactions, such as acylation, alcohol dehydration [33] and the Bronsted centers of zeolites are the active sites; in the other reactions, such as isomerization [33] Lewis centers are the active sites. The present study was performed to understand the role of Bronsted and Lewis acid sites in the catalytic activity of zeolites. It will be very fruitful and may allow tailoring zeolite for specific reaction. For this work transition metal /zeolite have been prepared by solution ion exchange and solid state ion exchange for increasing Bronsted and Lewis acid sites, respectively. Activities of these catalysts have been tested in following reactions:

5.1. Interamolecular Acylation

The result of intramolecular reaction for 4-phenylbutyric acid and phenoxyacetic acid show that the yield of these reactions increasing when microwave solid state ion exchange zeolite was used. Then Lewis acid sites are the active centers in the intramolecular acylation reaction. The following mechanism proposed for this reaction:

In the first step, acid located to the transition metal in zeolite (Lewis sites) after this hydroxyl group bonded to Lewis site and acylium ion separated from zeolite. The acylium ion is considerably more stable than ordinary carbocations since every atom has not octet of electron. In the last step, acylium ion attached to aromatic ring and produced ketone.

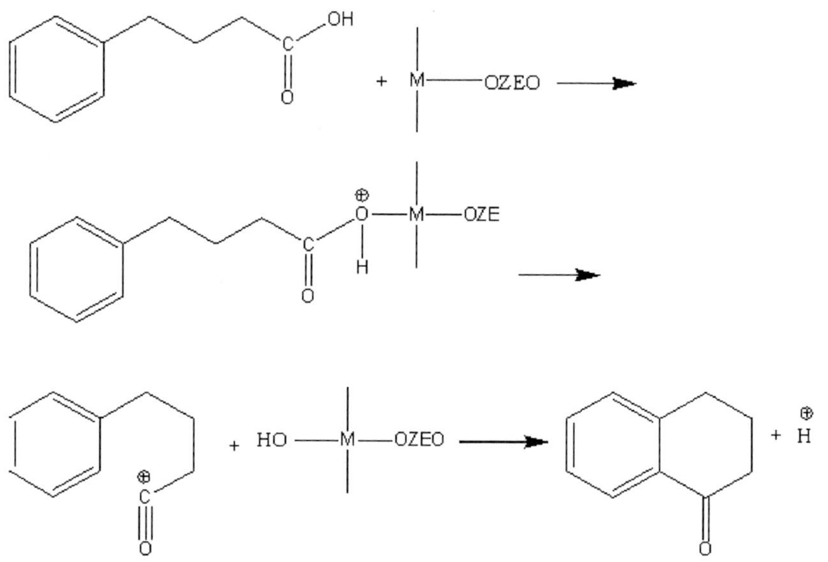

5.2. Esterification

The results for the esterification of benzyl alcohol with acetic acid, isoamyl alcohol with acetic acid and butanoic acid show that these reactions catalyze with Bronsted site with the following mechanism. The yield of these reactions increased when using ion exchange zeolite in the aqueous solution conditions [7].

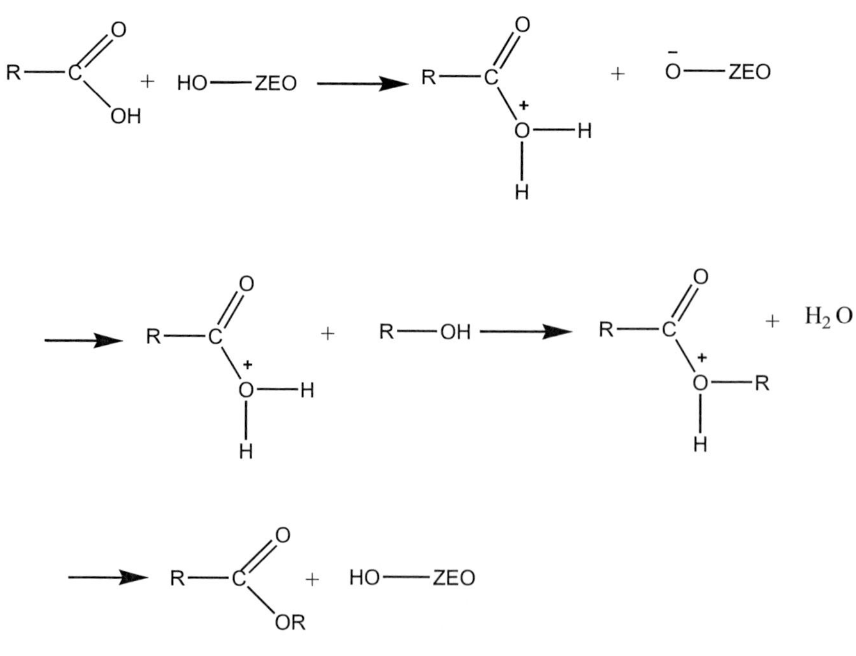

5.3. Diels-Alder Reactions

Diels-Alder reaction is an important tool in the synthesis of cyclic compounds. Several inorganic compounds, such as silica gel [34], alumina [35], clays [36], mesoporous [37] and zeolites [38] have been used as catalysts in Diels-Alder reactions. The yield for these reactions was higher when aqueous solution ion-exchanged metal zeolite was used. It appears that these reactions are catalyzed with Bronsted sites. Since these reactions are catalyzed with acid, the acidity is strongly linked to the kind of cation and will vary with the degree of cation exchange. For the small ions, such as Zn^{2+}, Cu^{2+}... etc, high degrees of ion exchange are observed. The following mechanism may be proposed for the Diels-Alder reaction in the present study.

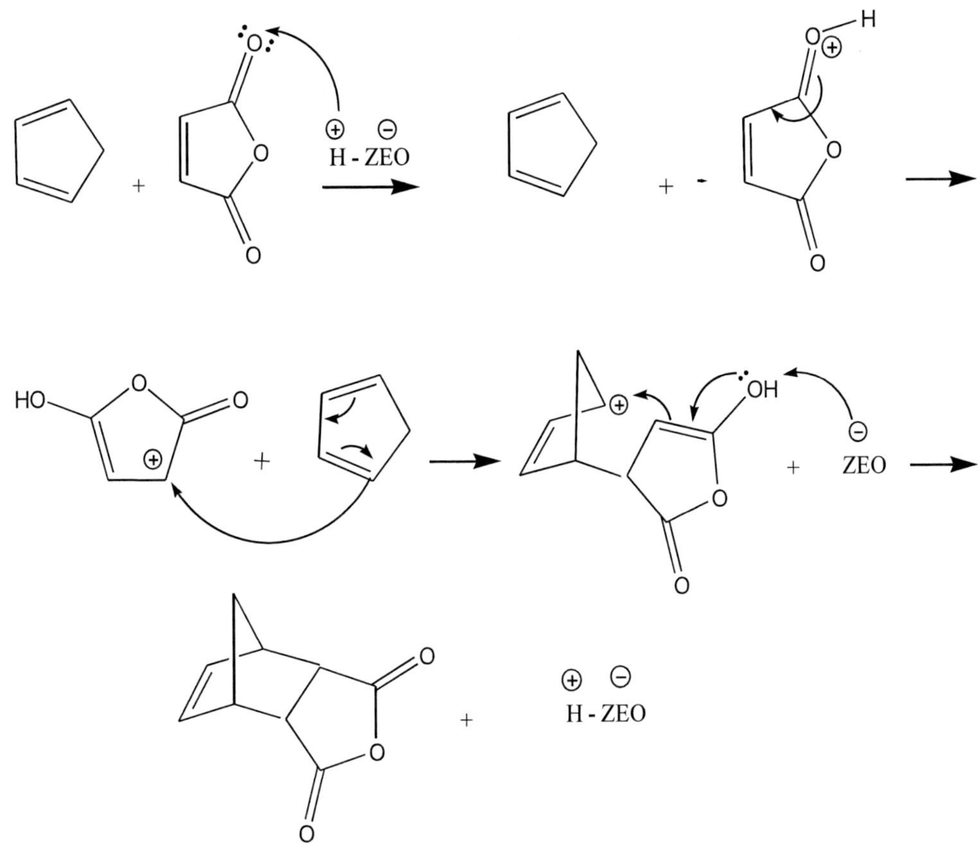

The predominant frontier molecular orbital interaction in a Diels-Alder reaction involves the diene HOMO and dienophile LUMO. The mode of action of the Bronsted acid appears to involve coordination of the carbonyl group of the dienophile. The interaction has the effect of lowering the energy level of the LUMO of the dienophiles, enabling it to be more efficient in overlapping with the HOMO of the diene. Therefore, the reaction proceeds with great efficiency. Results were observed for these Diels-Alder reactions show that there is high endo-selectivity. These results were observed for some of the Diels-Alder reactions with a variety of zeolites [39].

5.4. Biginelli Condensation Reaction

3,4-Dihydropyrimidin-2(1H)-ones (DHPMs) have received significant attentions due to their diverse range of therapeutic and pharmacological activities such as antiviral, antimitotic, anticarcinogenic, antihypertensive [40-44] and calcium channel modulators[45]. Moreover, several natural marine alkaloids with interesting biological activities contain the dihydropyrimidine-5-carboxylate moiety [46, 47]. The original method for the synthesis of 3,4-Dihydropyrimidin-2(1H)-ones, which was firstly reported by Biginelli in 1893, involves one-pot reaction of β-ketoester or β-diketone, arylaldehyde and urea [48]. This has led to the recent development of several improved reaction protocols by using different acid catalysts including inorganic solid acids such as nafion-H [49], HEU [54], zeolite [50], MCM-41 [51] and heterogeneous catalytic method. However, the most of these protocols seem to be mainly concerned with the enhancing of the yields reaction. Very rarely attention has paid to the role of different used inorganic solid acids as catalyst in the mechanism of the Biginelli reaction.

The results of Biginelli reaction for benzaldehyde, 2-methoxy benzaldehyde and 2-chloro benzaldehyde show the zeolite, which was used as catalyst in these reactions, was ion exchanged by microwave in the solid state and aqueous solution. The catalyst activated for 2 h and reactions were carried out to 3-4 times. In each case, yield was decreased about 5-6%. The results for concentration of cations in metal/ zeolite after reaction were also similar to those before reported. The yield of these reactions was high when solid ion-exchanged metal zeolite under microwave irradiation was used. On the other hand the water molecule bonded to the zeolite decreases at high temperature and consequently the number of Lewis acid sites increase. Therefore, the nature of the acid sites is related to the type of metal centers in addition to thermal pretreatment. The ion exchange in solid state under microwave irradiation increases the Lewis sites [52], which may be formed through the following mechanism:

Condensation of aldehyde, urea and ethyl acetoacetate forms an intermediate which is stabilized by framework M and all Lewis sites zeolite. The intermediate cylices connect to zeolite and forms product. It seems that created hydrogen bond between OMe group of 2-methoxy benzaldehyde and the framework zeolite is a factor, which causes the yield of reaction increased [53].

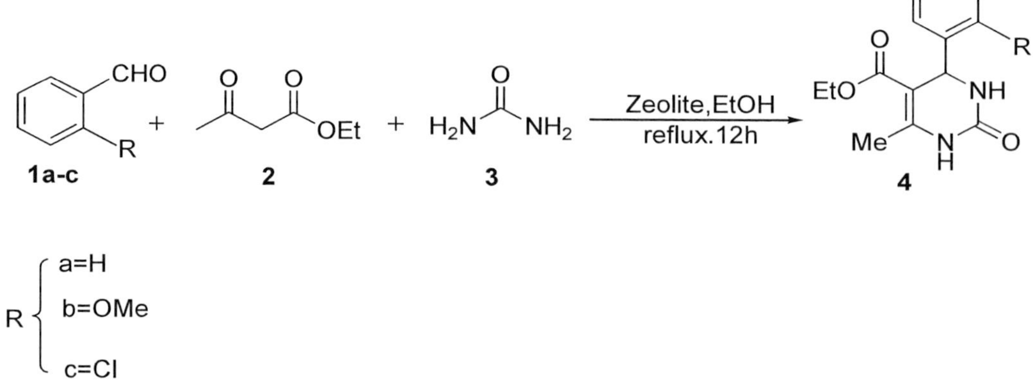

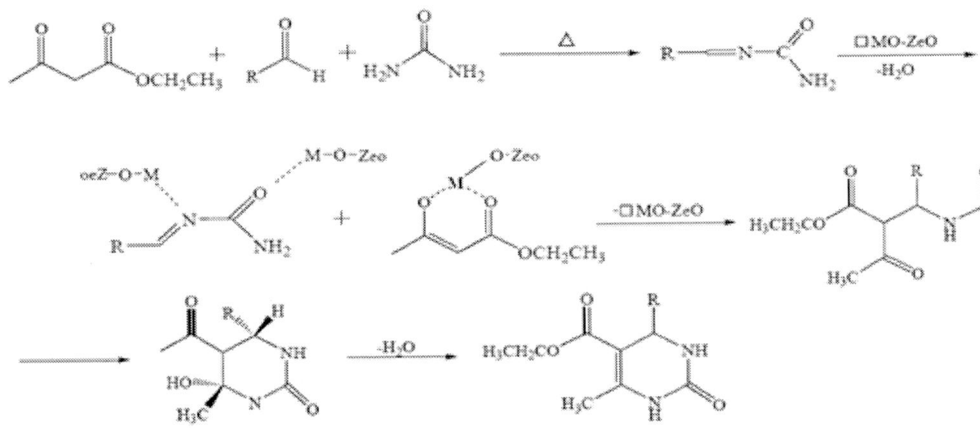

5.5. Conversion of Acids to Benzimidazoles

A large variety of benzimidsoles were reported to exhibit chemotherapeutic [54,55] and biological effects [56,57]. Substituted benzimidazoles have commercial applications in veterinarian medicine as anithelmintic agents and in diverse human therapeutic areas such as treatment of ulcers and as antihistaminic [58]. Recently, It has been found that two groups of benzimidazoles, namely the 5, 6-dinitro and 2-trifiuoromethyl derivatives to be promising candidates for antimicrobial drugs [59].Conversion of acids to benzimidazoles reaction between 4-methyl-1, 2-phenylenediamine and 3-nitrobenzoic acid. 2-Aryl substituted benzimidazole was considered and the results show that the yield for this reaction was higher when solid state ion-exchanged metal zeolite was used. It appears that reactions are catalyzed by Lewis sites. These Lewis acid sites may be formed through the following mechanism.

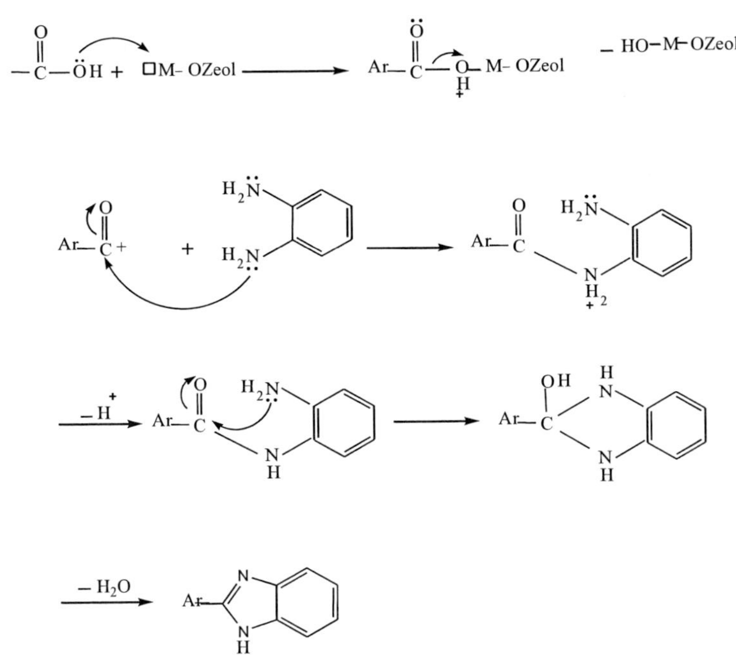

Interestingly, in the present study, we found that by using hydrated zeolite ion-exchanged under microwave condition, the yield of the reaction decreased by 10 %. Apparently, in the microwave solid state ion exchange the water bonded to the zeolite decreases at high temperature and consequently the number of Lewis acid sites increases. Therefore, the nature of the acid site is related to the type of metal centers in addition to thermal pretreatment [60,61].

CONCLUSION

Various transition metals can be dispersed and ion exchanged onto the surface of zeolite under microwave irradiation via the solid state interaction of salts and zeolite. The DSC, DRS, XRD and FTIR spectra of pyridine adsorbed on the activated metal/zeolite provided evidence for the Lewis acid interaction. The results obtained from this study demonstrate that in the microwave solid state ion exchange method the Lewis acid sites increase and Bronsted acid sites decrease. The increase of Lewis sites can be related to metal bonded to zeolites. Also, the nature of cations and the method of exchange was influenced the catalytic activities of the transition metal cation exchanged zeolites. The good correlation between the yields of the some reaction in the presence of different transition metal/zeolite ion exchange in solid state with microwave irradiation suggests that the active centers are mostly Lewis sites.

REFERENCES

[1] J. B. Uytterhoven, J. B. Chirstner, W. K. Hall, The hydrogen held by solids. VIII. The decationated zeolites *J. Phys. chem.* 69 (1965)2117-26

[2] A. Croma, Inorganic Solid Acids and Their Use in Acid-Catalyzed Hydrocarbon Reactions, *Chem. Rev.* 95(1995)559-614

[3] F. R. Ribeiro, A.E.Rodriguse, L.D.Rollmann, C.Naccahe, Zeolite: Science and Technology, NATO ASI series, *Martinus Nijhoff publishers* 1983

[4] T. Frising, P. Leflaive. Extraframework cation distributions in X and Y faujasite zeolites: A review, *Microporous and Mesoporous Materials* 114 (2008) 27–63.

[5] I. Braun, G. Schulf-Ekloff, M. Bocksette, D. Wohrle, Microwave-assisted crystallization inclusion of coumarin and azo dyes in alpo$_4$-5 molecular sieves. *Zeolites* 19 (1997) 128-132.

[6] S. Collin, Cundy, Microwave Techniques in the Synthesis and Modification of Zeolite Catalysts. A Review, *Collect. Czech. Chem. Comm.* 63 (1998) 1699-1723

[7] M. Zendehdel, M. Kooti, M. M. Amini, Dispersion and Solid State Ion Exchange of VCl_3, $CrCl_3.6H_2O$, $MnCl_2.4H_2O$ and $CoCl_2.6H_2O$ onto the Surface of NaY Zeolite Using Microwave Irradiation *J. Pores Mater.* 12 (2005) 143-149

[8] M. Kooti, M.Zendehdel, M.Mohammadpour amini, Esterification and Intramoleculare Acylation Reactions with Transition Metal/Zeolites, *J. Inclusion Phenom. Macrocyclic Chem.* 42 (2002) 265-269

[9] M. Zendehdel, N. Fouroghfar, Z.Gaykani, Diels-Alder Reaction with Transition Metal/Zeolites *J. Inclusion Phenom. Macrocyclic Chem.* 53 (2005) 47-49.

[10] X. Li, X. Zhang, L. Lei, Preparation of CuNaY zeolites with microwave irradiation and their application for removing thiophene from model fuel. *Separation and Purification Technology* 64 (2009)326-331

[11] M.D.Romero, G.Ovejero, M.A.Uguina, A.Rodriguez, J.M.Gomez, Fast removal of the acid properties in the NaX zeolite by ion-exchange under microwave heating. *Catalysis Communications* 5 (2004)157-160

[12] J. A. Heuser, W. U. Spendel , A. N. Pisarenko ,C. Yu , M. J. Pechan , G. E. Pacey, Formation of surface magnetite nanoparticles from iron-exchanged zeolite using microwave radiation. *J Mater Sci* 42 (2007) 9057–9062

[13] D.Yin, D. Yin, The dispersion and solid-state ion exchange of $ZnCl_2$ onto the surface of NaY zeolite using microwave irradiation. *Microporous and Mesoporous Materials* 24 (1998) 123–126

[14] H. Firstar, U. Hatje, Investigations on the solid-state ion exchange of Ni^{2+}, Cu^{+} and Zn^{2+} ions into zeolite Y using EXAFS techniques, *Solid State Ionics* 101-103 (1997) 425-430

[15] M. Zendehdel, M. M. Amini, M. Emami Khansairi,Acidic Properties of CoCl2·6H2O and NiCl2·6H2O,Microwave Ion Exchanged NaY Zeolite, *Asian J. of Chemistry,* 19, 3, (2007) 1802-1808.

[16] K. Tanabe, Solid acid and Bases, *Academic Press,* NewYork 1978.

[17] J. C. Vedrine, A. Auroux, P. Dejaifive, V. Ducarme, H. Hoser, S. Zhou, Catalytic and physical properties of phosphorus-modified ZSM-5 zeolite *J. Catal.* 73 (1982)147-160.

[18] J. Datka, E. Tużnik, Hydroxyl groups and acid sites in NaZSM-5 zeolites studied by i.r. spectroscopy. *Zeolites* 5 (1985)230-232.

[19] N. Y. Yopsoe, K. Pederson, E. G. Derouane, Infrared and temperature-programmed desorption study of the acidic properties of ZSM-5-type zeolites. *J. Catal.* 70 (1981) 41-52.

[20] M. Maache, A. Janin, J. C. Lavalley, J. F. Joly, E. Benazi, Acidity of zeolites Beta dealuminated by acid leaching: An FT-IR. study using different probe molecules (pyridine, carbon monoxide. *Zeolite* 13 (1993) 419-426.

[21] M. Akcay, *J. Mol. Struct.* 694, 1-3 (2004) 21.

[22] A. Corma ,V. Fones an F.Ray, Quinoline as a probe molecule for determination of external Brönsted and Lewis acidity in zeolites. *Zeolite* 13 (1993) 56-59.

[23] P. Tyhjala, Tuula, T. Palkk Anen, *J. Mol. Catal. A : Chemical* 110 (1996)153.

[24] A. Martucci, M.Sacerdoti, G. Cruciani, C. Dalconi, In situ time resolved synchrotron powder diffraction study of mordenite. *Eur. J. Mineral. 15*(2003) 485–493

[25] Zendehdel M., Cruciani G. Dondi M Appraisal of microwave-assisted ion-exchange in mordenite by crystal structure analysis. *J. Pores Mater*

[26] B. Jasiewicz, E. Sikorska, I.V. Khmelinskii, B. Warzajtis, U. Rychlewska, W. Boczon, M. Sikorski, Spectroscopy and structure of spartine and 2-methyl spartine dichloride metal complexes, *Journal of Molecular Structure* 707 (2004) 89-96

[27] S. I. Mostafa, S. Ikedab, B. Ohtani, Transition metal Schiff-base complexes chemically anchored on Y-zeolite: their preparation and catalytic epoxidation of 1-octene in the suspension and phase boundary systems, *Journal of Molecular Catalysis A: Chemical* 225 (2005) 181–188

[28] S.Nanjundan, C.S. Jone Selvamalar, R. Jayakumar synthesis and characterization of poly (3-acetyle -4-hydroxyphenyl acrylate) and its Cu(II) and Ni(II) complexes, *European polymer journal* 40 (2004)2313-2321

[29] M.A. Zanjanchi, A. Ebrahimian, Studies on the solid-state ion exchange of nickel ions into zeolites using DRS technique, *Journal of Molecular Structure* 693 (2004) 211–216

[30] A.N. Pesteryakov, V.P. Petranovskii, A. Kryazhov, O. Ozhereliev, N. Pfander, A. Knop-Gericke, Study of copper nanoparticles formation on supports of different nature by UV-Vis diffiuse reflectance spectroscopy. *Chemical physics Letter* 385(2004)173-176

[31] 30.B. Jasiewicz, E. Sikorska, I.V. Khmelinskii, B. Warzajtis, U. Rychlewska, W. Boczon, M. Sikorski, Spectroscopy and structure of spartine and 2-methyl spartine dichloride metal complexes, *Journal of Molecular Structure* 707 (2004) 89-96

[32] N. Rane, M. Kersbulck, R.A. van Santen, E.J.M. Hensen, Cracking of n-heptane over Bronsted acid sites and Lewis acid sites in ZSM-5 zeolite, *Microporous and Mesoporous Materials* 110 (2008) 279–291

[33] E. BabuÊrek, J. NovaÂkova, Isomerization of n-butane over acid zeolites Role of Broensted and Lewis acid sites, *Applied Catalysis A: General* 185 (1999) 123-130

[34] W. Holderich, M. Hesse, F. Naumann, Zeolite:catalysts for organic syntheses. *Angew. Chem. Int. Engl* .27 (1988) 226-246

[35] M. Conrads, J. Mattay and J. Runsuik: *Chem. Ber.* 122, (1989) 2207.

[36] V. V. Veselovsky, A. S.Gybin, A. V.Lozanova, A. Moiseenkov, W. A.Smite and R.Caple: Dramatic acceleration of the Diels-Alder reaction by adsorption on chromatography adsorbents, *Tetrahedron Lett.* 29, (1988) 175-178.

[37] P. Laszlo and J. Lunchetti: Catalysis of the diels-alder reaction in the presence of clays *Tetrahedron Lett.* 25, (1984) 1567-1570.

[38] T.Kugita, M.Ezawa, T.Owada, Y.Tomita, S.Namba, N.Hashimoto and M.Onaka: MCM-41 as a highly active catalyst for Diels–Alder reaction of anthracene with p-benzoquinone, *Micropor and Mesopor. Mat.* 44, (2001) 531-536.

[39] S. Ghosh and N. BlaudHole-catalyzed and photoassisted hole-catalyzed pericyclic reactions on zeolites: *J. Catal.* 95, (1985) 300-304.

[40] N. Murthy and C.N.Pillai: Diels-Alder Reactions Catalyzed by Zeolites *Synth. Commun.* 21, (1991) 783-791

[41] L. Weber, *Drug. Disc. Today,* 7, (2002) 143-147.

[42] A. Domling, Recent advances in isocyanide-based multicomponent chemistry. *Curr. Opin. Chem. Biol.,* 6, (2002) 306-313.

[43] C. O.Kappe, Biologically Active Dihydripyrimidones of the Bignelli-Type. A Literature Survey. *Eur. J. Med. Chem.,* 35, (2000) 1043-1052.

[44] C. O . Kappe, 4-Aryldihydropyrimidines via the Biginelli Condensation: *Aza-Analogs of Niffedipine-Type Calcium Channel Modulators Molecules,* 5, (1998) 1-9.

[45] L. E.Overman, M. H. Rabinowitz, P. A. Renhowe, Coordinated Photoinduced Electron and Proton Transfer in a Molecular Triad. *J. Am. Chem. Soc,* 117, (1995) 1657-1658.

[46] B. B.Snider, Z. Shi, Biomimetic synthesis of (.+-.)-crambines A, B, C1, and C2. Revision of the structure of crambines B and C1, *J. Org. Chem.,* 58, (1993) 3828-3839.

[47] P.Biginelli, Biginelli reaction, *Gazz. Chim. Ital.,* 23, (1893) 360-416..

[48] Lin, HX. Zhao, QJ. Xu, B. Wang, XH, A green synthesis of dihydropyrimidinones by Biginelli reaction over Nafion-H catalyst. *Chinese Chemical Letters* 18 (5), (2007) 502-504

[49] M. Tajbakhsh, B. Mohajerani, M. M. Heravi and A. N. Ahmadi, Natural HEU type zeolite catalyzed Biginelli reaction for the synthesis of 3,4-dihydropyrimidin-2(1H) one derivatives. *Journal of Molecular Catalysis A: Chemical,* 236(1-2), (2005) 216.

[50] V. Radha Rani, N. Srinivas, M. Radha Kishan, S. J. Kulkarni and K. V. Raghavan, Zeolite-catalyzed cyclocondensation reaction for the selective synthesis of 3,4-dihydropyrimidin-2(1H)-ones. *Green Chemistry,* 3, (2001) 305-306.

[51] V. R. Choudhary, V. H. Tillu, V. S. Narkhede, H. B. Borate and R. D. Wakharkar, Microwave assisted solvent-free synthesis of dihydropyrimidinones by Biginelli reaction over Si-MCM-41 supported $FeCl_3$ catalyst, *Catalysis Communications,* 4(9)(2003) 449-453.

[52] H. Adrienn, H. Zoltán, V. Ilona, Convenient One-Pot Heterogeneous Catalytic Method for the Preparation of 3,4-Dihydropyrimidin-2(1H)-ones. *Synthetic Communications,* 36(1) (2006) 129-136

[53] U. Lohose, R.Bertran, K.Jancke, L.Kurzawski, B.Parlitz, E. Lofler, and Schreier, Acidity of aluminophosphate structures. Part 2.—Incorporation of cobalt into CHA and AFI by microwave synthesis, *J. Chem. Soc. Faraday Trans.* 91, (1995) 1163-1172

[54] M.Zendehdel, A.Mobinikhaledi,A.Asgari, Zeolite an Efficient Catalyst for the Biginelli Condensation Reaction, *J. Incl. Phenom. Macrocycl. Chem.* (2008) 60:353–357

[55] N. E. Sidorina and V. A. Osyanin, Novel Approach to Synthesis of Imidazo[2,1-b][1,3]benzoxazines 42, (2004) 1406–1407.

[56] K. Kubo, Y. Kohara, E. Imamia, Y. Sugiura, Y. Inada, Y. Furukawa, K. Nishikawa, T. Naka, Nonpeptide angiotensin II receptor antagonists. Synthesis and biological activity of benzimidazolecarboxylic acids, *J. Med. Chem.,* 36 (1993) 2182-2195.

[57] H. Afaf, E1-Masry, H. H. Fahmy, S. H. Ali Abdelwahed, Synthesis and Antimicrobial Activity of Some New Benzimidazole Derivatives, *Molecules,* 5, (2000) 1429-1438.

[58] M. Rizk, The antischistosomal activity of some benzothiazole derivatives Egypt. *J. Pharm. Sci.,* 34, (1993) 243-252.

[59] A. A. Spasov, I. N. Yozhitsa, L. I. Bugaeva, V. A. Anisimova, Rearrangement strategy for the syntheses of 2-amino anilines. *Pharm. Chem.* J. 33, (1999) 232-43.

[60] J. Z. Stefanska, R. Gralewska, B. J. Starosciak, Z. Kazimierczuk, Antimicrobial activity of substituted azoles and their nucleo- sides. *Pharmazie.* 54, (1999) 879-884.

[61] M. Zendehdel, A. Mobinikhaledi, F. Hasanvand jamshidi. Conversion of Acids to Benzimidazoles with Transition Metal/Zeolites, *J. Incl. Phenom. Macrocycle. Chem.* 59 (2007) 41–44

In: Zeolites: Synthesis, Chemistry and Applications
Editors: Moisey K. Andreyev and Olya L. Zubkov
ISBN: 978-1-61942-861-4

Chapter 8

ZEOLITES FROM HOME-MADE ATRANES

Sujitra Wongkasemjit and Dacho Kunnakorn
The Petroleum and Petrochemical College, and the Center for Petroleum, Petrochemicals, and Advanced Materials, Chulalongkorn University, Pathumwan, Bangkok, Thailand

ABSTRACT

Zeolites, mainly consisting of silicon, aluminum, and oxygen atoms that connect in three-dimensional frameworks, are three-dimensional microporous or mesoporous materials. They are widely used in many applications, such as catalysts, catalyst supports, membranes, etc. Zeolites can be classified in terms of framework type, such as AFI and MFI; or in terms of Si:Al ratio, such as LTA (NaA), ZSM-5; or even in terms of hydrolysis agent, such as LTA (NaA) and KA. The synthetic zeolites can be obtained from sol-gel process followed by a hydrothermal treatment, such as conventional heating, microwave heating or electrophoretic heating technique, to form crystalline-structured materials. The obtained synthetic zeolites are varied, depending on the production technique and raw materials used in synthesis. Many chemical compounds, used as raw materials for zeolite synthesis, produce different properties in the obtained zeolite, leading to different results in applications. In this chapter, home-made atrane materials (silatrane and alumatrane), used as silica and alumina sources, respectively, are shown to be distinguishable raw materials for synthesizing many types of zeolites since they provide good thermal and moisture stability, as compared to commercially available materials. The applications, properties, and performances of those synthetic zeolites are illustrated, as well.

INTRODUCTION

Zeolites are hydrated, crystalline tetra-aluminosilicates, in a thermodynamically stable phase, connected with oxygen atoms (Si-O-Al). Each oxygen atom is shared with silicon and aluminum atoms, which require positively charged alkaline metals to balance the negative charges in the framework. The common positive-charged alkaline metals (cations) employed in zeolite synthesis are Na^+, K^+, Li^+, and alkaline earth ions (Davis *et al.*, 1992). Synthetic zeolites can be synthesized by using alkali-metal hydroxides (or reactive forms of silica and alumina) and water. The addition of some surfactant (organic molecules) in a zeolite synthesis route can enhance the framework or structure of the synthetic zeolites. The mechanisms of zeolite formation are very complicated due to the difference of chemical reactions, equilibrium, and solubility variations that occur throughout the heterogeneous synthesis mixture during the crystallization process (Davis *et al.*, 1992). A brief understanding of the zeolite synthesis and its many applications is herein discussed. Furthermore, a better understanding of the zeolites synthesized from other raw materials, especially atranes compounds, suitable for the zeolite synthesis, is clarified. The atrane precursors are introduced and prove to be good candidates for zeolite formation in a novel route.

ZEOLITES: SYNTHESIS TECHNIQUESAND RAW MATERIALS

Since the 1980s, many researchers have observed and contributed to the development of zeolites for many applications, such as a refinery catalysts, catalyst supports, adsorbents, iron exchangers, etc., (Sathupanya *et al.*, 2002). Most of zeolites used in early development were derived from both nature and synthesis (Davis *et al.*, 1992). Although there are many types of natural zeolites, such as mordenite, erionite, chabazite, and clinoptilolite, their Si:Al ratio is very low, affecting the strength of the zeolites and their toleration to solvents (Davis *et al.*, 1992). On the other hand, the crystalline phase of synthetic zeolites can be easily and successfully prepared via sol-gel and hydrothermal treatment processes. The hydrothermal process for the crystalline zeolite formation consists of conventional heating, microwave heating, and electrophoretic techniques.

The conventional heating process is the easiest technique for synthesizing zeolites and provides a higher yield of crystals in any given period of time. When compared to the microwave heating technique, however, conventional heating requires a much longer time for crystallization for the zeolite formation but also yields a wider range of particle size distribution and an overall larger particle size.

The microwave heating technique is widely used in many applications for both organic and inorganic syntheses. It is not only a very fast and energy-efficient technique for zeolite formation, but also prevents other side reactions, owing to its unique nature for certain interactions. The energy transfer from microwave to material occurs either through resonance or relaxation, resulting in a rapid heating and causing simultaneous, abundant nucleation, and fast dissolution of gel. With those advantages, the use of the microwave heating process, thus provides a higher crystallization rate of the zeolite formation in a very short period of time, leading to small particle size with narrow particle-size distribution and high purity (Sathupanya *et al.*, 2002; Sathupanya *et al.*, 2003). In other words, the microwave heating technique provides the following advantages over the conventional heating technique: a higher heating rate, a selective heating of materials with no energy wasted in a bulk heating of sample, a greater control of the heating process, a reduction of the equipment size and waste, and no direct contact between the heating source and the heating materials (Jones *et al.*, 2002).

Generally, the starting materials used for zeolite synthesis are a mixture of silicon and aluminum compounds (such as silica and alumina), alkaline metal cation, organic molecules, and water (Kuanchertchoo *et al.*, 2006; Feigen *et al.,* 1999). The main parameters which affect the phase and the performance of zeolites are the synthesis mixture, the synthesis time, and the temperature during the synthesis process. The crystallization process is thermally activated and allows a high yield of crystals to be achieved in an acceptable period of time. The crystallization period of zeolites can be subdivided into two steps: a nucleation of new population of zeolite crystals and a growth of the existing population of crystals (Kuanchertchoo *et al.*, 2006). However, Wongkasemjit's research grouphas developed new methods for zeolite production using synthetic silatrane and aluminatrane as precursors (so called "atrane" precursors). The atrane precursors provide a remarkable property that is moisture-stable and that cannot be found in commercially available silicon and aluminium alkoxides. This special property easily facilitates the synthesis process, likely leading to zeolite products with better physical properties, such as small particle size with narrow particle-size distribution and high purity. Moreover, due to a greater ability to construct inorganic-organic micelles of these atrane precursors, the hydrolysis rate can be retarded, resulting in easier handling and better thermodynamic stability in an aqueous-base system. Nevertheless, they can also be formed complexes with metal ion to increase a possibility for expanding its coordinated sphere. Therefore, the synthetic atrane precursors (silatrane and aluminatrane) could be suitable for use as precursors in the zeolites synthesis.

Analcium (ANA), possessing ability in matrix storage for radioactive materials (Anchell *et al.,* 1994), can be produced using aluminum sulfate/sodium metasilicate as precursors and cetyltrimethylammonium bromine (CTAB) as a template (Wiersema *et al.*, 1996; Yokomori *et al.*, 1998; Huang *et al.,* 2001; Dimitrijevic *et al.,* 1995). Na-P1 (GIS), mostly used as molecular sieve, can be synthesized from natural clay (Kaolinite) by crystallizing at around 70–80 °C for various reaction times (1–60 days). However, the obtained product shows low purity (Rees *et al.,* 1993; Baccouch *et al.,* 1998). Suthupunya *et al.* used silatrane and alumatrane as silica and alumina precursors, respectively, to synthesize both ANA and GIS zeolites via microwave heating technique and found that both zeolites showed uniform particle sizes with narrow particle-size distribution, as compared to ANA and GIS zeolite synthesized using silica and alumina precursors. In addition, a milder condition, such as lower synthesis temperature and time, was needed to synthesize both ANA and GIS zeolites using those atranes.

ZSM-5 or MFI is also used in many petrochemical industry applications, especially as a catalyst in reforming processes, because seems to be a selective catalyst, cracking only linear hydrocarbons while preserving branched alkanes. It was found that the uniformity of pore size can enhance specific interaction with selected hydrocarbons. ZSM-5 or MFIis commonly synthesized using silica and alumina as precursors and tetra-propylammonium bromide (TPA) as a template. De Moor *et al.* (De Moor *et al.,* 1999; De Moor *et al.,* 2000) studied the MFI synthesis using tetraethoxysilane (TEOS) as a silica precursor and TPA as a template to observe the formation of MFI nanoblocks. The MFI formation observed in this reaction must be conducted under controlled humidity, which makes it more difficult to study when using TEOS since this precursor is extremely susceptible to hydrolysis by water (Phiriyawirut *et al.,* 2003; Phiriyawirut *et al.,* 2003). Therefore, in 2003, Phiriyawirut *et al.* used silatrane and alumatrane as a silica and alumina precursors, respectively. They compared two different templates, viz. tetrabutylammonium hydroxide (TBAOH) and TPA, to study the MFI zeolite formation, and found that the complete conversion of silatrane precursor to MFI zeolite crystals was achieved when using longer aging times. However, the longer aging and heating times provided the better conversion of atrane precursor to MFI zeolite (Phiriyawirut *et al.*, 2003). Furthermore, Kritchayanon *et al.* (Kritchayanon *et al.*, 2006) successfully synthesized Fe-loaded MFI zeolite using silatrane as precursor and TPA as a template via microwave technique. The synthesized Fe-MFI zeolites were used as a catalyst in CO oxidation and gave high oxidation reaction results. Vanadium loaded on MFI zeolite was also synthesized using silatrane precursors via hydrothermal synthesis technique (Phiriyawirut *et al.*, 2005). The silatrane precursor provides a good potential for preparing VS-1 zeolite with high vanadium loading and satisfactory dispersion of vanadium in the MFI flamework. VS-1 zeolite was used as a catalyst in oxidative dehydrogenation of propane and the obtained results were acceptable.

Another type of zeolite being used in many applications, such as household products, aquaculture, agriculture, water treatment, etc., is Sodium A zeolite (NaA or LTA). It can be used as absorbance, ion exchange and size selectivity materials. Wongkasemjit *et al.* (Sathupanya *et al.*, 2003; Phiriyawirut *et al.*, 2003) developed a method to produce a small, uniform NaA zeolite using silatrane and alumatrane as precursors via sol-gel process and microwave heating technique. Nanosize NaA zeolite was successfully synthesized with uniform and narrow particle-size distribution (0.1–0.2 μm NaA zeolite crystals).A much shorter reaction time for NaA zeolite synthesis was also observed when using microwave heating technique as compared to hydrothermal heating or conventional heating techniques. The moisture absorption ability of the synthesized NaA zeolite using atrane precursors was found to be considerably higher than that of commercial NaA zeolites (Kuanchertchoo *et al.*, 2006).

Furthermore, due to hydrolytic stability of atrane precursors for the zeolites synthesis, metal-loaded zeolite was also interesting to study. A high purity of metal-loaded zeolite with controllable cation-loading and a highly uniform structure was successfully synthesized via our novel route and home-made atrane precursors (Sathupunya *et al.*, 2004). Using LiOH instead of NaOH, for instance, Li-zeolites were directly synthesized by using silatrane and alumatrane precursors via sol-gel process and microwave heating technique. Many types of Li-zeolite were obtained by controlling the synthesis parameters and conditions (Sathupunya *et al.*, 2004). Sathupunya *et al.* also successfully synthesized K-H zeolite (potassium-loaded zeolite) using silatrane and slumatrane as precursors and KOH as a base. A high crystallinity and pure K-H zeolite was obtained. The new zeolite structure with narrow particle-size distribution was suitable forusing as a drying agent that could absorb moisture up to 156% (1.565g H_2O/g of dried solid).This newly discovered zeolite was named "PPC-ZM-1"("Petroleum and Petrochemical College-Zeolite Synthesized by Mathavee-1"). PPC-ZM-1 has a flower-petal-like shape.

Synthesis of Ni-supported K-H zeolite was studied using atrane precursors (Athiya *et al.*, 2007). The Ni-supported KH zeolite synthesized by using the conditions of Athiya *et al.* provided a better activity of CO_2 reforming of methane due to better properties of the synthesized catalyst, in terms of morphology of the Ni-supported synthetic KH zeolite.

The silatrane precursor was also successfully used to synthesize mesoporous SAPO-5 with AFI-type formation (Utchariyajit *et al.*, 2010). Interestingly, the mesoporous SAPO-5 product contains both classic, single-crystal SAPO-5 and 1–10 μm mesoporous matrix nanostructure.

Zeolite: Chemistry of Zeolite from Atrane Precursors

As described above, many types of zeolites can be synthesized using home-made atranes (silatrane and alumatrane) as starting materials instead of commercially available, but expensive, silica and alumina precursors. The atrane precursors not only provide remarkable moisture stability, resulting in a slowing of the hydrolysis rate during the sol-gel process, but also are easily synthesized from commercially available and inexpensive raw materials. Moreover, the synthesized zeolites show good characteristics and properties. Varying the Si:Al ratio (by varying the silatrane to alumatrane ratios) can change the structure, the morphology, and the stability of the zeolites to solvent and environment. A change in the alkali cations also affects the structure of the amorphous gel to be transformed to crystalline aluminosilicate, and the stability of the synthesized products under hydrothermal conditions (Sathupunya *et al.*, 2004; Singh *et al.*, 1998). The Si:Al ratio, the type of zeolite, and the structure or morphology of the synthesized zeolite obtained from atrane precursors are summarized in Table 1.

Table 1. Synthetic zeolites from atrane precursors

Type of zeolite	Si:Al	Raw materials	Structure / morphology	Ref. No.
SAPO-5 zeolite	<1	silatrane and alumatrane	AFI type structure	Utchariyajit *et al.*, 2010
NaA zeolite	~1	silatrane and alumatrane	cubic form of building unit cells structure	Sathupunya *et al.*, 2004
			nano-size of NaA zeolite particle	Kuanchertchoo *et al.*, 2006
Li-zeolite	Varies	silatrane and alumatrane	Si:Al = 0.25-0.50 ; EDI product (cubic form structure)	Sathupunya *et al.*, 2004
			Si:Al = 1 ; ABW product (needle-like structure)	
			high Li loaded ; FAU product (unit cell structure)	
K-H	~2	silatrane and alumatrane	flower-petal-like KH zeolite	Sathupunya *et al.*, 2004
			Ni-supported on flower-shaped KH zeolite	
ANA (anacium)	4	silatrane and alumatrane	distorted T6-ring which distorted 4-rings in between	Sathupunya *et al.*, 2002
GIS (Gismondine, Na-P1)	4	silatrane and alumatrane	Periodic building unit (PBU) connected through T4-rings	Sathupunya *et al.*, 2002
ZSM-5	>10	silatrane and alumatrane	MFI crystal structure	Phiriyawirut *et al.*, 2003
			Fe-loaded MFI crystal structure	Kritchayanon *et al.*, 2006
VS-1 zeolite	>10	silatrane	high Si content zeolite with MFI structure	Phiriyawirut *et al.*, 2005

For example, mesoporous SAPO-5 zeolite was synthesized using atrane precursors with 0.2:1Si:Al ratio by microwave heating technique. The product obtained indicates an AFI type with a well-formed hexagonal outer surface. Increasing the silicon content (or increasing the Si:Al ratio) in the system affected the structure of the obtained zeolite by providing a long, hexagonal, rod-like AFI structure. Incorporation of Si atoms in the AFI-type framework influenced the crystal growth in the c-axis (Utchariyajit *et al.,* 2010). Moreover, at very high silicon content-loaded conditions, an amorphous impurity occurred along with the hexagonal plate-like crystals (Utchariyajit *et al.,* 2010).

NaA zeolite is a well-known zeolite in the separation application field. It can be synthesized by mixing silica, alumina, alkali-metal hydroxide, and water. Sathupunya *et al.* (Sathupunya *et al.,* 2003) successfully synthesized NaA (LTA) zeolite particles using silatrane and alumatrane as silica and alumina precursors via sol-gel process and microwave heating technique. The obtained NaA zeolite showed a crystalline aluminosilicate and sharp edge cubic form building unit with a sodalite cage (β-cage or T24 unit) connection in which the Si:Al:Na-ratio was 1:1:1.25. The thermal stability of the synthesized NaA zeolite was up to 600 °C with no mass loss of carbon residue, no phase-transformation, and lower agglomeration of zeolite particles, as compared to commercial NaA zeolite particles. In 2006, Kuanchertchoo *et al.* successfully synthesized uniform and nano-sized NaA zeolite using silatrane and aluminatrane via sol-gel process and microwave technique. The composition of Si:Al:Na was found around 1:1:3–6; and a narrow size distribution of 0.1–0.2 μm with uniform and homogeneous NaA zeolite particle was obtained. Moreover, using the same conditions of the NaA zeolite particle synthesis, NaA zeolite membrane was also fabricated using silatrane and alumatrane precursors (Kuanchertchoo *et al.*, 2009). However, the NaA zeolite membrane fabrication required the use of NaA zeolite particles synthesized using atrane precursors as seeding particles. A uniform and dense columnar NaA crystallayer was formed, and the morphology of the NaA zeolite membrane was improved. The crystals on the support surface were found to be uniform and homogeneous with 1–3 μm size. The XRD pattern also confirmed the size of the crystal using the characteristic peak of the NaA zeolite membrane. As a result, the overall membranes showed better orientation than membranes synthesized from commercial NaA zeolites.

As described previously, Li-zeolites were directly synthesized using atrane precursors via sol-gel process and microwave heating technique. The perfect crystalline zeolite with EDI type zeolite (unit cell structure) was obtained by using the Si:Al ratio around 1.2:1. As the Si:Al ratio was increased to 1.33:1, the ABW type zeolite with needle-like zeolite was obtained. Furthermore, as the amount of Li-loading increased, the unit cell structure and crystal morphology of the Li-zeolite was also changed from ABW to FAU type zeolite. K-H zeolite provides another example of how morphology changes with a change in reaction parameters. The KH zeolite with flower-petals-like crystalline morphology was obtained using 2:1Si:Al ratio (Sathupunya *et al.,* 2004). The obtained KH zeolite showed a crystalline and highly pure potassium aluminosilicate, and also provided a very narrow particle-size distribution. In addition, very fine and uniform needle-like single crystals tended to aggregate and formed polycrystal KH zeolites. When Ni was loaded onto KH zeolite via microwave heating technique (Kaengsilalai *et al.,* 2007), dog-bone and flower-shaped Ni-loaded KH zeolite was obtained.

To synthesize ZSM-5 or MFI zeolite, known as a selective catalyst for petrochemical industry, especially in the reforming processes, the pore structure must be stabilized by the use of organic molecules template. In 2003, Phiriyawirut *et al.* used atrane precursors and varied the Si:Al ratio to synthesize MFI. They found that as alumatrane content increased (or the Si:Al ratio decreased) the shape of the MFI zeolite became rounder, and MFI crystal formation was also impeded. However, a lower level of amorphous phase can be achieved by increasing the heating time or the aging time. The templates studied also played important roles. In this case TPA was more efficient than TBA for the MFI zeolite synthesis. Most importantly, the complete conversion of silatrane precursor to MFI crystals was successfully achieved with increases in the aging time or heating time (Phiriyawirut *et al.*, 2003). When loading iron onto MFI (Kritchayanon *et al.*, 2006), Fe-loaded MFI zeolite obtained from the use of atrane precursors showed a cubic MFI shape; however, the crystal size seemed to decrease, and the crystal shape become rounder and less homogeneous with an increase in the iron content loaded onto the MFI zeolite. Up to this point, the atrane precursors have shown a great potential for use as raw materials in synthesizing many types of zeolites and metal-loaded zeolites with impressive morphology, as compared to commercial raw materials.

ZEOLITE: APPLICATIONS

According to a variety of properties (i.e. absorption, ion exchange, hydrophobic /hydrophilic characteristics of surface, size selectivity properties, and meso- or micro-porous structure) zeolites or crystalline aluminosilicates can be used in many applications, such as catalysts, catalyst supports, water treatment, separation, refinery industries, etc. As mentioned before, many types of zeolites synthesized by atrane precursors provide many advantages and show remarkable performances in those applications.

In terms of separation technology, the well-known zeolite used in the separation technologyis NaA zeolite in the form of either powder or membrane. Especially in liquid-liquid separation technology, NaA zeolite membrane is used to separate water from water-organic mixture using a pervaporation system. The widest use of NaA zeolite membrane is in water-ethanol separation using the pervaporation system. One of the reasons is due to the 0.4 nm pore size of the NaA zeolite. According to the kinetic diameters of water and ethanol of 0.3 and 0.44 nm, respectively, the size of the water molecule is lower than 0.4 nm while that of the ethanol molecule is higher. Thus, the NaA zeolite membrane is chosen to be applied for separating water from ethanol-water solution.

In the Wongkasemjit research group, the NaA zeolite membrane is synthesized using NaA zeolite crystal seeds obtained from atrane precursors (Kuanchertchoo *et al.,* 2009: Kunnakorn *et al.,* 2011). The obtained NaA zeolite membrane can be synthesized using either microwave or autoclave (or conventional) heating technique, resulting in a good performance, as confirmed by the obtained value of the separation factor higher than 10,000 with the total water flux as high as 2 $kg/m^2/h$. Wongkasemjit *et al.* also discovered that the NaA zeolite membrane can be synthesized using commercial silica and alumina; however, the NaA zeolite seeds used before membrane fabrication have to be synthesized using atrane precursors since the synthesized seeds are in the nano-size range which is suitable to be used as nuclei for growing membrane.

The resulting membrane also provides good performance in terms of both the total water flux and the separation factor. All results obtained from the NaA zeolite membranes synthesized using either atrane or commercial silica and alumina precursors are summarized in Table 2. As can be seen from Table 2, our NaA zeolite membranes showed good performance for separating water from water-ethanol mixture using pervaporation system, providing a separation factor higher than 10,000. Moreover, we also studied NaA zeolite membrane synthesis via electrophoretic technique and found that the membrane showed moderate performance for water-ethanol separation via pervaporation system. However, the obtained membrane from this technique showed better uniform and dense columnar NaA crystallayer than the other synthesis techniques (microwave and autoclave heating techniques). Thus, to obtain better morphology of the zeolite crystals and membrane, electrophoretic route could be the choice (Kuanchertchoo *et al.*, 2009).

Table 2. Performance of NaA zeolite membrane from different synthesis techniques and raw materials

	Synthesized technique	Total water flux (kg/m^2/h)	Separation factor	References
NaA zeolite membrane	Conventional heating technique, seeding technique with NaA zeolite crystal from atrane precursors	~ 2.82	> 10,000	Kunnakorn *et al.*, 2011
	Microwave heating technique, seeding technique with NaA zeolite crystal from atrane precursors	~ 0.8	> 10,000	Kunnakorn *et al.*, 2011
	Conventional heating technique, seeding technique with NaA zeolite crystal from commeicial silica and alumina precursors	~ 0.6	> 10,000	Ahn *et al.*,2006
	Hydrothermal synthesis	~1.67	> 10,000	Kondo *et al.*, 2010

In terms of catalyst technology, many types of zeolites are used as catalysts in many industrial applications, such as LTA-type, MFI-type, zeolite Y, ZSM-5, etc. The most common uses of zeolites in industrial applications are in catalytic cracking process, gas absorption, and water purification. The applications of the zeolites synthesized from atrane precursors were also investigated. For example, Fe-loaded MFI zeolite was used as catalyst for CO oxidation reactions, giving 62.3% of CO conversion at a temperature of 450 °C (Kritchayanon *et al.,* 2006). This result was coincident with works by Malero *et al.* and Lobree *et al.*, who synthesized Fe-loaded MFI zeolite using commercial precursors.

Ni-supported KH zeolite synthesized from atrane precursors was used as a catalyst for CO_2 reforming of methane (CH_4).As mentioned previously, the synthesized KH zeolite from atrane precursors contained flower and dog-bone shapes with little amount of disorder in the

KH zeolite morphology. The activity of both Ni/flower-shape KH and Ni/dog-bone-shape KH for CO_2 reforming showed a higher ratio of H_2/CO than the Ni/disordered KH zeolite. As compared to the conventional catalyst (γ-alumina) for CH_4 reforming, the Ni/synthesized KH zeolite showed a higher H_2/CO ratio than the Ni/γ-alumina. Moreover, the carbon deposition/a coke formation for Ni supported on both flower and dog-bone-shaped KH was lower than that for the disordered KH zeolite. The synthesized Ni-loaded KH zeolite also showed better performance for CH_4 reforming than the conventional catalyst.

Therefore, the synthesized Ni-loaded KH zeolite, due to its high hydrogen production, should be a promising catalyst to be used in fuel processors for fuel cell applications (Kaengsilalai *et al.,* 2007). VS-1 zeolite is another catalyst synthesized from atrane precursors and used for oxidative dehydrogenation of propane (Phiriyawirut *et al.*, 2009). The propane conversion increased with an increase in the vanadium content in VS-1 zeolite. Tetrahedral vanadium within the zeolite crystal structure is believed to be responsible for selective oxidative dehydrogenation, whereas extrinsic vanadium or polymeric species is responsible for the over-oxidation, selectively giving high carbon monoxide and carbon dioxide (Phiriyawirut *et al.*, 2009).

CONCLUSION

Many types of zeolites were successfully synthesized from atrane precursors (silatrane and alumatrane) via hydrothermal heating techniques. The synthetic zeolites from atrane precursors showed good morphologies and good properties. Various Si:Al ratios of atrane precursors generated different types of zeolites with different morphologies and properties. Synthesized zeolites from atrane precursors could be used in many applications, especially, in separation technology and catalyst application technology.

ACKNOWLEDGMENTS

The authors would like to thank the National Research Council of Thailand, the Center for Petroleum, Petrochemicals and Advanced Materials (PPAM), Chulalongkorn University, Thailand, the Reverse Brain Drain Project, National Science and Technology Development Agency, Ministry of Science and Technology (Thailand). Special thanks goes to Mr. John M. Jackson for English proof-reading.

REFERENCES

Ahn, H., Lee, H., Lee, S.B., and Lee, Y., (2006) Pervaporation of an aqueous ethanol solution through hydrophilic zeolite membranes, *Desalination*, 193, 244-251.

Anchell, J.L., White, J.C., Thompson, M.R., and Hess, A.C., (1994) An AB-inito periodic hartree-fock study of group IA cations in ANA-type zeolite, *Journal of Physical Chemistry*, 98, 4463.

Baccouch, A., Srasra, E., and El Maaoui, M., (1998) Preparationo f Na-P1 and sodalite octahydrate zeolites from Interstratified Illitesmectite, *Applied Clay Science,* 13, 255.

Davis, M.E., and Lobo, R.F., (1992) Zeolite and molecular sieve synthesis, *Chemistry of Materials,* 4,756-768.

De-Moor, P., Beelen, T.P.M., and Santen, R.A., (1999) In situ observation of nucleation and crystal growth in zeolite synthesis. A small-angle X-ray scattering investigation on Si-TPA-MFI, *Journal of Physical Chemistry B.,* 103, 1639.

De-Moor, P., Beelen, T.P.M., Santen, R.A., Beck, L.W., and Davis, M.E., (2000) Si-MFI crystallization using a "Dimer" and "Trimer" of TPA studied with small-angle X-ray scattering, *Journal of Physical Chemistry B.,* 104, 7600.

Dimitrijevic, R., and Dondur, V., (1995) Synthesis and characterization of $KalSiO_4$ polymorphs on the SiO_2-$KAlO_2$ join 2 the end-member of ANA type of zeolite framework, *Journal of Solid State Chemistry,* 115, 214.

Feigen, E.J.P., Martens, J.A., and Jacobs, P.A., (1999) Hydrothermal zeolite synthesis, *In Preparation of Solid Catalysts, Wiley-VCH: Weinheim.*

Huang, L.M., Chen, X.Y., and Li, Q.Z., (2001) Synthesis of microporous molecular sieve by surfactant decomposition, *Journal of Materials Chemistry*, 11/2, 610.

Jones, D.A., Lelyveld, T.P., Mavrofidis, S.D., Kingman, S.W., and Miles, N.J., (2002) Microwave heating applications in environmental engineering (a review), *Resources, Conservation and Recycling,* 34, 75-90.

Kaengsilalai, A., Luengnaruemitchai, A., Jitkarnka, S., and Wongkasemjit, S., (2007) Potential of Ni supported o KH zeolite catalysts for carbon dioxide reforming of methane, *Journal of Power Sources,* 165, 347-352.

Kondo, M., and Kita, H., (2010) Permeation mechanism through zeolite NaA and T-type membranes for practical dehydration of organic solvents, *Journal of Membrane Science,* 361, 223-231.

Kritchayanon, N., Thanaboodeekij, N., Jitkarnka, S., Jamieson, A.M., and Wongkasemjit, S., (2006) Synthesis of Fe-loaded MFI zeolite using silatrane as precursor and its CO activity, *Applied Organometallic Chemistry,* 20, 155-160.

Kuanchertchoo, N., Kulprathipanja, S., Aungkavattana, P., Atong, D., Hemra, K., Rirksomboon, T., and Wongkasemjit, S., (2006) Preparation of uniform and nano-sized NaA zeolite using silatrane and alumatrane precursors, *Applied Organometallic Chemistry,* 20, 775-783.

Kuanchertchoo, N., Kunnakorn, D., Kulprathipanja, S., Aungkavattana, P., Atong, D, Hemra, K., Rirksomboon, T., and Wongkasemjit, S., (2009) Oriented NaA zeolite membrane formation and performance using seeding and electrophoretic techniques, *Advanced Metal and Meal Oxide Technology, Transworld Research Network.*

Kunnakorn, D., Rirksomboon, T., Aungkavattana, P., Kuanchertchoo, N., Atong, D., Kulprathipanja, S., and Wongkasemjit, S., (2011) Performance of sodium A zeolite membranes synthesized via microwave and autoclave techniques for water-ethanol separation: Recycle-continuous pervaporation process, *Desalination,* 269, 78-83.

Kunnakorn, D., Rirksomboon, T., Aungkavattana, P., Kuanchertchoo, N., Atong, D., Kulprathipanja, S., and Wongkasemjit, S., (2011) Optimization of synthesis time for high performance of NaA zeolite membranes synthesized via autoclave for water-ethanol separation, *Desalination,* 280, 259-265.

Phiriyawirut, P., Jamieson, A.M., and Wongkasemjit, S., (2005) VS-1 zeolite synthesized directly from silatrane, *Microporous and Mesoporous Materials,* 77, 203-213.

Phiriyawirut, P., Jitkarnka, S., Jamieson, A.M., and Wongkasemjit, S., (2009) Oxidative dehydrogenation of propane over home-made VS-1 zeolite, *Advanced Metal and Meal Oxide Technology, Transworld Research Network.*

Phiriyawirut, P., Magaraphan, R., Jamieson, A.M., and Wongkasemjit, S., (2003) MFI zeolite synthesis directly from silatrane via sol-gel process and microwave technique, *Materials Science and Engineering A,* 361, 147-154.

Phiriyawirut, P., Magaraphan, R., Jamieson, A.M., and Wongkasemjit, S., (2003) Morphology study of MFI zeolite synthesized directly from silatrane andalumatrane via the sol-gel process and microwave heating, *Microporous and Mesoporous Materials,* 64, 83-93.

Rees, L.V.C., and Chandrasekhar, S., (1993) Hydrothermal reaction of kaolinite in presence of fluride ions at Ph-Less-Than-10, *Zeolites,* 13,535.

Sathupunya, M. Gulari, E., and Wongkasemjit, S., (2004) Microwave preparation of Li-zeolite directly from alumatrane and silatrane, *Materials Chemistry and Physics,* 83, 89-95.

Sathupunya, M., Gulari, E., and Wongkasemjit, S., (2002) ANA and GIS zeolite synthesis directly from alumatrane and silatrane by sol-gel process and microwave technique, *Journal of the European Ceramic Society,* 22, 2305-2314.

Sathupunya, M., Gulari, E., and Wongkasemjit, S., (2003) Na-A (LTA) zeolite synthesis directly from alumatrane and silatrane by sol-gel microwave techniques, *Journal of the European Ceramic Society,* 23, 1293-1303.

Sathupunya, M., Gulari, E., Jamieson, A., and Wongkasemjit, S., (2004) Microwave-assisted preparation of zeolite K-H from alumatrane and silatrane, *Microporous and Mesoporous materials,* 69, 157-164.

Singh, R., and Putta, P.K., (1998) Stabilization of natural Faujasite zeolite: possible role of alkaline earth metal ions, *Microporous and Mesoporous Materials,* 21, 103-109.

Utchariyajit, K., and Wongkasemjit, S., (2010) Effect of synthesis parameters on mesoporous SAPO-5 with AFI-type formation via microwave radiation using alumatrane and silatrane precursors, *Microporous and Mesoporous Materials,* 135, 116-123.

Wiersema, G.S., and Thompson, R.W., (1996) Nucleation and crystal growth of analcimefrom clear aluminolicate solution, *Journal of Materials Chemistry,* 6/10, 1693.

Yokomori, Y., and Idaka, S., (1998) The crystal structure of analcime, *Microporous and Mesoporous Materials,* 21/4-6, 365.

In: Zeolites: Synthesis, Chemistry and Applications
Editors: Moisey K. Andreyev and Olya L. Zubkov
ISBN: 978-1-61942-861-4

Chapter 9

APPLICATION OF ZEOLITE CONTAINING ROCKS IN BERRY CROP GROWING

M. N. Kuznetsov*[*], *E. V. Leonicheva, S. M. Motyleva, T. A. Roeva and L. I. Leont'eva

SSI All Russian Research Institute of Horticultural breeding, Orel, Russia

ABSTRACT

The possibilities of application of zeolite-containing rock from Khotynetz deposit (Orel region, Russia) have been investigated in the field and laboratory experiments for the purpose of microelement composition improvement of berry crops: raspberry, gooseberry and black currants. Zeolite (2…30 t/ha) was applied in the grey forest soil, which had gross content of heavy metals 2…6 times higher than background level.

The application of zeolite-containing rock led to the lowering of the contents of mobile forms Pb, Ni, Zn and Cu in the soil by 30…50% depending on the metal and zeolite dose. The reduction of heavy metal compound supply within the reach for plants changed the leaf and fruit microelement composition of berry crops. The character of changes depended on a zeolite dose, plant species, metal nature and its physiological significance. The contents of toxic Pb and Ni in fruit were reduced by 20…70% depending on the crop. The lowing of mobile forms Zn and Cu in the soil by 30…50% did not lead to a deficiency of those biogenic elements for berry crops.

The application of mineral fertilizer effected the action of the zeolite-containing rock as a deactivator of heavy metals in the "soil-plant" system. On $N_{90}P_{90}K_{90}$ background the content of Cu in black currant berries was higher, but the contents of Pb, Ni and Zn were lower than in the variants without fertilizers.

The application of the zeolite-containing rock increases the highest possible hygroscopic soil humidity and accordingly the wilting point of plants. Zeolite application in doses upwards of 16 t/ha positively reduces the supply of productive moisture that may lead to the lowing of berry crop productivity. Possible deterioration of plant water provision is a limiting factor in a choice of doses of zeolite-containing rocks for the purpose of solving other land-reclamation measures.

[*] E-mail: info@vniispk.ru.

Keywords: Zeolite contenting rock, heavy metals, berry crops, soil medium fertility, supply of productive moisture

INTRODUCTION

An application of natural mineral absorbents is considered to be a promising method of up-to-date ecologicalized agriculture. Natural zeolite is one of the most popular absorbents suggested for the improvement of the circuit of nutrient elements and detoxication of soil pollution. Russia possesses large resources of zeolite containing rocks (ZCR). Their deposits are both in the European part of the country and Siberia. The investigations concerning the possibility of the application of a zeolite containing material in the plant-growing have been carrying out for the last 30 years. By nowadays a sufficient amount of data has been collected on the application of zeolites from different deposits for various plant cultures [1]. For all this, a scientific-methodical basis of selection of optimum ZCR dozes has insufficiently been elaborated taking into account both various soil-and-climate conditions and specific characters of ZCR in different deposits.

The goal of our investigation was to study a possibility of application of zeolite containing rocks from Khotynetz deposit (Orel region) for the purpose of reducing of heavy metal (HM) entering berry plants when growing berry crops on the grey forest soil in real conditions of heightened HM contents in the soil.

MATERIAL AND METHODS

The accumulation of micro elements in leaves and fruit of berry plants (black currant variety “Kipiana”; raspberry variety “Sputnitza”; gooseberry selection ELS24-15-21) with the application of zeolite containing rock was studied in the micro field experiments started in 2004 in orchards of the All Russian Fruit Crop Breeding Research Institute (Orel, VNIISPK).

Agricultural lands of VNIISPK have been used in fruit-growing for more than 50 years. HM contents in the soil and fruit production have been noticed to exceed the admitted levels for the last decades [2, 3]. In the soil of our field experiments the gross contents of HM exceeded the regional background level stated for grey forest soils of Orel region: Pb and Ni – by 2 times, Zn – by 3 times, Cu by 6 times, respectively (Table 1). MAC_{gross} for Cu was also exceeded.

ZCR from the Khotynetz deposit of Orel region was used in the experiments. The mineralogical composition of ZCR was the following: clinoptilolite 34%, mordenite 4%, crystoballite 28%, calcite 5%, montmorillonite 12%, mica 1%, and quartz 16%. The size of ameliorant particles was 0, 01 – 0, 005 mm. The schemes of the experiments included increasing dozes of the zeolite containing rocks (2 – 30 t/ha) on the background $N_{90}P_{90}K_{90}$. ZCR was used once when planting in the soil layer 0 – 20 sm.

Table 1. HM contents in the soil of experimental plot of VNIISPK used in fruit-growing for more than 50 years (mg/kg of soil)

Index	Pb		Ni		Zn		Cu	
	0… 20 sm	20… 40 sm	0… 20 sm	20… 40 sm	0… 20 sm	20… 40 sm	0… 20 sm	20… 40 sm
Gross content	26,0	20,7	29,0	31,5	42,5	45,5	72,6	58,0
MAC_{gross} [4]	65,0		40,0		110		66,0	
$Background_{gross.}$ [5]	12,7		15,0		13,4		12,3	
Content of mobile compounds	0,08	0,05	0,28	0,23	15,0	13,0	0,56	0,45
$MAC_{mob..}$ [6]	6,0		4,0		23,0		3,0	

The soil of the experiment plot was grey forest medium loamy. The agrochemical characteristics of the soil were the following: pH_{KCL}=5,4±0,15; mobile compounds of phosphorus 43,78±3,7 mg/100g of soil; mobile compounds of potassium 29,7±5,3 mg/100g; exchangeable Ca^{2+} 15,7±0,4 mmol+/100 g; exchangeable Mg^{2+} 5,8±0,5 mmol+/100 g; humus contents 3,7±0,14%.

Table 2. Contents of lead mobile forms in the soil with ZCR application in the experiment with gooseberry (ppm)

Variants (A)	Years of investigation (B)		$\overline{X}$ according to factor A $LSD_{0,05}$= 0,03
	2005	2006	
Layer 0 – 20 sm			
Control	0,15	0,19	0,17
$N_{90}P_{90}K_{90}$	0,27	0,26	0,26
$N_{90}P_{90}K_{90}$ + ZCR 8 t/ha	0,07	0,06	0,06
$N_{90}P_{90}K_{90}$ + ZCR 16 t/ha	0,10	0,08	0,09
$N_{90}P_{90}K_{90}$ + ZCR 24 t/ha	0,08	0,08	0,08
$\overline{X}$ according to factor B $LSD_{0,05}$= 0,03	0,13	0,13	
Layer 20 – 40 sm			$\overline{X}$ according to factor A $LSD_{0,05}$= 0,02
Control	0,23	0,23	0,23
$N_{90}P_{90}K_{90}$	0,21	0,21	0,21
$N_{90}P_{90}K_{90}$ + ZCR 8 t/ha	0,19	0,19	0,19
$N_{90}P_{90}K_{90}$ + ZCR 16 t/ha	0,19	0,17	0,19
$N_{90}P_{90}K_{90}$ + ZCR 24 t/ha	0,19	0,17	0,19
$\overline{X}$ according to factor B $LSD_{0,05}$= 0,03	0,20	0,19	

Table 3. Contents of nickel mobile forms in the soil with ZCR application in the experiment with gooseberry (ppm)

Variants (A)	Years of investigation (B)		$\overline{X}$ according to factor A $LSD_{0,05}$= 0,16
	2005	2006	
Layer 0 – 20 sm			
Control	0,64	0,61	0,63
$N_{90}P_{90}K_{90}$	1,18	1,10	1,11
$N_{90}P_{90}K_{90}$ + ZCR 8 t/ha	0,58	0,56	0,67
$N_{90}P_{90}K_{90}$ + ZCR 16 t/ha	0,79	0,67	0,83
$N_{90}P_{90}K_{90}$ + ZCR 24 t/ha	0,83	0,64	0,81
$\overline{X}$ according to factor B $LSD_{0,05}$= 0,11	0,80	0,71	
Layer 20 – 40 sm			$\overline{X}$ according to factor A $LSD_{0,05}$= 0,15
Control	0,86	0,91	0,79
$N_{90}P_{90}K_{90}$	1,12	1,10	1,07
$N_{90}P_{90}K_{90}$ + ZCR 8 t/ha	0,58	0,56	0,61
$N_{90}P_{90}K_{90}$ + ZCR 16 t/ha	0,53	0,52	0,60
$N_{90}P_{90}K_{90}$ + ZCR 24 t/ha	0,94	0,84	0,96
$\overline{X}$ according to factor B $LSD_{0,05}$= 0,09	0,81	0,78	

The HM contents (Pb, Ni, Zn and Cu) in the plant samples were determined by high performance liquid chromatography. Heavy metal gross content in the soil was determined by x-ray fluorescence spectroscopy, the content of their mobile compounds was determined by HELC in the extract with acetate-ammonium buffer (pH 4,8).

Agrochemical indices, water content and maximum hydroscopic moisture of the soil were determined according to the standard methods. The results were processed by statistical methods in Microsoft Excel.

RESULTS AND DISCUSSION

The dynamics of mobile forms of Pb, Ni, Zn and Cu in the soil of micro field experiments with gooseberry during two vegetative periods is shown in Tables 2 – 5. The contents of mobile forms of all studied elements were positively reduced with the zeolite application both in the layer 0-20 sm where the ameliorant was applied, and also in the lower layer 20-40 sm. That could be related to a vertical dislocation of mobile HM compounds in the soil profile under the ascending movement of soil water owing to capillary forces.

The decline of HM mobility occurred in the first year of ZCR application (2005). There were no reliable differences between the contents of mobile forms Pb, Ni, Zn and Cu in the soil in the first and second periods of the vegetation (Tables 2-5).

Table 4. Contents of zinc mobile forms in the soil with ZCR application in the experiment with gooseberry (ppm)

Variants (A)	Years of investigation (B)		$\overline{X}$ according to factor A $LSD_{0,05}$= 0,54
	2005	2006	
Layer 0 – 20 sm			
Control	5,95	5,73	5,84
$N_{90}P_{90}K_{90}$	3,16	4,34	3,75
$N_{90}P_{90}K_{90}$ + ZCR 8 t/ha	3,14	3,48	3,31
$N_{90}P_{90}K_{90}$ + ZCR 16 t/ha	6,12	5,90	6,01
$N_{90}P_{90}K_{90}$ + ZCR 24 t/ha	2,56	2,11	2,33
$\overline{X}$ according to factor B $LSD_{0,05}$= 0,42	4,18	4,31	
Layer 20 – 40 sm			$\overline{X}$ according to factor A $LSD_{0,05}$= 1,5
Control	5,06	5,14	5,10
$N_{90}P_{90}K_{90}$	4,22	4,91	4,56
$N_{90}P_{90}K_{90}$ + ZCR 8 t/ha	3,54	3,12	3,33
$N_{90}P_{90}K_{90}$ + ZCR 16 t/ha	4,53	3,57	4,05
$N_{90}P_{90}K_{90}$ + ZCR 24 t/ha	1,82	1,91	1,86
$\overline{X}$ according to factor B $LSD_{0,05}$= 0,95	3,83	4,13	

Table 5. Contents of copper mobile forms in the soil with ZCR application in the experiment with gooseberry (ppm)

Variants (A)	Years of investigation (B)		$\overline{X}$ according to factor A $LSD_{0,05}$= 0,30
	2005	2006	
Layer 0 – 20 sm			
Control	1,492	1,463	1,477
$N_{90}P_{90}K_{90}$	1,023	1,272	1,148
$N_{90}P_{90}K_{90}$ + ZCR 8 t/ha	1,301	1,213	1,257
$N_{90}P_{90}K_{90}$ + ZCR 16 t/ha	1,487	1,259	1,373
$N_{90}P_{90}K_{90}$ + ZCR 24 t/ha	1,375	1,137	1,256
$\overline{X}$ according to factor B $LSD_{0,05}$= 0,23	1,335	1,268	
Layer 20 – 40 sm			$\overline{X}$ according to factor A $LSD_{0,05}$= 0,32
Control	1,309	1,311	1,310
$N_{90}P_{90}K_{90}$	1,113	1,255	1,184
$N_{90}P_{90}K_{90}$ + ZCR 8 t/ha	1,380	1,301	1,341
$N_{90}P_{90}K_{90}$ + ZCR 16 t/ha	0,866	0,881	0,873
$N_{90}P_{90}K_{90}$ + ZCR 24 t/ha	1,667	1,363	1,531
$\overline{X}$ according to factor B $LSD_{0,05}$= 0,20	1,267	1,228	

Two probable reasons of reducing of the content of mobile HM may be designated. The first reason is the absorption of metal cations directly by zeolite. The second reason is the intensification of absorption and cation exchange process as a result of increase in total surface of some particles. The increase of aggregate part with size less than 2 mm under ZCR application is shown in Figure 1. This may indirectly indicate the zeolite peptizing influence on the soil colloids which fasten together small structural separate parts.

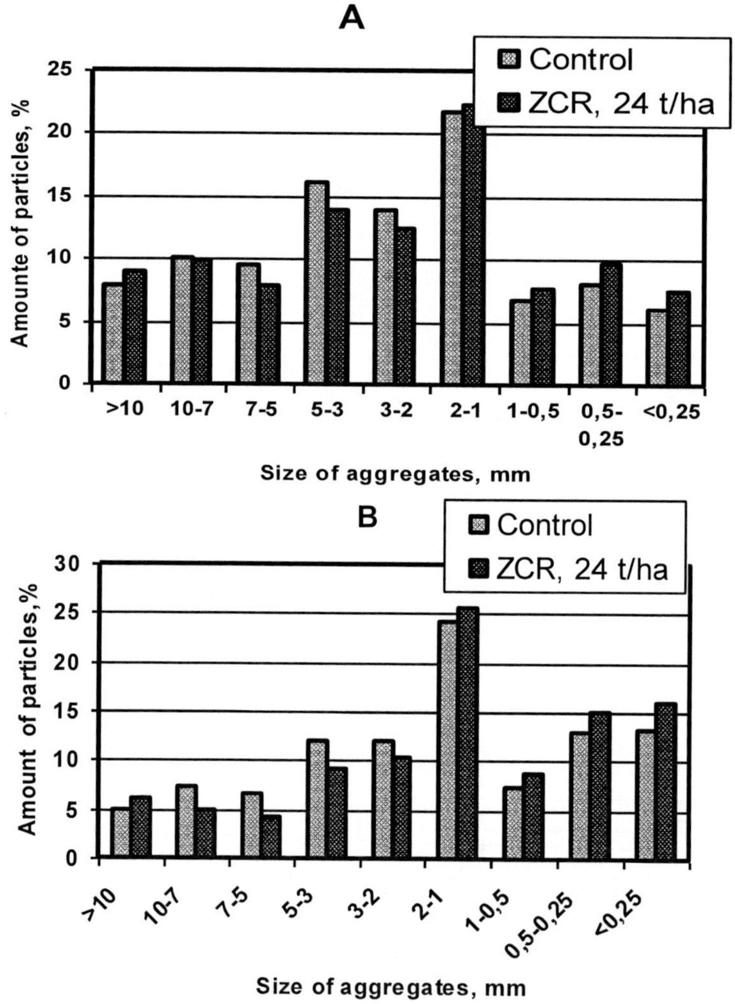

Figure 1. Analysis of soil macro structure of experiment plots (according to Savinov method): A – gooseberry, B – black currant.

In all our investigations the application of ZCR led to the content reduce of mobile forms Pb, Ni, Zn and Cu by 30 – 50 % depending on a metal kind and ameliorant doze. The contents of mobile HM forms were reduced in proportion to the increase of ZCR doze up to 8-10 t/ha, but further rise of dozes did not proved a significant reduce of HM mobility in conditions of our experiments (Figure 2).

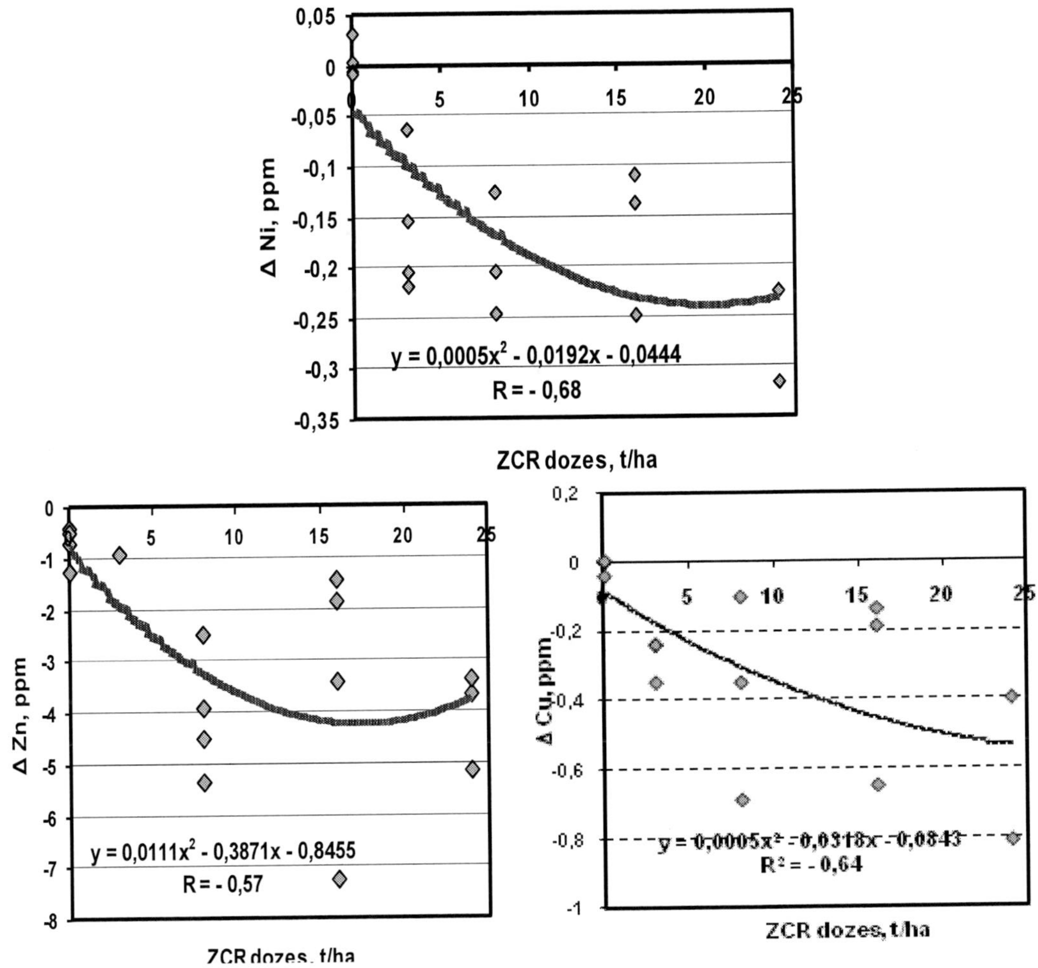

Figure 2. Theoretical curves of HM content change with application of increasing ZCR dozes of the experiment with black currant.
*ΔHM (ΔPb, ΔNi, ΔCu and ΔZn, respectively, ppm) – difference between final and initial contents of mobile forms of each element.

The efficiency of ZCR as an inactivator of HM in the system "soil-plant" in berry crop cultivation significantly depended on the agrochemical background, on which the ameliorant was applied, and chemical features of the element. Figure 3 demonstrates the changes of the content of HM mobile forms (ΔPb, ΔNi, ΔCu and ΔZn) in comparison with the initial level 2 years after ZCR application. The content of mobile Ni and Cu was reduced more in the variants with ZCR application without mineral fertilizers than on $N_{90}P_{90}K_{90}$. The agrobackground influence on the efficiency of ZCR as an inactivator of mobile Pb and Zn depended upon the doze of the ameliorant. The significant influence of the agrobackground on the inactivation of mobile HM compounds by zeolite might be bound up with changes of balance conditions in the soil absorption complex. As a result, the competition of cations for absorbing places was intensified.

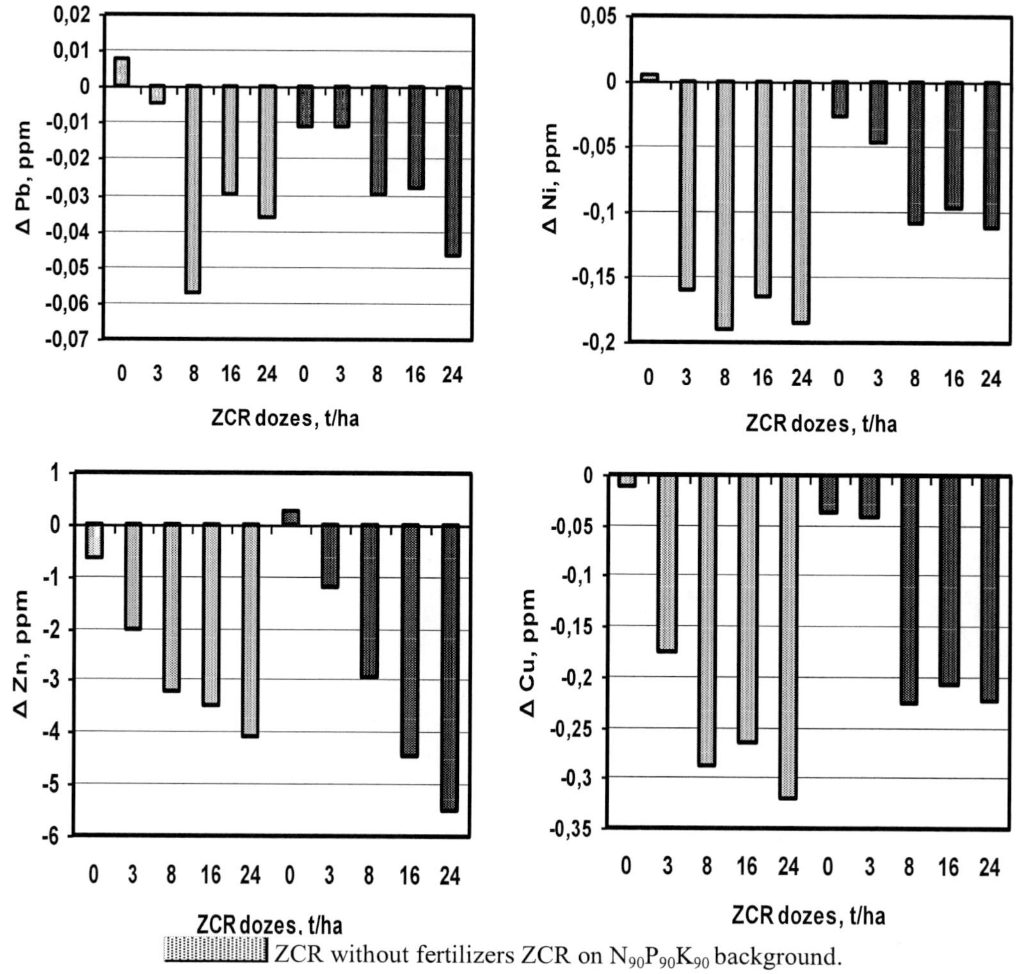

Figure 3. Agrobackground influence on HM mobility in the soil with application of increasing ZCR dozes (2004, 2006).

To understand the affect of the zeolite on HM entering a plant, it is necessary alongside with the estimation of resources of mobile forms in the soil to determine the content of HM in the plant itself, a leaf of which is the most important indicative organ.

HM content in different organs of a plant is determined by a complex of factors: the total content of accessible forms of the element for the plant in the soil; the soil-chemical conditions, which influence the concentration of the element in the soil solution; physiological importance of the element for plant determining mechanisms of the element entering and its transportation to the vegetative and generative organs.

The reducing of reserves of microelement compounds accessible for plants affected the microelement composition of berry crop leaves. It is known that microelement entering to the plants may be passive (according to a concentration gradient) and active (against the gradient of concentration with energy input). The presence of various ways of HM entering to the plants suggests the presence of two leading factors of formation of the element chemical composition of plants: genetic and ecological ones [7].

ZCR application in increasing dozes led to really lower content of Pb and Ni in leaves of black currant, raspberry and gooseberry (Figure 4). These toxic elements generally enter plants by a passive way, and in all studied cultures the contents of Pb and Ni in leaves change in compliance with the content of accessible forms in the soil. The indices of the correlation between Ni content in leaves and Ni mobile forms in the soil were the following: 0, 7 for black currant; 0, 9 for raspberry and 0, 5 for gooseberry, respectively. The following correlation between Pb content in leaves and soil was observed: r_{Pb} = 0,5 for raspberry and r_{Pb} = 0,82 for gooseberry.

ZCR doze increasing over 2 t/ha in the experiment with raspberry and over 8 t/ha in the experiments with black currant and gooseberry did not cause the subsequent significant reduce of Pb and Ni contents in leaves (Figure 4).

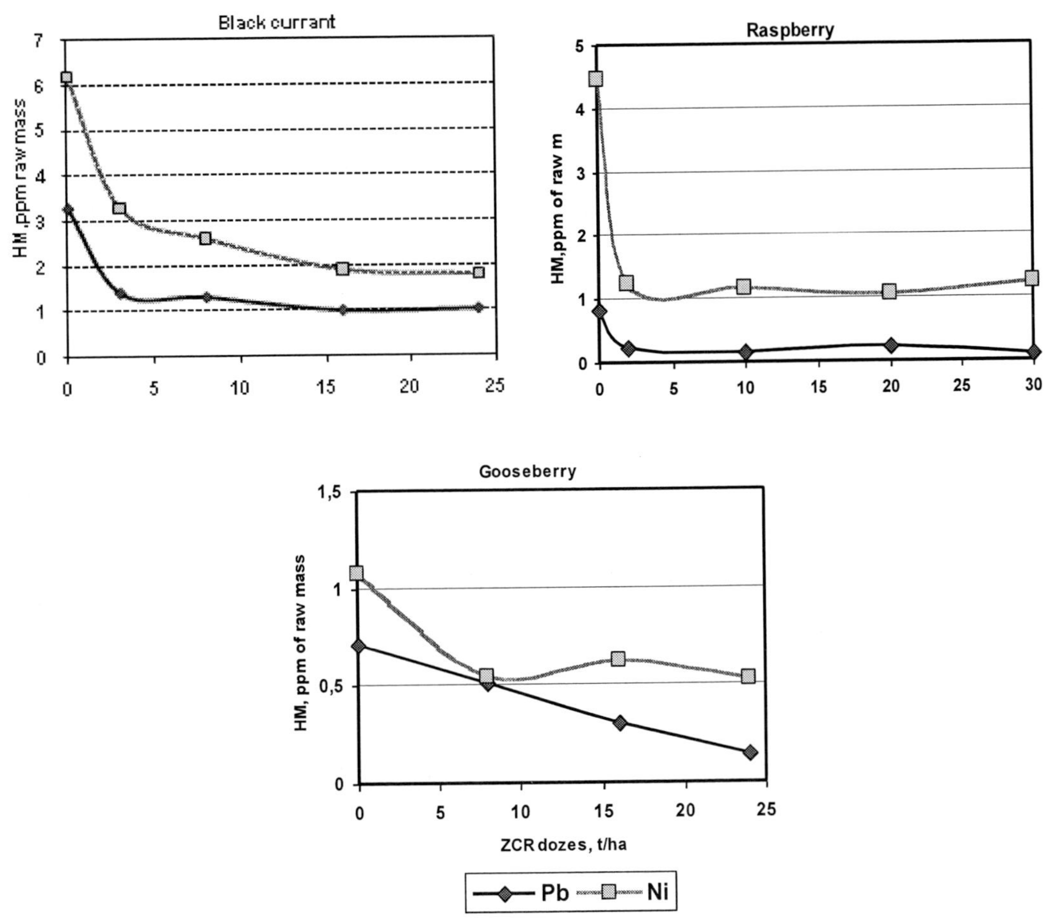

Figure 4. Pb and Ni contents in leaves of berry crops (mg/kg of raw mass) with application of increasing dozes of ZCR (2005-2007).

The accumulation of biogenic elements Zn and Cu in leaves of berry crops differed from the accumulation of toxic Pb and Ni. If Pb and Ni in the leaves of all studied crops submitted to a common regularity (Figure 4), the change of Zn and Cu contents in leaves with the application of higher dozes of ZCR was specific for each crop (Figure 5).

Low contents of biogenic elements in leaves were typical for raspberry and black currant with the application of small dozes of ZCR (2 and 3 t/ha, respectively). With the subsequent increasing of dozes the content of Zn and Cu increased in the leaves of those two crops, that was especially noticeable in raspberry, and with ZCR application 20-30 t/ha it could exceed the control (Figure 5).

The maximum content of Zn in gooseberry leaves was observed with ZCR application 16 t/ha, the maximum content of Cu was in the variant 8 t/ha.

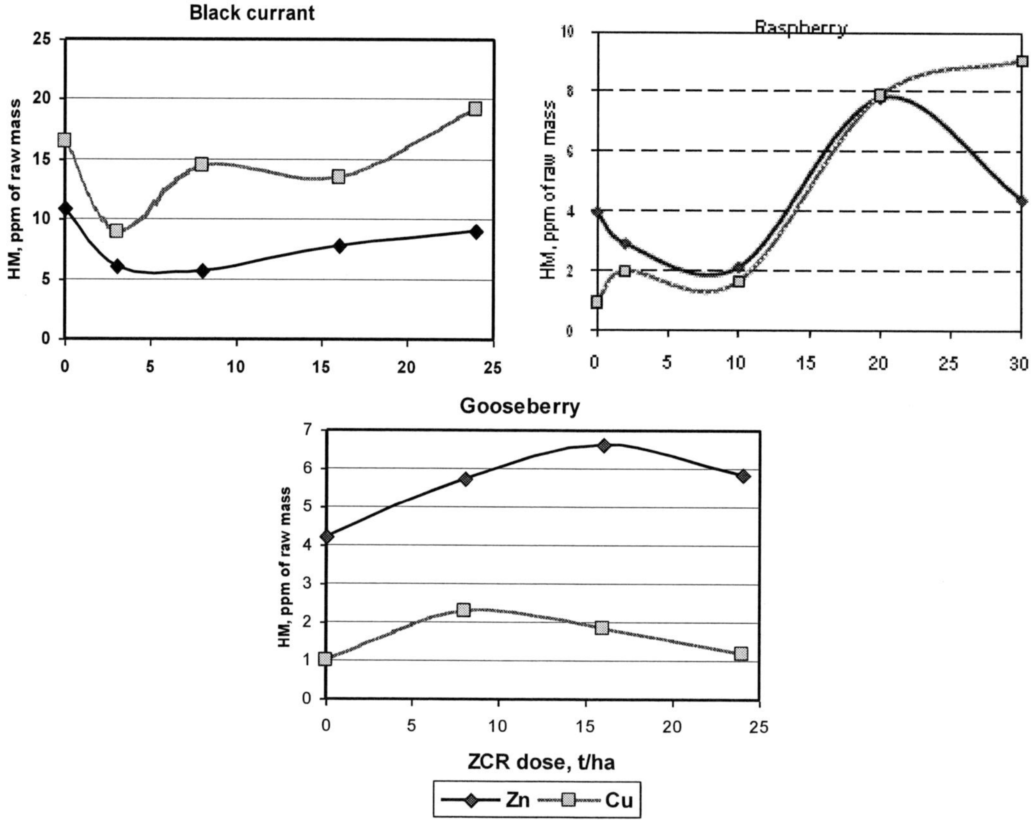

Figure 5. Zn and Cu contents in leaves of berry crops (mg/kg of raw mass) with application of increasing dozes of ZCR (2005-2007).

The increasing of biogenic microelement content in leaves of berry crops while the reserve of their excessive compounds in the soil had reduced might be indicative of using the mechanisms of active metabolic uptake by plants.

Thus, the content of toxic microelements in leaves of berry crops was mainly determined by an ecological factor – the content of excessive forms of those elements in the soil. Going of biogenic microelements to the leaves of berry plants to a considerable extent reflected the action of genetic controlling. The selective specific absorption of Zn and Cu ions was reserved in spite of exceeding regional background concentration of Zn in the soil by 3 times and of Cu by 12 times. The reducing of the content of mobile forms of Zn and Cu in the soil by 30-50 % with application of increasing dozes of ZCR did not lead to the deficiency of those biogenic elements.

The content of metals in berry crop fruit in conditions of our investigations did not exceed standards admitted in Russia. At the same time, the presence of toxic Pb and Ni in berries and the change of Zn and Cu part in comparison with the medium background level [8] were indicative of the effect of the environmental anthropogenic factors on the mineral composition (Table 6).

The forming of the microelement fruit composition was also under the influence of both ecological and genetic factors.

Table 6. Microelement content in fruit of berry crops (ppm of raw mass) with application of increasing dozes of ZCR (2005-2007)

Berry crop	Variants	Pb	Ni	Zn	Cu
Black currant	Control	0,025	0,15	0,48	0,26
	$N_{90}P_{90}K_{90}$	0,021	0,20	0,58	0,35
	$N_{90}P_{90}K_{90}$ + ZCR 3 t/ha	0,018	0,07	0,15	0,18
	$N_{90}P_{90}K_{90}$ + ZCR 8 t/ha	0,025	0,05	0,23	0,20
	$N_{90}P_{90}K_{90}$ + ZCR 16 t/ha	0,023	0,04	0,34	0,25
	$N_{90}P_{90}K_{90}$ + ZCR 24 t/ha	0,015	0,15	0,39	0,40
	$LSD_{0,05}$	0,01	0,10	0,09	0,10
	Medium background level [8]	-	-	0,8	0,8
Raspberry	Control	0,016	0,35	1,85	0,26
	$N_{90}P_{90}K_{90}$	0,014	0,34	1,35	0,24
	$N_{90}P_{90}K_{90}$ + ZCR 2 t/ha	0,010	0,26	1,49	0,19
	$N_{90}P_{90}K_{90}$ + ZCR 10 t/ha	0,007	0,21	1,07	0,17
	$N_{90}P_{90}K_{90}$ + ZCR 20 t/ha	0,003	0,21	1,23	0,18
	$N_{90}P_{90}K_{90}$ + ZCR 30 t/ha	0,003	0,19	1,64	0,18
	$LSD_{0,05}$	0,005	0,14	0,11	0,03
	Medium background level [8]	-	0,06	2,0	1,7
Gooseberry	Control	0,053	0,10	0,84	0,16
	$N_{90}P_{90}K_{90}$	0,051	0,15	0,80	0,13
	$N_{90}P_{90}K_{90}$ + ZCR 8 t/ha	0,027	0,08	0,21	0,09
	$N_{90}P_{90}K_{90}$ + ZCR 16 t/ha	0,027	0,04	0,19	0,26
	$N_{90}P_{90}K_{90}$ + ZCR 24 t/ha	0,006	0,05	0,05	0,13
	$LSD_{0,05}$	0,010	0,01	0,13	0,08
	Medium background level [8]	-	-	0,9	1,3
MPC for fruits and berries [9]		0,4	0,5	10,0	5,0

Pb and Ni contents in fruit with ZCR application were reduced by 20-70% depending on a crop (Table 6). For all this, the appropriateness of Pb and Ni going to fruit of berry crops was similar to those elements going to leaves (Figure 4, Table 6). That was confirmed by a reliable correlation between Pb and Ni contents in fruit and leaves of raspberry (r_{Pb} = 0,83 and r_{Ni} = 0,76), gooseberry (r_{Pb} = 0,96 and r_{Ni} = 0,94) and black currant (r_{Ni} = 0,77).

The change of Zn and Cu contents in fruit of berry crops with the application of increasing ZCR dozes was specific for each crop (Table 6).

The appropriateness of Zn and Cu going to berries of black currant was similar to Zn and Cu going to leaves (Figure 5, Table 6). The indices of the correlation between the content of microelements in leaves and berries were the following: $r_{Cu} = 0,\ 8$; $r_{Zn} = 0,5$.

Zn and Cu concentration in black currant berries was also connected with the content of mobile forms of those elements in the soil. The reliable indices of the correlation between Zn and Cu contents in the soil and berries were the following: $r_{Zn} = 0,\ 78$; $r_{Cu} = 0,\ 73$. Thus, the accumulation of biogenic microelements in black currant berries occurred with the participation of both genetic factor and ecological factor.

In raspberry berries the minimal content of Zn and Cu was noticed in the variant with application of ZCR 10 t/ha on the background $N_{90}P_{90}K_{90}$. With the application of higher dozes of ameliorant the content of Zn and Cu in fruit increased, though not so intensively as in leaves.

Gooseberry berries in the variants with ZCR application contented reliably less Zn than in the control variant. A high positive correlation (r = 0,91 - 0,94) was noticed between the content of mobile forms of Zn in the soil layer 20-40 sm and the content of that element in gooseberry berries.

With ZCR application 16 t/ha on $N_{90}P_{90}K_{90}$ background the content of Cu in fruit of gooseberry was reliably higher relative to the background. In other variants gooseberry fruit contented Cu on the level of the background and control. That is indicative of mainly genetic controlling of Cu entering gooseberry fruit in the environmental conditions of our experiments.

The observed specific peculiarity of Zn and Cu going to fruit and leaves of raspberry, gooseberry and black currant shows the primary role of a genetic factor in the nutrient supply of berry crops with vitally important microelements. The reduce of the content of mobile Zn and Cu forms by 30 – 50% with the application of increasing dozes of ZCR did not lead to the deficiency of those biogenic elements.

It is known that the application of mineral fertilizers is an efficient method of enforcing of physiological barrier functions of plants with respect to HM on the border "root – over ground organs" [11]. ZCR influence on Pb, Ni, Zn and Cu going to black currant plants significantly depended upon the agrochemical background.

The content of toxic elements of Pb and Ni in black currant leaves was considerably less when ZCR was applied on the fertilized background. With the application of ZCR on the background of a full mineral fertilizer the content of biogenic Zn and Cu in leaves was reducing in the variants with dozes 3 and 8 t/ha and increasing with subsequent doze increasing.

The peculiarities of the microelement composition of black currant berries grown with application of ZCR on different agrobackground are shown in Figure 6. The content of Pb in berries on the background of mineral fertilizers was lower in all variants of the experiment than on the unfertilized background and did not reliably change with ZCR application. On the mineral agrobackground the reducing of Ni content under the effect of ZCR was more expressed with 3...16 t/ha application. In the variant 24 t/ha black currant berries on $N_{90}P_{90}K_{90}$ background contented more Ni than on the unfertilized background. With ZCR application in dozes 3...24 t/ha on $N_{90}P_{90}K_{90}$ background the content of Cu in black currant berries was higher and Zn content was lower than in the variants with similar dozes of ZCR without fertilizers.

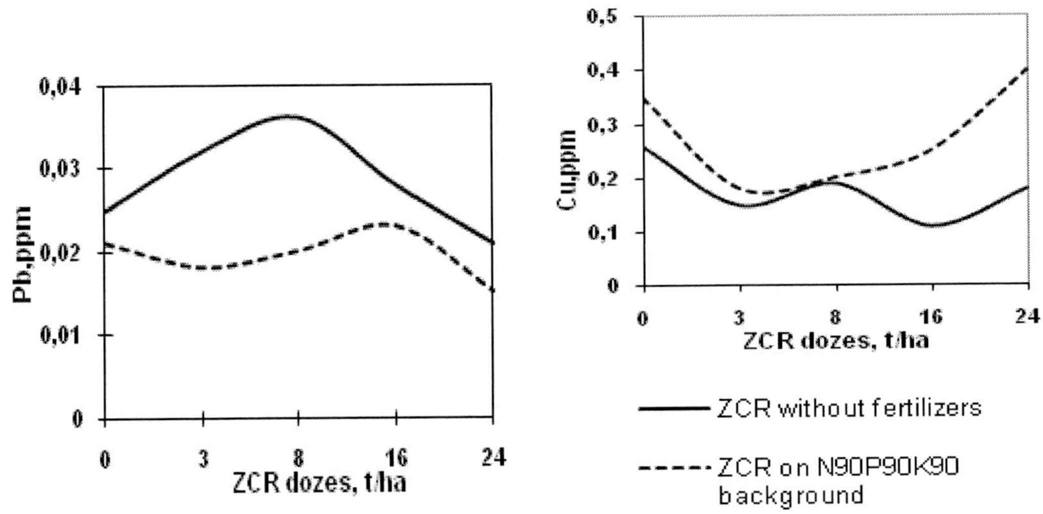

Figure 6. Agrobackground influence on HM content in black currant berries with application of increasing dozes of ZCR (2006-2007).

Zeolite influence upon the water regime in the system "soil – plant" may become a limiting factor when zeolite containing rocks are used for the purpose of optimization of mineral nutrition of plants and as inactivators of heavy metals.

It is known that minerals of zeolite group possess a high absorption capacity relative to water. Therefore, it may be suggested that water fixed with a zeolite containing rock is strongly retained. The content of strongly retained water is characterized with an index of maximum hygroscopic humidity (Wm.h., %). Our laboratory experiment revealed a linear dependence of maximum hygroscopic humidity value on a ZCR doze (Figure 7)

Humidity of plant withering increases in proportion to the value W_{mh}: from 8, 65% on the control to 9, 90% with ZCR application, 24 t/ha. Thus, a part of soil humidity inaccessible for plants increased with application of zeolite containing rocks.

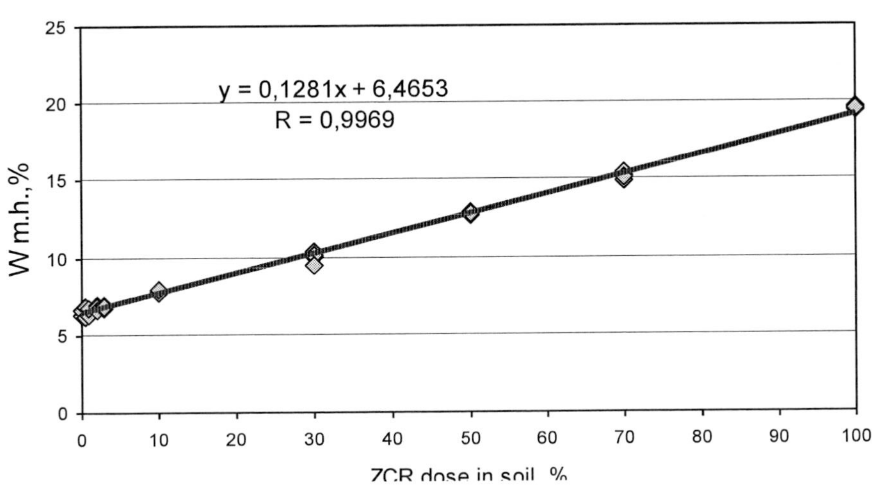

Figure 7. Dependence of maximum hygroscopic humidity of soil on ZCR doze.

Table 7. Total soil humidity (%) and reserves of productive water (m^3/ha) with application of increasing dozes of zeolite containing rocks (2010)

	May		June		July		August	
	Total humidity	Reserves of productive water	Total humidity	Reserves of productive water	Total humidity	Reserves of productive water	Total humidity	Reserves of productive water
Precipitation sum, mm	10,1		35,8		39,9		10,7	
Layer 0 – 20 sm								
Control	16,86	195,40	10,47	38,41	11,25	61,88	7,66	-
ZCR 3 t/ha	16,63	176,73	9,77	16,88	9,25	14,68	6,63	-
ZCR 8 t/ha	16,44	169,51	8,81	6,41	10,05	22,54	6,83	-
ZCR 16 t/ha	15,67	113,14	10,68	19,13	10,25	16,53	6,21	-
ZCR 24 t/ha	17,01	150,73	10,83	19,72	11,57	35,60	6,40	-
LSD $_{0,05}$		39,30	1,6	18,16	2,09	24,61	0,54	
Layer 20 – 40 sm								
Control	15,98	197,29	9,91	29,74	12,50	90,86	11,2	60,18
ZCR 3 t/ha	15,45	165,47	10,81	48,00	12,25	82,56	10,30	35,76
ZCR 8 t/ha	15,68	156,11	9,23	12,78	11,00	46,73	8,00	-
ZCR 16 t/ha	15,73	141,41	11,51	46,10	12,50	67,35	8,86	-
ZCR 24 t/ha	16,13	134,57	11,68	38,45	12,00	44,28	9,56	-
LSD $_{0,05}$		37,24					1,15	

The total content of humidity in the soil, which is determined by a standard method of hot drying under the temperature 105°C, includes both accessible and inaccessible, strongly retained water for plants.

The determination of the reserves of productive humidity gives a more distinct picture of provision of the plants with water. We calculated the reserves of the productive water in the extremely draught vegetative period in 2010 (Table 7). We used the data of Roeva T.A. [11], which show the deterioration of soil density with application of ZCR from Khotynetz deposit.

The sum of precipitation from May till August 2010 was by several times lower than on the average for many years. As a result at that period there was a deficiency of productive humidity in soil layers at 0-20 and 20-40 sm. With application of ZCR the content of accessible humidity for plants was significantly lower even in those variants where the total humidity was higher than the control.

The probable deterioration of plant water supply may become a limiting factor in the selection of dozes of zeolite containing rocks for solving some other problems of melioration.

Conclusion

Our investigations have shown that zeolite containing rocks of the Khotynetz deposit may be used for the purpose of reducing HM entering berry crop fruit. ZCR use on the soils having

HM content level several times higher than on the background efficiently reduce toxic elements going to the plants but does not result in the deficiency of biogenic elements.

In conditions of our experiments the increasing of ZCR dozes upwards of 8-10 t/ha was not efficient. The mineralogical composition and features of a zeolite containing rock, characteristics of the soil, the agrobackground, and the level of the pollution and chemical peculiarities of each metal, and specific peculiarity of cultivated crops should be taken into account while selecting dozes of zeolite containing materials for the purpose of melioration of soils polluted by HM.

The zeolite containing rock application reliably increases the part of strongly retained water in the soil and this may result in the deterioration of plant water supply in drought periods.

ACKNOWLEDGMENT

This work has been done under the financial support of the Russian Fund of Fundamental Investigations: project № 11-04-97572 р-центр-а.

REFERENCES

[1] Loboda, B.P. Zeolite containing mineral raw material use in plant-growing// *Agrochemistry,* - 2000, № 6, p. 78-91.

[2] Motyleva, S.M. Peculiarities of HM (Pb, Ni, Zn, Fe, Cu) content in fruit, berries and precipitation relative to variety estimation for use in breeding: specialty 06.01.05 "Breeding and Seed-growing", 03.00.04 *"Biochemistry": synopsis of thesis.* – Orel, 2000. – p. 23: Bibliogr.: p. 23.

[3] Gromova, V.S. The influence of continuous application of mineral fertilizers on the agroecological characteristics of apple orchard soil and fruit// *Fruit-growing and Berry-growing of Russia.* M. VSTISP, 1995. – p. 153-157.

[4] Tentative allowable concentrations (TAC) of heavy metals and arsenic in soils (Addendum №1 to the list of MAC and TAC № 6229-91): Hygienic standards. HS 2.1.7.020-94 / M.: *Publ. Goskomsanepidnadzor of Russia.*- 1995.

[5] Regionally underground content of chemical substances in soils of Orel region. – Orel: *State committee for environmental protection in Orel region.*, 1999.

[6] List of MAC and TAC of chemical substances in soil. Special edition / M.: *Publ. Goskomsanepidnadzor of Russia.*- 1991.

[7] Ilyin, V.B. Heavy metals in the soil plant system. – Novosibirsk: *Nauka,* 1991. – 150 p.

[8] Chemical composition of food products. Reference book edited by I.M. Skurikhin, M.N. Volgarev. M. 1987.

[9] Bespamyatnov, G.P., Krotov, Yu. A. Maximum Permissible concentrations of chemical substances in the environment. – *L.: Chemistry,* 1985. P. 309.

[10] Solovyeva, Yu.B., Lebedeva, L.A., The influence of consequences of different systems of fertilization on the protective physiological plant functions on the soddy podzolic

soil polluted by heavy metals// *Ecological Agrochemistry*. – M.: edited by MGU, 2008. – P. 66-81.

[11] Royeva, T.A. Ameliorant application for reducing heavy metals entering berries of black currant: specialty 06.01.07 *"Fruit-growing": candidate dissertation.* – Orel, 2008. – 23 p. Bibliogr.: p. 22-23.

In: Zeolites: Synthesis, Chemistry and Applications
Editors: Moisey K. Andreyev and Olya L. Zubkov
ISBN: 978-1-61942-861-4

Chapter 10

Adsorption Equilibrium and Kinetics of Fluoride on Two Synthetic Zeolites

Lucy Mar Camacho[1], Shuguang Deng[2] and Arely Torres[2]
[1]Center for Inland Desalination Systems, University of Texas at El Paso, El Paso, Texas, US
[2]Department of Chemical Engineering, New Mexico State University, NM, US

Abstract

Synthetic 4A and 5A zeolites were studied to explore the feasibility of applying these adsorbents for fluoride removal from drinking water. The zeolites were characterized with SEM/EDS and N_2-adsorption/desorption for their chemical and pore textural properties. The adsorption isotherms were correlated with the Langmuir and Freundlich adsorption equations. The fluoride adsorption isotherms on synthetic zeolite 4A and 5A followed the Freundlich model. The zeolite 5A adsorbent showed the highest fluoride adsorption capacity. It adsorbed up to 0.1 and 71 mg/g when the fluoride equilibrium concentration in water was 4.0 and 514 mg/L, respectively. Zeolite 5A adsorbed seven times more fluoride than zeolite 4A at the same initial concentration.

A pseudo-second order model and an intraparticle kinetic model fitted well the adsorption kinetic data for both synthetic zeolite adsorbents. It was found that both external and intraparticle diffusions contribute to the rate of removal of fluoride by the zeolites. From the experimental data obtained in this work, it was found that synthetic zeolite 5A has potential adsorption properties for fluoride removal from drinking water.

1. Introduction

Fluoride is a naturally occurring contaminant in water, air, plants, and animals. It is added to drinking water supplies at approximately 1 mg/L to prevent dental decay. Excess of fluoride in water can cause denser bones and skeletal damage. Excessive fluoride in groundwater is found in many regions worldwide. Fluoride Levels exceed 1.5 mg/L in some regions of the southwest United States [1]. It also exceeds 6 mg/L in many groundwater wells in the North-central part of Mexico [2]. The maximum contaminant level (MCL) in drinking water in USA is 4.0 mg/L. The maximum acceptable concentration of fluoride in drinking water established by the World Health Organization is 1.5 mg/L. Treatment methods to remove fluoride from drinking water include precipitation, membrane processes, ion-exchange and adsorption processes [3, 4]. The main disadvantages of these technologies are the generation of unwanted chemicals and waste disposal issues, the low economic viability, high maintenance cost, fouling, scaling, and membrane degradation [5]. Ion-exchange and adsorption can reduce fluoride concentrations to values below the permissible limits. Adsorption is a fast and inexpensive purification method. A variety of adsorbents materials have been used for removing fluoride from drinking water. They include biopolymer chitosan, coconut shell carbon, activated red mud, aligned carbon nanotubes, amorphous alumina supported on carbon nanotubes, and alumina cement granulates [6, 7, 8]. Synthetic zeolites are potential candidates for fluoride adsorption due to their porous structure and ion-exchange capacity. They are crystalline, three-dimensional molecular sieves made up of silicon and alumina atoms. Positive ions, most commonly sodium, calcium, or potassium are added in order to balance the molecule. The specific cation chosen influences the pore diameter and the adsorptive properties of the molecule sieve. Zeolite types 4A and 5A are the most common forms of molecule sieve. Zeolite 4A is the sodium form of molecular sieves. It will adsorb those molecules having a critical diameter of less than four angstroms. Zeolite 5A is the calcium form of molecular sieves. It will adsorb those molecules having a critical diameter of less than five angstroms. The purpose of the present chapter is to present results on the fluoride adsorption capacity and pH dependency of two synthetic zeolites available in the market. Results on characterization regarding surface area, chemical and pore structural properties are also presented.

2. Materials and Methods

2.1. Characterization of Adsorbent Materials

The chemical composition and pore textural properties of the synthetic zeolites 4A and 5A were determined by scanning electron microscopy with energy dispersive spectroscopy (SEM/EDS) (Hitachi, S-3400N) and N_2 adsorption and desorption (Micromeritics® ASAP 2020) analyses. Prior to SEM/EDS analysis the adsorbent samples were coated with a gold film to minimize charging and improve resolution. The pore textural properties including BET specific surface area, pore volume and pore size distribution were obtained by analyzing the nitrogen adsorption and desorption isotherms with built-in software in ASAP-2020.

2.2. Batch Equilibrium and Kinetic Adsorption Experiments

Batch adsorption experiments were conducted to determine the fluoride adsorption equilibrium and kinetic properties of the two adsorbents. Fluoride standard solutions with initial fluoride concentrations ranging from 1 to 1000 mg/L were prepared by diluting a certain amount of a fluoride standard stock solution of 10,000 mg/L with deionized water to a final volume of 150 mg/L. The standard stock solution was prepared by dissolving 22.11 g of reagent grade sodium fluoride (Fisher Scientific) into 1000 ml of deionized water. The pH of the standard solutions was adjusted to 5.5 with 0.10 M NaOH or 0.1M HCl to simulate drinking water conditions. For every experiment, 150 mL of the standard solution were mixed with one gram of the adsorbent material in a 200 ml high-density polyethylene plastic bottle. The bottle was closed tightly and placed on an automatic shaker (Lab-line Orbit No. 359) set at a shaking speed of 100 rpm for 48 hours. After removing the bottles from the shaker, solution samples of 20 ml were taken, filtered with 0.45 μm nylon sterile filters, mixed with 2 ml of a total ionic strength adjustment buffer (TISAB III), and analyzed for final pH and fluoride concentration with a combined pH and ion-meter (Accumet Excel XL25, Fisher Scientific). A duplicate was run for each of the different solution concentrations. Fluoride concentration and pH data for equilibrium analysis were collected at the beginning and at the end of each experiment. Concentration and pH data for kinetic experiments were collected initially every 15 minutes and then with increasing intervals while maintaining both the initial concentration and volume of the feed sample solutions constant. The latter was achieved by replacing 20 ml of the solution with another 20 ml of a new fluoride solution which concentration was determined by conducting a mass balance calculation.

3. Results and Discussion

3.1. Adsorbent Properties

The Nitrogen adsorption and desorption isotherms and the corresponding BJH pore size distributions [9] of the two synthetic zeolite adsorbents are plotted in Figure 1A and 1B, respectively. The desorption hysteresis on the two adsorbent in Figure 1A suggests that zeolite 5A has a wider range of pore sizes than zeolite 4A. The increase in the amount of fluoride adsorbed by zeolite 5A at higher pressures is due to an increase of the pore diameter as the pressure is raised. Figure 1B shows that there is a significant difference in the BJH pore size distribution of zeolite 4A and 5A. Zeolite 4A has a mesoporous structure with a median pore size of 157 Å, while zeolite 5A has a median microporous size of 23 Å and a media mesoporous size of 250 Å.

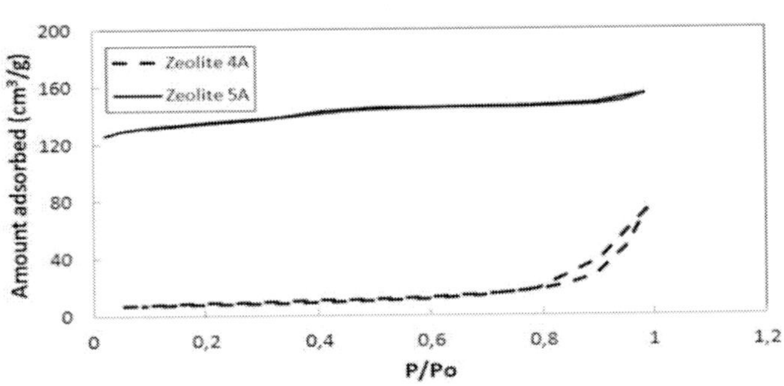

(a)

Figure 1. (Continued).

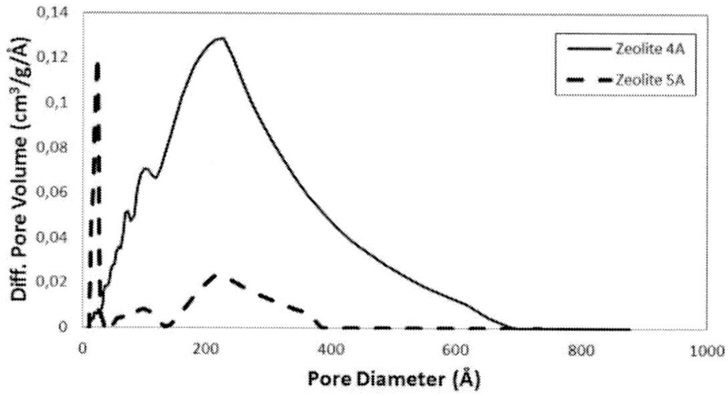

(b)

Figure 1. (A) Nitrogen adsorption/desorption isotherms and (B) BJH pore size distributions on two synthetic zeolite adsorbents.

The BET specific surface area of zeolite 4A is 28.6 m^2/g; almost fifteen times less than the BET specific surface area of zeolite 5A (412.06 m^2/g) but almost two times more than the specific surface area of natural clinoptilolite zeolite (18 m^2/g). The values of BET specific surface area, median BJH pore size and pore volume for the two adsorbents are presented in Table 1.

Table 1. Pore textural properties for synthetic 4A and 5A zeolite adsorbents

Property	Zeolite 4A	Zeolite5A
BET surface area (m^2/g)	28.60	412.06
BJH pore volume (cm^3/g)	0.1111	0.060
BJH average pore size (Å)	157.504	23.221

(a) (b)

Figure 2. (A) SEM image of 4A zeolite and (B) 5A zeolite (Scale bar 5µm).

Amorphous crystal clusters with open channel network were observed in the SEM of zeolite 4A and 5A.

The image of the inner structure of zeolite 5A was not well defined even though the sample was coated with a gold film. Energy dispersive electroscopic mapping (not included) identify the presence of sodium in synthetic zeolite 4A and calcium in zeolite 5A.

3.2. Adsorption Equilibrium

Figure 3A displays the fluoride adsorption equilibrium data obtained on the two synthetic zeolite adsorbents at initial feed concentrations ranging from 1 to 1,000 mg/L. A pH of 5.5 was selected to conduct the equilibrium experiments. From the two adsorbent materials, zeolite 5A showed the highest adsorption capacity. It adsorbed up to 0.1 and 71 mg/g when the fluoride equilibrium concentration in water was 4.0 and 514 mg/L, respectively. Zeolite 5A adsorbed seven times more fluoride than zeolite 4A at the same initial concentration.

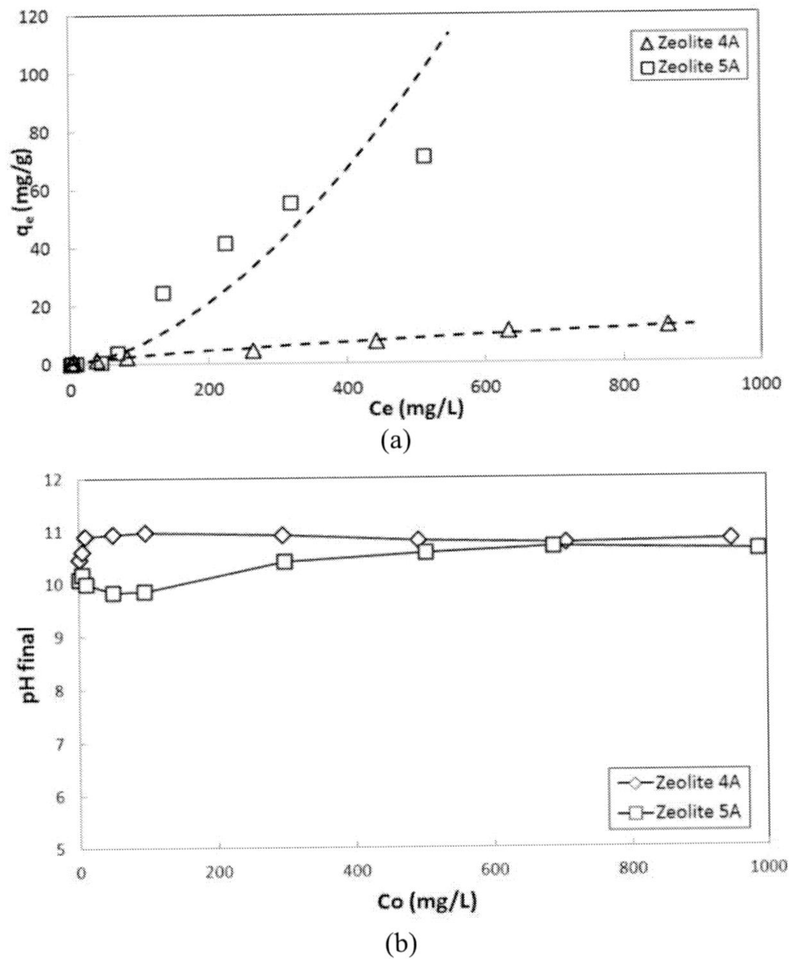

Figure 3. (A) Fluoride adsorption isotherms on two synthetic zeolites adsorbents at 25°C; (B) Effect of initial fluoride concentration on the equilibrium pH.

The main factor that may contribute to the high adsorption capacity of fluoride on the zeolite 5A is the ability of the adsorbent to adjust to the pH of the solution at low and high concentrations (Figure 3B) as well as the better pore textural and surface properties.

Freundlich and Langmuir isotherm models were applied to correlate the fluoride adsorption equilibrium data obtained in this work. The estimated adsorption isotherm parameters are presented in Table 2. The Freundlich [10] and Langmuir [11] isotherms are two-parameter models that provide useful information to predict the adsorption capacity of the adsorbent materials at given feed concentrations. For both models, q_m (mg/g) is the amount of fluoride adsorbed per weight of adsorbent at equilibrium, and C_e (mg/L) is the equilibrium concentration of the adsorbate in the solution. In the Freundlich model (Table 2), K is the Freundlich constant representing the relative adsorption affinity of the adsorbent toward the adsorbate molecules, and n represents the heterogeneity of the adsorbent. In the Langmuir model (Table 2), b (l/mg) is the Langmuir constant that characterizes the adsorbent-adsorbate interaction, and a (mg/g) represents the monolayer adsorption capacity. The Freundlich adsorption isotherm fitted well the fluoride adsorption on the zeolite 4A (R^2 = 0.982) and 5A (R^2 = 0.956). The non-linearity of the adsorption isotherms of zeolite 4A is evident because n>1.

Table 2. Freundlich and Langmuir isotherm constants for fluoride adsorption on synthetic zeolites adsorbents at pH of 5.5 and 25°C

Adsorbent	Freundlich Isotherm $q_e = k*C_e^{1/n}$			Langmuir Isotherm $q_e = a*b*C_e/(1+b*C_e)$		
	K (L/g)	1/n	R^2	a (mg/g)	b (L/mg)	R^2
4A	0.1314	0.6728	0.9824	14.7275	0.0036	0.7536
5A	0.0029	1.6765	0.9565	2.4370	0.0027	0.3102

3.3. Adsorption Kinetics

Adsorption kinetic studies were conducted to obtain essential information related to optimum parameter conditions in the application of the synthetic zeolite adsorbents for water treatment systems.

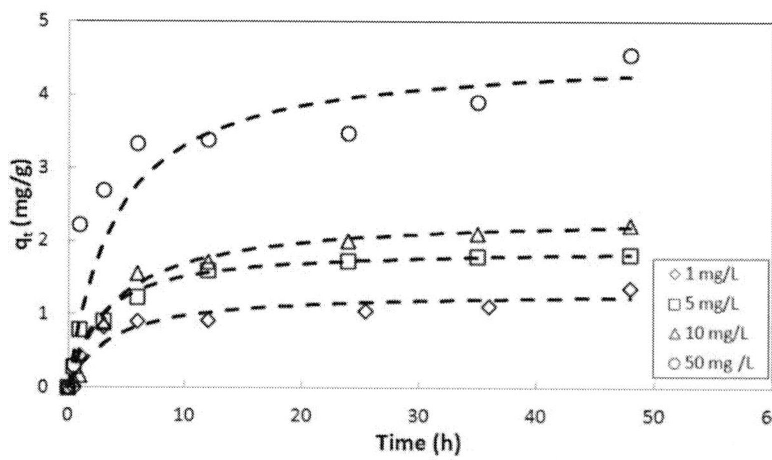

(a)

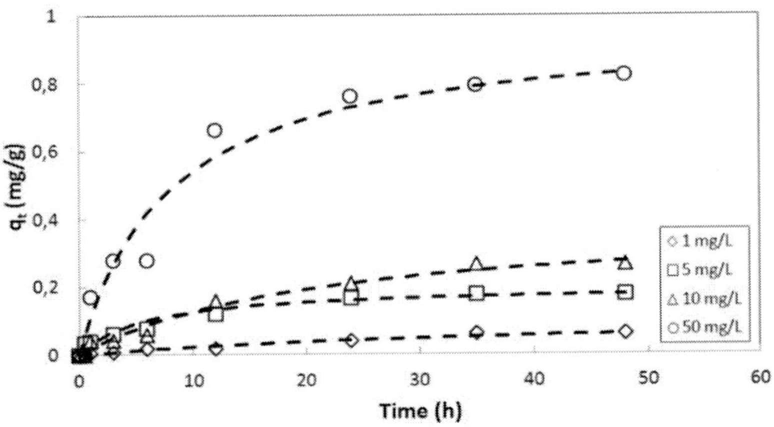

(b)

Figure 4. (A) Pseudo-second order plots for fluoride adsorption on 4A zeolite and (B) 5A zeolite at low concentrations, pH of 5.5, and 25°C (symbols are experiments data and dashed lines are model predictions).

Figure 4A and 4B show the adsorption kinetics of fluoride on zeolite 4A and 5A at low concentrations (1 mg/L - 50 mg/L). Similar plots for high concentrations (100 mg/L – 1,000 mg/L) are shown in Figure 5A and 5B. It was observed that the time required to obtain equilibrium increased with the increase of initial feed concentration for the two adsorbents.

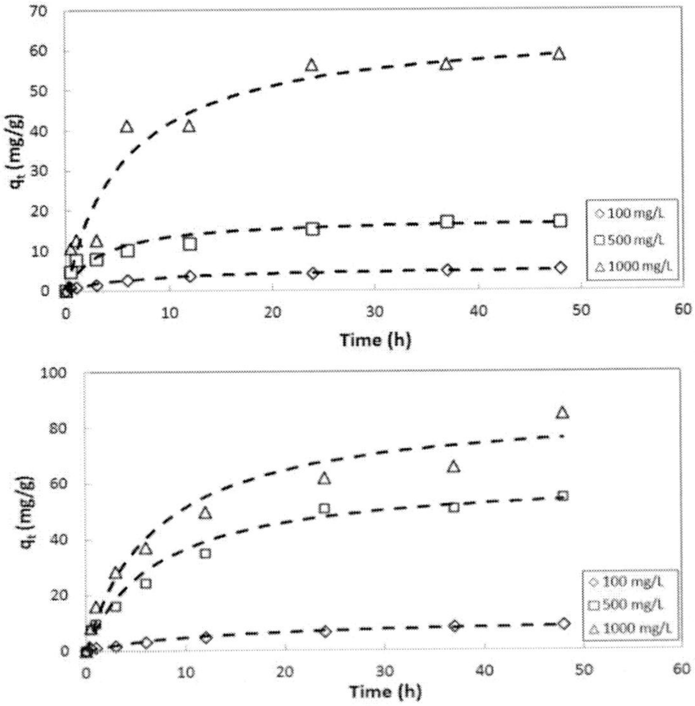

Figure 5. (A) Pseudo-second order plots for fluoride adsorption on 4A zeolite and (B) 5A zeolite at high concentrations, pH of 5.5, and 25°C (symbols are experiments data and dashed lines are model predictions).

The same increase in equilibrium time by increasing the initial fluoride concentration was observed when conducting kinetic adsorption studies of fluoride on sol-gel-modified activated alumina adsorbents [12]. Zeolite 4A required 10 and 20 hours to equilibrate solutions with initial fluoride concentrations of 1 mg/L and 50 mg/L, while zeolite 5A required 10 and 30hours to reach equilibrium at the same initial concentrations. At concentrations of 1,000 mg/L the equilibrium for zeolite 4A and 5A was not reached yet after 40 hours (Figures 5A and 5B). This may attributed to a saturation of the active sites readily available for fluoride adsorption

A pseudo-first order model [13], a pseudo-second order model [14], and an intraparticle diffusion models [15] were used to correlate the adsorption kinetics of fluoride on the two synthetic zeolite adsorbents. Even though adsorption onto the solid surface of the adsorbent is a more general assumption, intraparticle diffusion in the zeolite adsorbents can also be considered.

The mathematical equations representing the models are given as:

Pseudo-first order equation: $$q_t = q_e(1 - e^{k_1 t}) \quad (1)$$

Pseudo-second order equation: $$q_t = \frac{q_e^2 k_2 t}{1+q_e k_2 t} \quad (2)$$

Intraparticle diffusion model: $$q_t = k_p t^{1/2} \quad (3)$$

In equation (1), (2) and (3), q_t (mg/g) is the adsorption amount at a given time t (h), q_e (mg/g) is the adsorption equilibrium capacity. In the equations, k_1 (h^{-1}), k_2 (g/mg·h), and k_p (mg/g·$h^{1/2}$) represent the adsorption rate constants of the pseudo-first order, pseudo-second order adsorption reactions, and the intraparticle diffusion rate constant, respectively. The linear forms of the pseudo-first and pseudo-second order equations are:

$$\ln(q_e - q_t) = \ln(q_e) + k_1 t \quad (4)$$

$$\frac{1}{q_t} = \frac{1}{q_e} + \frac{1}{q_e^2 k_2}\frac{1}{t} \quad (5)$$

By plotting log(q_e-q_t) versus time in equation (4) a straight line is obtained. The pseudo-first rate constant k_1 is obtained from the slope of the plot. From equation (5) a plot of $1/q_t$ versus 1/t generates a straight line and the values of q_e and k_2 can be calculated from the slope and intercept of this line.

The slope of the linear plot of q_t versus $t^{1/2}$ in equation (3) gives the value of k_p. If the linear plot of equation (3) presents two or three plateaus it means that the kinetic adsorption process is complex. In this case the first slope (initial curved portion) represents the external surface sorption, the second slope (intermediate linear portion) the intraparticle diffusion, and the third slope (last portion of the linear plot) the equilibrium sorption. If the linear plot presents only one slope and this slope passes through the origin then intraparticle diffusion is the only rate-controlling step in the adsorption process [16].

The kinetic parameters obtained from the slope and intercept of the linear plots of the experimental data in the pseudo-first order and second-order models and the experimental adsorption capacities are given in Table 3. The pseudo-first order rate constant (k_1 = 0.04-0.12) and the correlation coefficient ($R^2 < 0.98$) for the concentrations tested were relatively low, suggesting a slow adsorption rate. The predicted amounts of fluoride adsorbed are smaller than the experimentally values.

Therefore it may be suggested that the adsorption of fluoride on the adsorbents tested does not follow the first-order rate model.

The near linear fits obtained by applying the pseudo-second order model with correlation coefficient values (R^2) ranging from 0.82 to 0.99 for most of the samples tested with the two adsorbents indicates that the observed kinetics behavior can be better approximated to the pseudo-second order kinetics as shown in Figure 4A and 4B, and Figures 5A and 5B for low and high concentrations, respectively.

For most of the initial concentrations tested the pseudo-second order rate constant k_2 was low, indicating that the rate of fluoride adsorption process is fast. The predicted amounts of fluoride adsorbed are similar to the experimental values for both adsorbents. It was evident that the rate constant decreased with the increase of initial feed concentrations. This can be attributed to the increased number of fluoride ions available for adsorption. The pseudo-second order fitting suggests that chemisorption is responsible for the fluoride adsorption on the two synthetic zeolites adsorbents.

Table 3. Kinetic constants for adsorption of fluoride on synthetic zeolite adsorbents at pH of 5.5 and 25°C

Initial fluoride concentration (mg/L)	Pseudo-first order			Pseudo-second order			Experimental
	k_1 $(hr)^{-1}$	q_e mg/g	R^2	k_2 (g/mg.hr)	q_e (mg/g)	R^2	q_e (mg/g)
4A							
1	0.0413	0.8968	0.723	0.2143	1.3284	0.974	1.3574
5	0.1206	1.3453	0.982	0.1952	1.9201	0.999	1.8144
10	0.0976	1.9099	0.924	0.1020	2.3843	0.998	2.2262
50	0.0445	2.7483	0.704	0.0550	4.6083	0.975	4.5582
100	0.0696	4.3921	0.986	0.0265	5.6561	0.992	5.0902
500	0.0871	12.5309	0.950	0.0170	17.8571	0.991	16.837
1,000	0.0934	47.4891	0.910	0.00263	65.7895	0.971	58.643
5A							
1	0.0408	0.0598	0.955	0.2392	0.1084	0.647	0.0599
5	0.1075	0.1669	0.985	0.8622	0.1983	0.984	0.1754
10	0.0647	0.2659	0.975	0.1179	0.3960	0.820	0.2659
50	0.0982	0.7671	0.975	0.1289	0.9709	0.966	0.8234
100	0.0747	9.0412	0.968	0.0066	11.0254	0.959	8.7347
500	0.0788	48.1779	0.932	0.0023	61.7284	0.986	54.6524
1,000	0.0398	68.8686	0.920	0.0017	86.9566	0.963	84.6374

By plotting q_t versus $t^{1/2}$ straight lines with three plateaus were obtained. This behavior is observed in Figure 6A and 6B, and in Figures 7A and 7B for low and high initial feed concentrations, respectively. The plateaus are more notorious for synthetic zeolite 4A, which are attributed to boundary layer diffusion, intraparticle diffusion and adsorption equilibrium, respectively [17]. The second linear portion of the lines did not pass through the origin meaning that both external and intraparticle diffusion contributed to the rate of removal of fluoride on the synthetic zeolites. The slope and intercepts of the plots increased by increasing the initial fluoride concentrations and the intraparticle diffusion is slow at low concentrations.

The increasing slope may be attributed to the effect of increasing concentration gradients and to the effect of external mass transfer resistance due to the boundary layer thickness.

Table 4 shows the intraparticle diffusion constants for the two zeolites. Synthetic zeolites 4A and 5A are porous materials therefore the main resistance to mass transfer is the diffusion of fluoride within the pores of the particles.

From the experimental observations and modeling it can be concluded that fluoride adsorption onto the synthetic zeolite adsorbents tested is a complex process. Both surface adsorption and intraparticle diffusion contribute to the rate-limiting step.

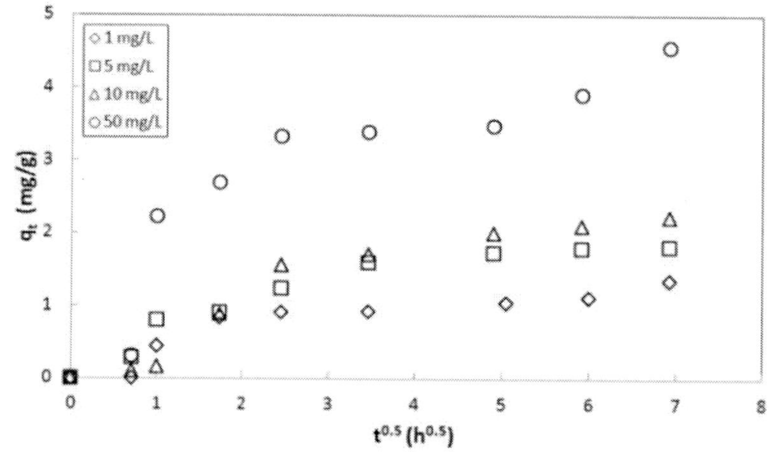

(a)

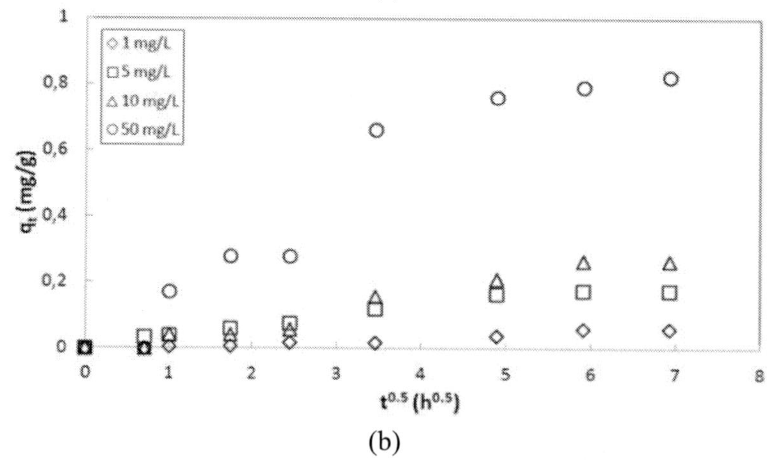

(b)

Figure 6. (A) Intraparticle diffusion plots for fluoride adsorption on synthetic 4A zeolite and (B) 5A zeolite at low concentrations, pH of 5.5 and 25°C.

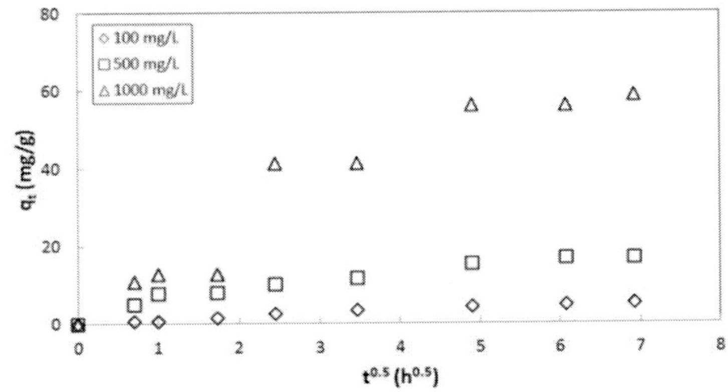

(a)

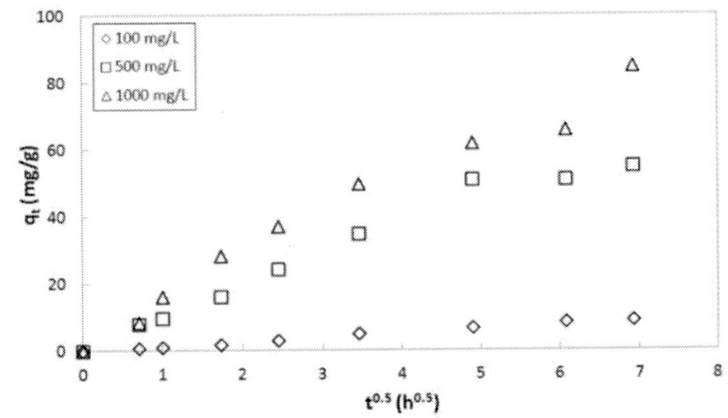

(b)

Figure 7. (A) Intraparticle diffusion plots for fluoride adsorption on synthetic 4A zeolite and (B) 5A zeolite at high concentrations, pH of 5.5 and 25°C.

Table 4. Intraparticle diffusion constants for adsorption of fluoride on synthetic zeolite adsorbents at pH of 5.5 and 25°C

	4A		5A	
	Slope	R^2	Slope	R^2
1	0.3267	0.8772	0.0108	0.925
5	0.4225	0.9181	0.0330	0.993
10	0.6380	0.8909	0.0567	0.957
50	0.7645	0.9898	0.1886	0.843
100	0.9262	0.9643	1.5777	0.988
500	2.0273	0.9749	10.689	0.998
1,000	6.3756	0.8301	11.618	0.983

CONCLUSION

Two synthetic zeolites were studied to explore the feasibility of applying these adsorbents for fluoride removal from drinking water. Zeolite 4A is characterized for being the sodium form of molecular sieves while zeolite 5A is the calcium form.

The following conclusions were drawn from this experimental study:

- The BET specific surface area of zeolite 4A is 28.6 m^2/g; fifteen times less than the BET surface area of zeolite 5A (412.06 m^2/g).
- From the two adsorbent, zeolite 5A appears to have the highest adsorption capacity. It adsorbed up to 0.1 and 71 mg/g when the fluoride equilibrium concentration in water at pH 5.5 was 4.0 and 514 mg/L, respectively. The alkalinity of this adsorbent is the factor that contributes to the high adsorption capacity of fluoride.
- The Freundlich adsorption isotherm represents well the fluoride adsorption on zeolite 4A and zeolite 5A adsorbents.
- The observed kinetic behavior for the two adsorbents is better explained by the pseudo-second order kinetic model. The low value of the pseudo-second order rate constant indicates that the rate of the fluoride adsorption process is fast.
- Both external and intraparticle diffusion contribute to the rate of removal of fluoride on the two synthetic zeolites adsorbents. The process of intraparticle diffusion is slow at low concentrations.

The experimental data obtained in this work demonstrated that synthetic zeolite 5A is a promising adsorbent for removing fluoride from drinking water. Its high surface area, high adsorption capacity, and rapid equilibrium time suggesting a fast rate of adsorption, makes this adsorbent material feasible for implementation of a fluoride removal system.

REFERENCES

[1] B.E. Yoldas. Transparent porous alumina, *Am. Ceramic Assoc.*, 54 (1975) 289.

[2] C. Diaz-Nava, M. T. Olguin, M. Solache-Rios. Water defluoridation by Mexican heulandites-clinoptilolite, *Sep. Sci. Technol.*, 37 (2002) 3109-3128.

[3] H.M.F. Freundlich. Uber die adsorption in lösungen, *Z. Phys. Chem.*, (1906) 385-470.

[4] I. Langmuir. The adsorption of gases on plane surfaces of glass, mica and platinum, *J. Am. Chem. Soc.*, 40 (1916) 1361-1403.

[5] L.M. Camacho, A. Torres, D. Saha, S. Deng. Adsorption equilibrium and kinetics of fluoride on sol-gel derived activated alumina adsorbents.

[6] M. Sarkar, A. Banerjee, P.P. Pramanick, A.R. Sarkar. Design and operation of fixed bed laterin column for the removal of fluoride from water, *Chem. Eng. J.,* 131 (2007) 329-335.

[7] M. Srimurali, A. Pragathi, J. Karthikeyan. A study on removal of fluorides from drinking water by adsorption onto low-cost materials*, J.* Environ. *Pollut.,* 99 (1998) 285-289.

[8] R.S. Sathish, N.S.R. Raju, G.S. Raju, G.N. Rao, K.A. Kumar, C. Janardhana. Equilibrium and kinetics studies for fluoride adsorption from water on zirconium impregnated coconut shell carbon, *Sep. Sci. Technol.*, 42 (2007), 769-788.

[9] S. Ayoob, A.K. Gupta, V.T.A. Bhat. A conceptual overview on sustainable technologies for the defluoridation of drinking water, *Crit. Rev. Environ. Sci. Technol.*, 38 (2008) 401-470.

[10] S. Ayoob, A.K. Gupta. Sorptive response profile of an adsorbent in the defluoridation of drinking, *Chem. Eng. J.*, 133 (2007) 273-281.

[11] S. Lagergren. Zur theory der sogennanten adsorption gelöster stoffe, S. *Kungliga Svenska Vetenskapsakademiens*, 24 (1898) 1-39.

[12] S.M. Maliyekkal, A.K. Sharma, L. Philip. Manganese-oxide-coated alumina: a promising sorbent for defluoridation of water, *Water Res.*, 40 (2006) 3497-3506.

[13] S.X. Teng, S.G. Wang, W.X. Gong, X. W. Liu, B.Y. Gao. Removal of fluoride by hydrous manganese oxide-coated alumina: Performance and mechanism, *J. Hazard. Mater.*, 168 (2009) 1004-1011.

[14] U.S. Department of Health and Human Services (USDHHS). Toxicological profile for fluorides, hydrogen fluoride and fluorine. September, 2003, Atlanta, 404 pp. http://www. atsdr.cdc.gov /toxprofiles/tp11.pdf.

[15] W. Rongshu, L. Haiming, N. Ping, W. Ying. Study of a new adsorbent for fluoride removal from waters, Water Qual. *Res. J.,* Canada. 30 (1995) 81-88.

[16] W.J. Weber Jr., J.C. Morris. Kinetics of adsorption on carbon from solution, *J. Sanit. Eng. Div. ASCE,* 89 (1963) 31-59.

[17] Y.S. Ho, G. McKay. Sorption of dye from aqueous solution by peat, *Chem. Eng. J.,* 70 (1998) 115-124.

In: Zeolites: Synthesis, Chemistry and Applications
Editors: Moisey K. Andreyev and Olya L. Zubkov
ISBN: 978-1-61942-861-4

Chapter 11

TRANSITION METAL IONS IN ZEOLITES: CASE OF IRON

Károly Lázár
Institute of Isotopes, Hung. Acad. Sci., Budapest, Hungary

ABSTRACT

Iron ions may occupy various positions in zeolites and their coordination and oxidation states may be changed during different treatments. An apparent distinction among their siting is the framework/extra-framework assignment. In situ Mössbauer spectroscopy is a sensitive and unique tool to analyze the siting of iron in zeolites. A few examples for the identification of iron located in framework or extra-framework sites is presented and their interconversions are correlated with the detected changes in their oxidation and coordination states. Finally, the formation of coupled framework/extra-framework iron pairs is also demonstrated.

INTRODUCTION

The primary building units in zeolites are the $[SiO_{4/2}]$ tetrahedra. In certain cases however, not only silicon but other, e.g. transition metal ions (V^{4+}, Ti^{4+}, Fe^{3+}) may also occupy the central tetrahedral (T) site in the center of tetrahedron surrounded by four oxygens. This isomorphous substitution may take place in a certain range of ionic radii ($35 < T_{radius} < 65$ pm) [1].

The charge neutrality in the primary building units is provided if the T ions are four-valent (Si^{4+}, V^{4+}, Ti^{4+}, Sn^{4+},...). In a large number of cases, the substituted T ion may also be trivalent (the most common example is Al^{3+}), resulting in the formation of negatively charged primary $\lfloor Al^{(3+)}O_{4/2}]^{(-)}$ tetrahedron. Its negative charge should be compensated by positive "extra-framework" (EFW) ions, which may easily be accommodated in the structural pores of zeolites (H^+, Na^+ etc.).

Most of transition metals may exist in different valency states (Fe^{2+}/Fe^{3+}, V^{5+}/V^{4+}, Ti^{2+}/Ti^{4+},....). The change in their valency may take also place in a limited extent on the ions located in the T-substituted, “framework” (FW) sites. Depending on the change in the ionic radius accompanying the valency change, the ions may be retained in the FW site, or, if the radius of the newly formed ion is outside of the 35 – 65 pm range, they may excange the FW position to EFW siting. Abundant literature is devoted for a more detailed general description of the properties of transition metals in zeolites [2,3].

All of the mentioned features can be related to iron. It may occupy FW sites in tetrahedral coordination, it is able for Fe^{2+}/Fe^{3+} valency change, and iron ions, or simple iron oxycompounds ($Fe^{(3+)}O^+$, $Fe^{(2+)}(OH)^+$, etc.), may play EFW charge compensating role as well.

A unique advantageous method for studying iron is Mössbauer spectroscopy. It can conveniently be applied to distinguish between the ferric and ferrous states since the value of a basic Mössbauer parameter, the isomer shift (IS), is directly correlated to the electron density, i.e. to oxidation state.

Furthermore, this technique provides information on the coordination state of ions as well, since the value of another principal and easily extractable parameter of spectra, the quadrupole splittig (QS), is correlated to the symmetry of the electric field around the iron nucleus, i.e. to the coordination of ligands. Fortunately, the ferrous and ferric regions do not overlap in an IS vs. QS representation. Thus, in most cases, both the oxidation and coordination states of iron can be determined simultaneously in ferrisilicates [4] providing a mean to extract unique structural informations.

In the overview, the possibilities of the application of the in situ Mössbauer technique are presented in a selection of examples. Examples of the introduction of ferrous iron to EFW sites are described. The FW substitution is demonstrated in different host matrices, and the conditions of displacement from FW to EFW siting are followed. The possibilities of the formation of FW-EFW dinuclear iron pairs are also considered. Exposure to various atmospheres (reductive, neutral and oxidative) may result in changes both in the coordination and oxidation states. Cycling among the treatments may lead to changes in the siting of iron as well. These processes are also illustrated.

EXTRA-FRAMEWORK IONS

Extra-framework ions are located in the so-called charge-compensating positions, where they neutralize the uncompensated charges generated by the isomorphous substitution of trivalent ions in the tetrahedral framework (T) sites. In most cases, the trivalent ion is Al^{3+} in the T sites. The charge compensation is usually provided by simple, common ions, like Na^+, NH_4^+, H^+ or H_3O^+, etc.

In certain cases, iron ions may also take part in the EFW charge-compensating processes. Various single-valent iron containing species may exist depending on the actual conditions: $[Fe^{(3+)}O]^+$, $[Fe^{(3+)}(OH)_2]^+$, $[Fe^{(2+)}(OH)]^+$, etc. The ligand sphere of the iron ions may be completed either by further neutral water molecules or by bonds to the oxygen atoms of the zeolite framework. Depending on the conditions, the local symmetry around the iron ions may vary – from low-coordination trigonal to octahedral [2,5].

Two examples are presented in the following paragraphs to illustrate the behavior of iron-conatining EFW species in cages of zeolites.

EFW Iron Ions Introduced by Solid State Ion Exchange (I) - FAU

Extra framework location can apparently be provided if iron ions are introduced to a zeolite framework in a separate secondary process. Usually, the secondary ion exchange can be performed in an aqueous solution, however, it can also be accomplished by mixing the solid state components and heating them afterwards. NH_4-form of zeolites can advantageously be used in this later procedure. For instance, processes described with the general scheme

$$NH_4\text{-FAU} + FeCl_2 \rightarrow \text{Fe-FAU} + NH_4Cl \text{ ; and } NH_4Cl \rightarrow HCl + NH_3$$

may proceed (FAU represents the faujasite zeolite cage, Y zeolite by earlier nomenclature). This process was followed in detail in a Si:Al = 2:1 faujasite zeolite. $FeCl_2.4H_2O$ was added in stoichiometric amounts, corresponding to Fe:Al = 0.5 : 1 ratio, then the components were intensively grinded. The formed mixture was put into an in situ Mössbauer cell – and spectra were sequentially recorded after heating the sample to different temperatures in vacuum (420, 520, 620 and 720 K). Parallel measurements were performed in a temperature-programmed mass spectrometer. Mössbauer spectra recorded after the treatments are shown in Figure 1.

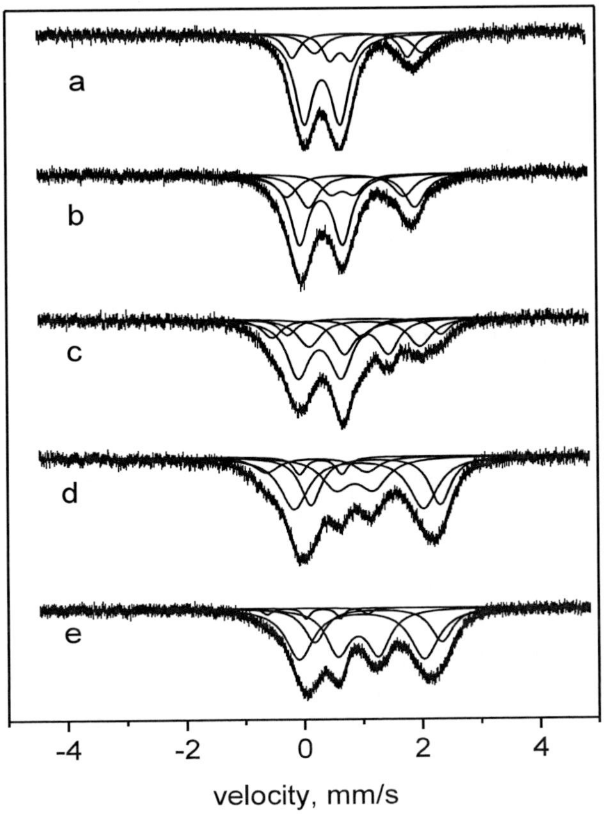

Figure 1. Sequential Mössbauer spectra recorded at the various stages of the $NH_4Y + FeCl_2 \rightarrow FeY + NH_3 + HCl$ process. After mixing the components a), and after heat treatments at 420 K, b), 520 K, c), 620 K, d) and 720 K, e).

Table 1. Mössbauer parameters extracted from spectra recorded at different stages of the $NH_4Y + FeCl_2 . 4H_2O \rightarrow FeY + NH_3 + HCl$ process (IS: isomer shift, related to metallic iron, mm/s; QS: quadrupole splitting, mm/s; RI: relative intensity, % spectral area)

Treatment	Component	IS	QS	RI
As mixed	$Fe^{3+}_{(octahedral)}$	0.37	0.63	57
	$Fe^{2+}_{(tetrahedral)}$	0.69	0.35	13
	$Fe^{2+}_{(octahedral)}$	0.83	1.95	9
	$Fe^{2+}_{(FeCl2.xH2O)}$	1.13	1.83	21
Evacuation 420 K	$Fe^{3+}_{(octahedral)}$	0.35	0.72	44
	$Fe^{2+}_{(tetrahedral)}$	0.72	0.36	11
	$Fe^{2+}_{(octahedral)}$	0.79	2.00	22
	$Fe^{2+}_{(FeCl.xH2O)}$	1.04	1.79	23
Evacuation 520 K	$Fe^{3+}_{(octahedral)}$	0.31	0.75	38
	$Fe^{3+}_{(trigonal)}$	0.25	1.53	11
	$Fe^{2+}_{(trigonal)}$	1.08	0.78	21
	$Fe^{2+}_{(octahedral)}$	1.06	2.60	9
	$Fe^{2+}_{(FeCl.xH2O)}$	1.06	1.90	21
Evacuation 620 K	$Fe^{3+}_{(octahedral)}$	0.30	0.72	5
	$Fe^{3+}_{(trigonal)}$	0.23	1.71	11
	$Fe^{2+}_{(trigonal)}$	0.88	0.62	21
	$Fe^{2+}_{(octahedral)}$	0.95	2.20	37
	$Fe^{2+}_{(octahedral)}$	1.23	2.19	26
Evacuation 720 K	$Fe^{3+}_{(octahedral)}$	0.33	0.58	2
	$Fe^{3+}_{(trigonal)}$	0.22	1.73	2
	$Fe^{2+}_{(trigonal)}$	0.92	0.68	30
	$Fe^{2+}_{(octahedral)}$	0.99	2.13	45
	$Fe^{2+}_{(octahedral)}$	1.27	2.16	20

During grinding on air, a part of the ferrous iron ions was oxidized to ferric state as spectrum of Figure 1a demonstrates (and data of decompositions of the corresponding Mössbauer spectra is collected in Table 1). A modest heating to 420 K results in some loss of water, a slight increase in the Fe^{2+} proportion is also detected. Further loss of water takes place in the heating to 520 K, a part of iron ions completely loses the water from their coordination sphere, the intensity of the low quadrupole splitting component increases. The subsequent heating to 620 K results in the most profound changes, most of the iron ions occupy the final positions in the zeolite framework. Further heating stabilizes these positions at 720 K. At this final stage, only 4 % of iron is found in Fe^{3+} state (originally, the Fe^{3+} proportion was 57 %). During heating, the removal of water was probaly connected with some removal of oxygen, (e.g. $2\ Fe^{(3+)}OH \rightarrow 2\ Fe^{(2+)} + H_2O + ½\ O_2$ may presumably proceed) and the ammonia formed from the decomposition of NH_4Cl may also contribute to the reduction of iron at higher temperatures. Further details are described in [6].

Various components were identified during the process, demonstrating that the oxidation and coordination states are changing in the iron-containing species in correspondence to the experimental conditions. In this example, the behavior of a low Si:Al zeolite case was shown. In partucular, the Al content was comparatively large, the amount of exchangeable NH_4 ions was also large, i.e. the solid state iron exchange with iron was facile.

EFW Iron Ions Introduced by Ion Exchange (II) - MFI

The following case presents an example in which the number of the exchangeable charge compensating sites is less with more than one order of magnitute in comparison to the previous case of FAU. In that instance $FeCl_2$ was introduced to MFI framework (or, using the earlier nomenclature, to ZSM-5). NH_4ZSM-5 was used with Si:Al = 25 ratio, the Fe^{2+} was introduced by impregnation from ethanolic solution. The amount of impregnated iron was low, c.a. 1/10th of the nominally available exchange capacity (at the Al sites) was filled up by iron counterions.

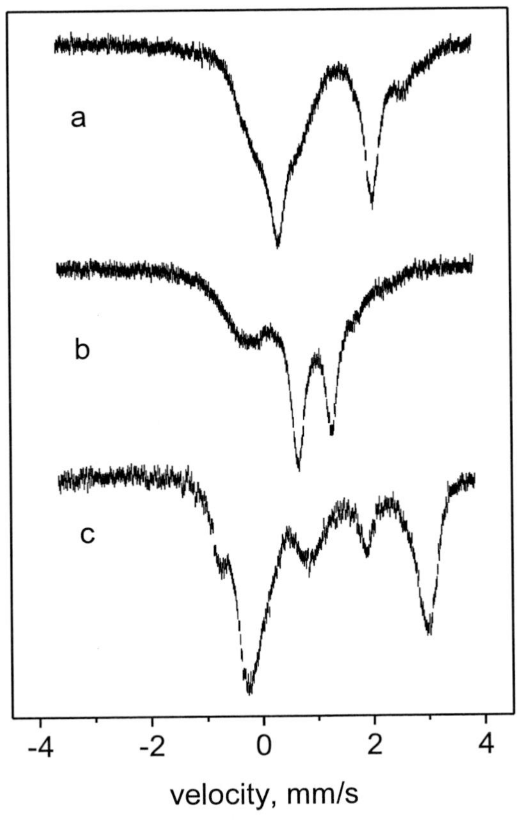

Figure 2. Mössbauer spectra of $FeCl_2$ + NH_4ZSM-5 sample. After impregnation of the $FeCl_2$ from ethanolic solution, a), after evacuation at 470 K, b), after storage on air with moisture, c).

Corresponding spectra are shown in Figure 2. The processes can be followed by identifying the main components and analyzing their interconversions, however, the exchange was not completely perfect. Figure 2a displays the 77 K spectrum of the impregnated sample. In this stage, iron ions do not occupy charge-compensating positions yet, the main component is the $FeCl_2$. H_2O ($IS_{77\,K}$ = 1.22 mm/s, $QS_{77\,K}$= 1.70 mm/s). The interaction with the ZSM-5 framework and the exchange of NH_4^+ ions to ferrous ones proceeds upon evacuation at 470 K. The characteristic component in the spectrum recorded during the treatment is the doublet of Fe^{2+} ions exhibiting a low quadrupole splitting ($IS_{470\,K}$= 0.99 mm/s, $QS_{470\,K}$ = 0.62 mm/s, Figure 2.b). This component probably originated from ferrous ions without coordinated water, probably in the form of $[Fe^{(2+)}(OH)]^-$, or $[Fe^{(2+)}Cl]^-$ attached closely in distorted cordination to the pore wall at the vicinity of tetrahedrally-substituted Al sites. It is worth to note that the exhcanged Fe^{2+} is not oxidized noticeably to Fe^{3+} in a few hour exposure to air at room temperature. Instead, sorption of moisture in the ligand sphere of Fe^{2+} can be deduced from the significant increase of the quadrupole splitting ($IS_{77\ K}$ = 1.38 mm/s, $QS_{77\ K}$ = 3.26 mm/s). Further details are described in [7].

This series of spectra demonstrates that extra-framework ions may occur in various environments and the changes in their coordinations are clearly reflected in the in situ Mössbauer spectra.

Extra Framework Ions – Other Examples

Further studies show that extra framework ions respond sensitively not only to changes in their coordination state but also to reduction-oxidation treatments. For example, varations in the principal Mössbauer parameters are correlated to the effects of a series of various repeated reduction-oxidation treatments in hydrogen and in N_2O on Si:Fe = 200 LTA zeolite [8].

Iron Ions Substituted to Framework Sites

Iron ions may occupy structure building framework sites in zeolites, too. The ionic radius of ferric ions is 63 pm. This size is 57% larger than that of the usual main constituent, Si^{4+} exhibits [1]. In contrast to this significant difference, a certain amount of ferric ions can be accommodated in the structure without structural damages. Incorporation of trivalent ferric ions results in the appearance of an extra, uncompensated charge in the $[FeO_{4/2}]^-$ tetrahedra, spanned by the four oxygen, similarly to the most common case mentioned previously when aluminium ions are substituted in T sites.

The synthesis of ferrisilicate analogue zeolites usually proceeds in hydrothermal conditions in neutral or, most probably, in alcaline media. The preferred form of iron ions in alcaline media is the hydrolyzed oligomeric state – i.e. incorporation of single ions is less favored in neutral or alcaline media. Thus, during synthesis of framework substituted zeolite analog ferrisilicates incorporation of iron may take place not only to FW but to EFW sites also. However, the synthesis conditions can be advantageously modified in order to promote the predominance of FW siting for iron ions [9].

The FW siting can be confirmed from the appearance either of one of the two characteristic features detectable in the Mössbauer spectra. The first sign can be the appearance of the so-called paramagnetic behavior, a phenomenon characteristic for ferric ions distributed in a few tenths of nanometer distance from each other. This phenomenon can be observed at low iron concentrations (Si/Fe < 100 atomic ratio). Namely, the direction of the internal magnetic field observed on separated iron ions does not change too fast – a magnetic splitting can be observed in the spectrum. The phenomenon is related to the low frequency of both the spin-spin coupling of neighboring pairs of ferric ions and spin-lattice coupling. As for the spin-spin interaction, the closer the iron ions are to each other, the faster is the flip-flop, the paramagnetic phenomenon disappears and the regular doublet is detected. The spins may flip-flop due to the interaction with the lattice, too. This latter interaction is temperature dependent, the higher the temperature, the faster the reversal is in direction of spins. The appearance of the paramagnetic behavior was detected on zeolites as early as in the 1990s [10], a few further examples have also been presented since then [11,12]. In the usual Fe-zeolite samples, the Si/Fe ratio is less than 100, the Fe-O- $(Si\text{-}O)_n$ -Fe distances are too short, the phenomenon of paramagnetism is not manifested. For illustration, a pair of spectra recorded on Fe-SOD (Si/Fe ~ 200) at 77 and 300 K are presented (Figure 3).

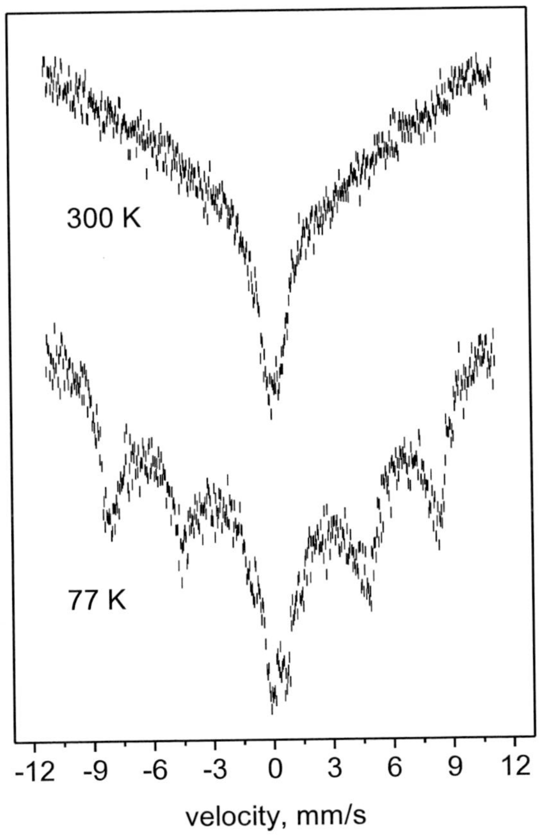

Figure 3. Temperature dependence of the the paramagnetic behavior of FW substituted Fe^{3+} ions reflected on the Mössbauer spectra recorded on an Fe-SOD (Si : Fe = 200) sample.

The sample was prepared by a recrystallisation of the parent zeolite including a step by removing the EFW Fe ions by complexation [13]. Correspondingly, the product contains iron ions in FW position. The 300 K spectrum displays a spectrum characteristic for Fe^{3+} ions with intermediate magnetic relaxation (Figure 3a). In the spectrum recorded at lower temperature (77 K), the magnetic ordering is more expressed in correspondence with the decrease of the rate of spin-spin relaxation among the interacting separated paramagnetic iron ions (Figure 3b). The other possibility to confirm the FW location of Fe^{3+} ions is to analyze their in situ Mössbauer spectra obtained under reducing/oxidizing conditions. Reduction/oxidation treatmens result in characteristic changes in spectra of EFW ions as was mentioned earlier. In contrast, resistance of EFW ions to reducing/oxidizing conditions is more expressed for the same treatments. In addition, the presence of Bronsted acidic sites ($O_3 \equiv Fe\text{-}(OH)\text{-}Si$) can be also deduced from Mössbauer spectra. These sites can only be formed on FW iron ions, where an acidic proton can be attached to compensate the charge introduced by the insertion of iron to framework T sites. The charge symmetry is severely distorted around the FW iron sites where Bronsted acidic centers are generated. Namely, the acidic proton is attached to one of the four oxygens surrounding the central ferric ion, resulting in the strong assymetry of charges. This assymmetry is reflected in the corresponding Mössbauer spectra, the related doublet of FW Fe^{3+} ions exhibit extra large quadrupole splitting (1.7 – 2.2 mm/s). If the charge compensating cations are not attached exclusively only to one oxgen, i.e. their diameter is larger than that of one single proton, the assymetry around the iron is less. In correspondence, the quadrupole splitting is usually less than 1.5 mm/s in these cases. Varieties of Mössbauer spectra recorded under different experimental conditions are presented for Fe-FER and Fe-MFI zeolites in which the previously described features are reflected [14,15].

Combination of Iron Ions, Formation of FW – EFW Pairs

Siting of iron ions in most of cases is neither exclusively FW nor is exlcusively EFW in the practice, usually a certain combination of them occurs. During synthesis, predominance of one characteristic siting can be provided, as the previously mentioned examples demonstrated. Then, during later treatmens, these positions may transform to each other. For instance, if cycling of reduction/oxidation treatments is performed on a ferrisilicate zeolite analogue with iron siting of FW dominace, a certain part of FW ions may change the siting to EFW one. Usually, a part of Fe^{3+} ions may leave the framework during reductive treatment, in reduced Fe^{2+} form. Since Fe^{2+} ions have larger ionic radii than Fe^{3+} have, they cannot be accommodated in tetrahedral T sites, instead, they remain in EFW positions. Once removed from the framework, they may act as charge-compensating counter-ions, e.g. in the form of $Fe^{(2+)}(OH)^{+}$. Furthermore, in that case, combined $Fe_{framework}$-O-$Fe_{extra\text{-}framework}$ pairs may be formed. One of the members of these pairs, that one located in EFW positions, experiences less spatial restrictions, i.e. it may change its size and coordination more freely then the other member, located in the FW sites. Hence, the EFW member may participate in $Fe^{2+} \leftrightarrow Fe^{3+}$ redox processes, or it may change simultaneously the coordination in its ligand sphere eg. in an $Fe^{(3+)}{}_{EFW}O^{+} + \frac{1}{2} H_2 \leftrightarrow Fe^{(2+)}{}_{EFW}(OH)^{+}$ or similar process, while the other one is retained in a confined tetrahedral FW ferric state.

These processes are illustrated in a series of spectra obtained on an Fe-MCM-22 sample exposed to cycled reduction-oxidation treatments (Figure 4). In the first series, at the first evacuation treatment, presence of the previously mentioned Bronsted acidic iron at the ($O_3 \equiv$ Fe-(OH)-Si) sites prevails, as the decomposition of the middle left spectrum in Figure 4 attests with dominating presence of the IS= 0.24 mm/s, QS = 1.92 mm/s component [16]. In the following stage, during a reducing treatment in hydrogen at 60 K, a part of the FW irons is removed to EFW Fe^{2+} positions (presence of Fe^{3+} component increases to 44 %). The original distribution of iron ions among the various positions can only be restored in part by a repeated calcination at 720 K on air. The amount of proton compensated ($O_3 \equiv$ Fe-(OH)-Si) sites decreases, and the presence of the $Fe_{framework}$-O-$Fe_{extra\text{-}framework}$ pairs can be deduced from the spectrum recorded after the second evacuation (Figure 4, middle right spectrum). This assigment is also supported by the subsequent spectrum, recorded after a repeated treatment in hydrogen (Figure 4, bottom right). In this spectrum, the presence of Fe^{2+} prevails, with ferrous iron located partly in $Fe^{(3+)}_{framework}$–O-$Fe^{(2+)}_{extraframework}$ pairs. Further experimental details and description of spectra with data of decompositions are discussed in [16].

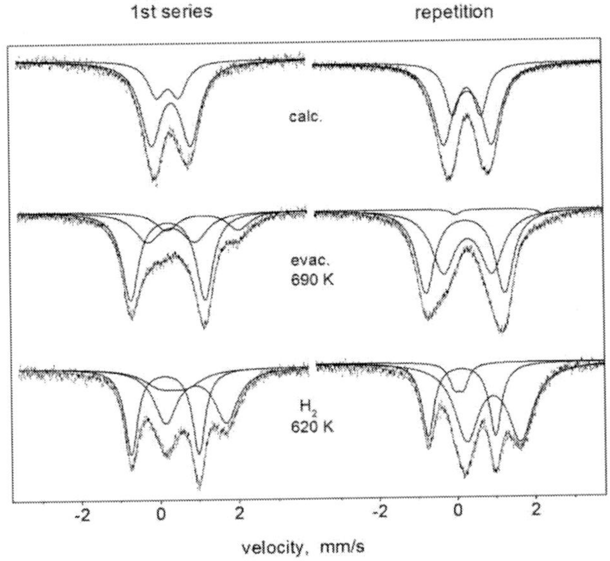

Figure 4. Partial FW → EFW displacement of iron ions during the stages of a reduction-oxidation cycling shown in sequential Mössbauer spectra recorded on an Fe-MCM-22.

References

[1] G. Perego, R. Millini, G. Belussi, in: Molecular Sieves, *Science and Technology*, (Eds. H.G. Karge, J. Weitkamp) Springer, 1998. Vol 5. p. 365.

[2] B. Wichterlova, J. Dedecek, Z. Sobalik, in: Catalysis by Unique Metal Ion Structures, (Eds. G. Centi, B. Wichterlova, A.T. Bell), *NATO Science Series*, (series II, Vol 13). Kluwer, 2001, p. 31.

[3] H.van Bekkum, E.M. Flaningen, P.A. Jacobs, J.C. Jansen (Eds.), Introduction to Zeolite Science and Practice, Elsevier, *Studies in Surface Science and Catalysis*, 137 (2001).

[4] R.G. Burns, *Hyperfine Interactions*, 91 (1994) 739.

[5] Z. Gao, L.V.C. Rees, *Zeolites*, 2 (1982) 215.

[6] Lázár, G. Pál-Borbély, H.K. Beyer, H.G. Karge, *J. Chem. Soc.*, Faraday Trans., 90 (1994) 1329.

[7] K. Lázár, N. Micheaud, R.M. Mihályi, G. Pál-Borbély, H.K. Beyer, *React. Kinet. Catal. Lett.*, 74 (2001) 289.

[8] Lázár, C. Vincent, P. Fejes, *Hyperfine Interactions* 187 (2008) 1.

[9] Ratnasamy, R. Kumar, *Catalysis Today*, 9 (1991) 329.

[10] R. Kumar, S.K. Date, E. Bill, A. Trautwein, *Zeolites*, 11 (1991) 211.

[11] K.A. Dubkov, N.S. Ovanesyan, A.A. Shteinman, E.V. Starokon, G.I. Panov, *J. Catal.*, 207 (2002) 341.

[12] J.B. Taboada, A.R. Overweg, P.J. Koyman, I.W.C.E. Arends, G. Mul, *J. Catal.*, 231 (2005) 56.

[13] P. Fejes, I. Kiricsi, K. Lázár, I. Marsi, A. Rockenbauer, L. Korecz, *Appl. Catal., A: General*, 242 (2003) 63.

[14] K. Lázár, G. Lejeune, R.K. Ahedi, S.S. Shevade, A.N. Kotasthane, *J. Phys. Chem. B* 102 (1998) 4865.

[15] K. Lázár, Á. Szegedi, G.Pál-Borbély, A.N. Kotasthane, P. Fejes, *Catal. Today* 110 (2005) 239.

[16] K. Lázár, Á. Szegedi, F. Martinez, R. Molina, P. Fejes, *Stud. Surf. Sci. Catal*, 158 (2005) 733.

In: Zeolites: Synthesis, Chemistry and Applications
Editors: Moisey K. Andreyev and Olya L. Zubkov
ISBN: 978-1-61942-861-4

Chapter 12

APPLICATIONS OF ZEOLITES IN CATALYTIC OLEFIN POLYMERIZATION

***Irina N. Meshkova*[1], *Tatyana M. Ushakova*[1], *Vitaly G. Grinev*[1], *Valentina A. Nikashina*[2] *and Lyudmila A. Novokshonova*[1]**

[1] Semenov Institute of Chemical Physics of Russian Academy of Sciences, Moscow, Russia

[2] Vernadsky Institute of Geochemistry and Analytical Chemistry of Russian Academy of Sciences, Moscow, Russia

ABSTRACT

The synthetic zeolites are characterized by a great variety of structures and compositions, as well as by the different adsorption ability to ions, clusters and the transition metal complexes. Therefore they are interesting as supports for heterogenization of the olefin polymerization catalysts. In addition the natural zeolites due to their considerable content in nature and low cost may be used as fillers for the polymer composite materials. The immobilized metalcomplex catalysts of olefin (ethylene and propylene) polymerization have been developed on the base of zeolites. In order to prepare these catalysts, the anchored aluminoxanes have been synthesized by reaction of aluminum alkyls ($AlMe_3$, $AlEt_3$ and $AliBu_3$) with water contained in intracrystalline cavities and pores of zeolites (NaY, NaZSM-5, HZSM-5, NH_4ZSM-5, mordenite and clinoptilolite-bearing tuff). These aluminoxanes, with zirconocenes, vanadium chlorides and other transition metals, form the heterogenized complexes that are active in the polymerization of ethylene and propylene without addition of other organoaluminum cocatalysts. The nature of zeolites used as a support, determines the content of zeolite water and affects the formation of anchored aluminoxanes and activity of immobilized catalysts. Polyolefins prepared with the immobilized on zeolites zirconocene catalysts demonstrate higher molecular mass and a wider molecular-mass distribution than those of corresponding homogeneous catalysts. The stereoselectivity of supported catalysts depends on the composition, structure of metallocenes and on the zeolite nature. The new composite materials on the base of ultrahigh-molecular-mass polyethylene UHMPE and natural zeolites were developed. The synthesis of composites was carried out by ethylene polymerization on the catalytically activated surface of zeolite fillers. Depending on the amount of zeolite filler in the UHMPE matrix such compositions can be used as wear-resistant, impact-resistant materials for production of units working in conditions of dry friction at high loading or as organomineral ion-exchange sorbents for water purification from radionuclides and heavy metal ions.

INTRODUCTION

Catalytic polymerization of olefins belongs to the key trends of the polymer chemistry and technology. Such polymers with most large-scale production as PE and PP are synthesized using metalcomplex catalysts. One of the important problems in the up-to-date metalcomplex catalysis of olefin polymerization is immobilization of highly active homogeneous metallocene catalysts activated by methylalumoxane (MAO) for application in the existing technological processes of polyolefin synthesis. Among the known methods of immobilized metalorganic catalysts preparing one of the simple method includes the formation of metalcomplexes with anchored alkylaluminoxanes, synthesized in the partial hydrolysis reaction of aluminum alkyls with the surface and inner bound water of inorganic and organic supports [1]. As well as vanadium precatalysts in the immobilized Ziegler–Natta systems [2], Zr-cenes form with the anchored alkylaluminoxanes ($support(H_2O)/AlR_3$) complexes, which are active in olefin polymerization without addition of other organoaluminum cocatalysts [3-6].

Table 1. Characteristics of the study synthetic zeolites

Zeolite	Si/Al ratio	Surface Area, m^2/g (BET)	Average Pore Diameter, Å (BJH)	Pore Volume*, cm^3/g (BJH)
NaY	6.0	726.39	55.00	0.0106
NaZSM-5	42.0	244.46	64.91	0.0211
Mordenite (Na form)	12.4	244.45	378.53	0.0235

*Volume of pores with Diameter between 17.00 and 3000.00 Å.

HZSM-5 (Si/Al = 17 and 34), NH4ZSM-5 (Si/Al = 32) and NaZSM-5 (Si/Al = 42)

Zeolites are of great interest as support for heterogenization of olefin polymerization catalysts owing to the diversity of their structural forms and compositions, their ability to adsorb ions, clusters, and complexes of transition metals, and also due to the high content of mobile zeolite water. In this chapter the effect of the nature of the zeolite support on the formation of anchored aluminoxanes, activity of immobilized zirconocene catalysts ($support(H_2O)/AlR_3/Zr\text{-}cene$) in olefin polymerization and properties of polymer products are examined. The used zeolites (NaY, NaZSM-5, HZSM-5, NH_4ZSM-5, mordenite and clinoptilolite-bearing tuff) are characterized by different topologies of the zeolite skeleton [7], Si:Al ratio and cations incorporated into the zeolite lattice (Table 1). On the zeolites surface the anchored aluminoxanes have been synthesized on the base of $AlMe_3$, $AlEt_3$ and $AliBu_3$, followed by formation of heterogenized complexes with zirconocenes Cp_2ZrCl_2 (**1**), *ansa*-zirconocenes rac-Et(Ind)$_2$ZrCl$_2$ (**2**), rac-Me$_2$Si(Ind)$_2$ZrCl$_2$(**3**), rac-[1-(9-η^5-Flu)-2-(5,6-

cyclopenta-2-Me-1-η^5-Ind)C_2H_4]$ZrCl_2$(**4**) and vanadium chlorides ($VOCl_3$, VCl_4). The prepared catalysts were used in the ethylene and propylene polymerization.

Formation of Anchored Aluminoxanes and Zeolite-(H_2O)/AlR_3/Zr-Cene Catalysts

Organoaluminum compounds (OACs) are characterized by an extremely high reactivity with respect to water. Under conditions of an uncontrolled hydrolysis of the OACs, a mixture of products composed of aluminoxane, hydroxylated organoderivatives of aluminum, polymer alkylaluminoxanes, and three-dimensional alumina are formed [8]. In the soft hydrolysis condition (low temperature and low water concentration, electronodonor solvent) the main product of reaction is alkylaluminoxane. The composition, molecular mass, and structure of alkylaluminoxane depend not only on temperature and the nature of solvent (inert or electron-donor) but also on the molar ratio of reagents (AlR_3 : H_2O).

Thus, at a molar ratio of AlR_3: H_2O = 2:1, the hydrolysis mostly yields tetramethylaluminoxane [9] and tetrabutylaluminoxane with molecular-mass (M) 240-285 [10] as the final products of hydrolysis. Polyisobutylaluminoxanes with more than 26 Al-O-Al units and M = (2.0 - 3.5) x 10^3 are produced when the synthesis is performed at an equimolar AlR_3 : H_2O ratio [10].

The reaction of $AlMe_3$ with water of the hydrated support (montmorillonite) at $AlMe_3$: H_2O = 1 : 1 in toluene gives oligomer methylaluminoxane with a structure similar to that of commercial aluminoxane (MAO). This was shown by temperature-programmed desorption measurements combined with mass spectrometric analysis [11].

The effect of zeolite nature on the formation of anchored aluminoxanes via the partial hydrolysis of aluminum alkyls under the action of zeolite water was examined. In Table 2 the results of zeolite dehydration – by thermal action and under the action of $AlMe_3$ – are given. According to derivatograms of the zeolite the content of zeolite water was determined. With the use of the McBain balance (the sensitivity of the balance quartz spring was 2.4 x 10^{-3} g/mm), the kinetics of desorption of zeolite water in vacuum at room temperature was studied. The amount of gas products from the partial hydrolysis of $AlMe_3$ methane, quantity of zeolite water, interacting with aluminiumalkyl, was established.

It was found that the content of mobile water in zeolite decreases with diminishing Al in its lattice (with increase of Si:Al ratio). All water adsorbed on the outer zeolite surface and contained within its channels and crystalline cavities interacts with aluminum alkyl (Table 2, columns 3 and 4). At room temperature in a vacuum, water is more readily removed from zeolites in the Na form (Table 2, column 5).

The kinetic curves of methane evolution on reaction of zeolite water with $AlMe_3$ at eguimolar$AlMe_3$: H_2O ratio are given in Figures 1a,b. The data show that the dehydration of the zeolites NaY and NaZSM-5 is at a much higher rate than that of other zeolites. The methane evolution completes for NaY and NaZSM-5 already after 3 and 5 min (Figure 1b), and after 20 to 30 min for HZSM-5 and NH4ZSM-5 (Figure 1a) [12]. For this reason, in the case of NaY and NaZSM-5 zeolites, the hydrolysis condition required for oligomer methylaluminoxane synthesis ($AlMe_3$: H_2O = 1:1) are actually realized.

Table 2. The results of zeolites dehydration by thermal treatment and using AlMe3

Zeolite	Si/Al ratio	Contents of H_2Oin zeolite by TGA,		Quantity of zeolite H_2O, take part in reaction with $AlMe_3$, mmol/g	Quantity of zeolite H_2O, desorbed (vacuum, 22^0C, hour), mmol/g
		mas%	mmol/g		
NaY	6.0	25.70	14.30	13.35	6.90
NaZSM-5	42.0	9.10	5.05	5.10	2.00
Mordenite (Naform)	12.4	13.9	7.70	8.00	1.75
HZSM-5	17.0	9.18	5.10	5.10	1.70
HZSM-5	34.0	4.40	2.44	3.00	1.90

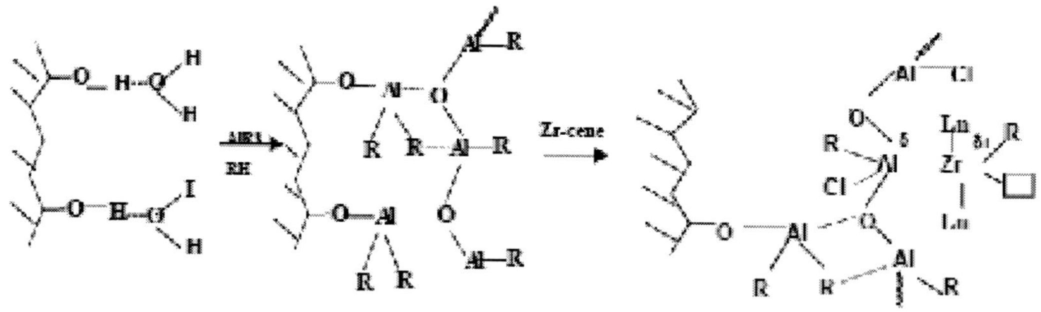

Scheme 1. Formation of anchored alkylalumoxanes and *zeolite-(H_2O)/AlR_3/Zr-cene* catalysts.

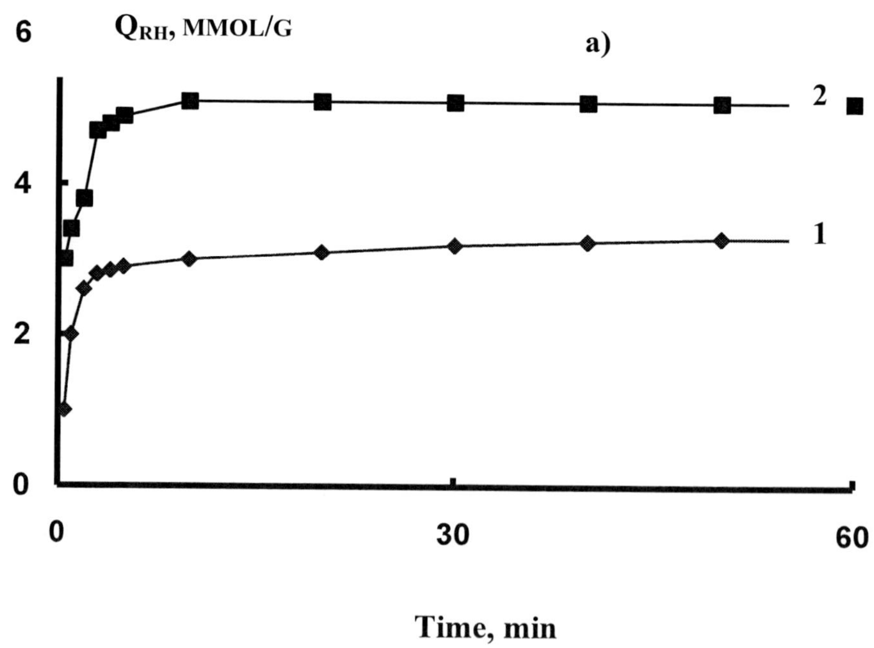

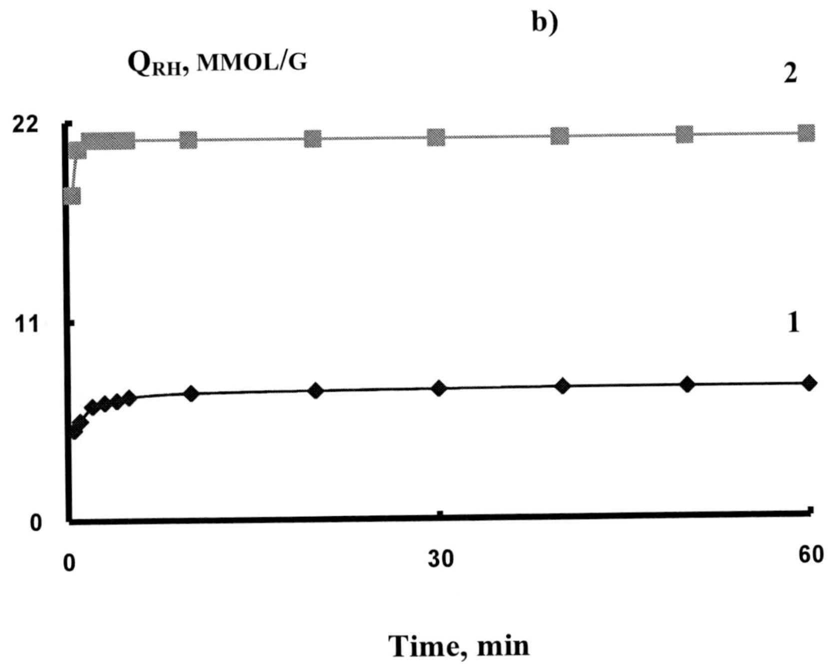

Figure.1a,b. Kinetic curves of methane release during the reaction of $AlMe_3$ with zeolite water in toluene at 22^0C (with disregard for the part of methane dissolved in toluene): (a) HZSM-5 (1) with Si : Al = 34 and HZSM-5 (2) with Si : Al = 17; (b) NaZSM-5 (1) and NaY (2).

According to the $AlMe_3$ and water adsorption isotherms obtained for NaZSM-5, previously heated in a vacuum at 300^0 C, $AlMe_3$ does not penetrate into the micropores of zeolite contrary water (Figure 2) [5]. Thus, heterogenized aluminoxanes and Zr-cene-aluminoxane complexes are formed in the secondary porous structure of zeolite and on its surface but not inside zeolite particles.

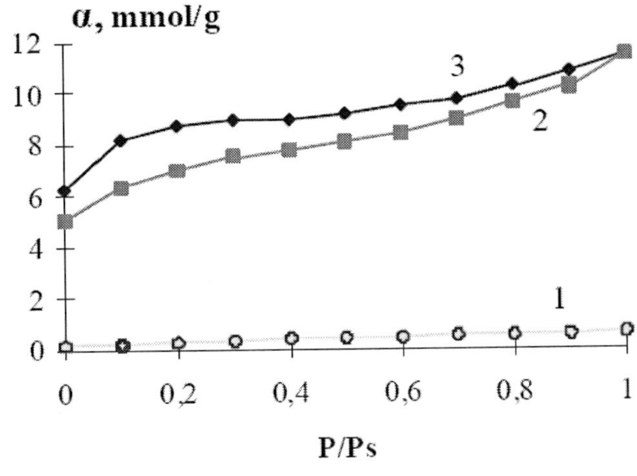

Figure 2. Isotherms of adsorption of $AlMe_3$ (1), adsorption (2) and desorption (3) of H_2O for NaZSM-5 zeolite heated in vacuum at 300^0 C.

Activity of Zirconocene Catalysts Immobilized on Zeolites and Kinetic Regularities of Ethylene and Propylene Polymerization on Zeolite(H_2O) /AlR_3 / Zr-Cene Catalysts

The study of the properties of the zirconocene catalysts show that the systems with zeolite supports NaY and NaZSM-5 are the most efficient in the polymerization of ethylene and propylene (Table 3 and 4). Their activity is close to the activity of the corresponding homogeneous catalyst. The activity of the catalysts on zeolite supports of the same type depends strongly not only on the Si : Al ratio but also on the nature of cations embedded into the zeolite lattice. Thus, when Na^+cation is replaced by H^+ or NH_4^+ in the ZSM-5 support, activity of the immobilized catalyst decreases drastically. As well known, AOCs interact with the compounds containing mobile hydrogen (water, alcohols, including organosilicon alcohols, ammonia, primary and secondary amines). In the synthesis of our catalysts, after elimination of zeolite water by aluminum alkyl, cations, incorporated to zeolite lattice, can react with heterogenized alkylalumoxane, resulting in a lowering of activity of the *zeolite(H_2O)/AlR_3/Zr-cene* catalytic system.

Kinetic regularities of olefin polymerization in the presence of zirconocene catalysts supported on zeolites are common for immobilized organometallic systems.

During the polymerization of ethylene and propylene, depending on the zirconocene type and concentrations of catalyst, the rate of the process increases or sharply decreases at the initial stage and then becomes steady-state (Figures 3 and 4).

An extremum dependence of the specific polymerization rate of olefin on the zirconium content in the zeolite support is observed (Figure 5). Thus, in ethylene polymerization with *NaZSM-5(H_2O)/$AlMe_3$ /Cp_2ZrCl_2* catalyst under the studied experimental conditions, the maximum values of Rp^{max} and Rp^{stat} are attained at a zirconocene concentration of $2.8x10^{-6}$mol/g zeolite which corresponds to a surface zirconocene concentration of $7x10^{-8}mol/m^2$ zeolite [13].

Table 3. Activity of catalysts zeolite(H_2O)/$AlMe_3$/2 and homogeneous system 2-MAO in ethylene polymerization

Zeolite	Ratio Si:Al	Zr- cene, μmol	OACs	OACs, mmol	$T_{pol,}$ ^0C	P_{C2H4}, bar	Activity, kg PE/mol Zr.bar.h
Homogeneous system	—	1.4		3.5	33	0.20	1800
		16.0	MAO	16.0	40	0.58	480
		2.1		6.8	74	0.36	1650
NaY	6.0	2.0	$AlMe_3$	8.9	50	0.40	1050
		6.7		14.8	50	0.40	860
NaZSM-5	42.0	2.6	$AlMe_3$	11.6	74	0.35	1180
		3.5		14.0	50	0.50	400
Mordenite (Na form)	12.4	3.3	$AlMe_3$	7.7	50	0.40	300
		8.8		14.0	50	0.40	620

Table 4. Activity of catalysts zeolite(H2O)/AlMe3/Zr-cene and homogeneous system Zr-cene-MAO in propylene polymerization

Zeolite	Si:Al ratio	Zr-cene	Zr-cene, μmol	OACs	OACs, mmol	T_{pol}, ^0C	P_{C3H6}, bar	Activity, kgPP/ molZr.bar.h
Homogeneous system	□	2	2.4	MAO	4.0	40	0.57	1115
NaY	6	2	3.4	$AlMe_3$	9.6	40	0.57	2150
NaZSM-5	42	2	2.6	$AlMe_3$	10.8	40	0.57	1050
HZSM-5	17	2	3.1	$AlMe_3$	8.0	40	0.57	405
HZSM-5	34	2	2.4	$AlMe_3$	7.5	40	0.57	220
NH_4ZSM-5	32.3	2	2.4	$AlMe_3$	8.0	40	0.57	210
NaY	6	4	4.8	$Al(iBu)_3$	4.1	50	0.68	1100
NaZSM-5	42	4	2.1	$Al(iBu)_3$	6.0	50	0.68	270

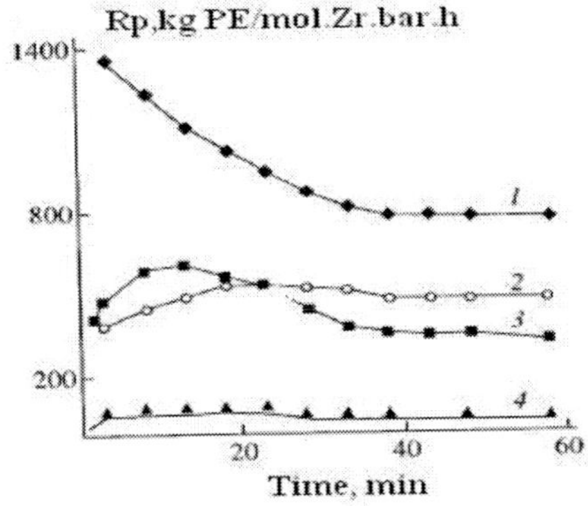

Figure 3. Rate-time profiles for ethylene polymerization with zeolite supported NaZSM-5(H_2O)/$AlMe_3$/**1**(2,3,4) and homogeneous **1**-MAO (1) catalysts. Polymerization conditions: P_{C2H4} = 0.46 bar; T_{pol} = 34^{0}C; amount of Zr to zeolite, wt.% : 0.024 (1), 0.028 (2), 0.16 (3); [Zr] = 14 10^{-2}mmol/l (4); [$AlMe_3$] mmol/l: 68(1), 60 (2), 90 (3), 70 (4); toluene.

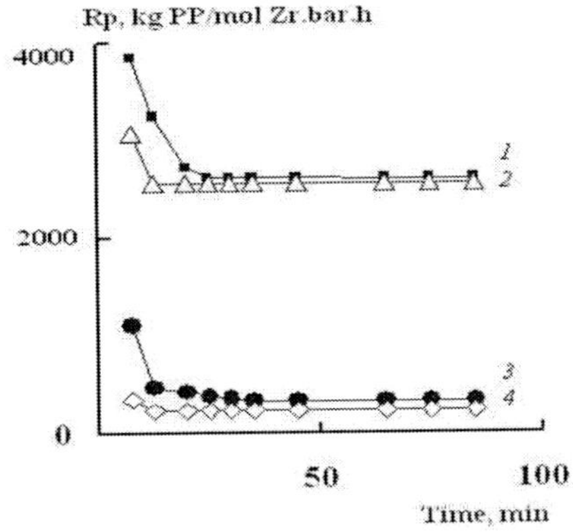

Figure 4. Rate-time profiles for propylene polymerization with zeolite(H_2O)/$AlMe_3$/**2** catalysts. Zeolite support: NaY (1), NaZSM-5 (2), HZSM-5 with Si:Al = 17 (3), and NH_4ZSM-5 (4). Polymerization conditions: [C_3H_6] = 0.26mol/l; T_{pol} = 40^0 C; amount of Zr-cene, μmol: 3.4 (1), 2.6 (2), 3.1 (3), and 2.0 (4); amount of $AlMe_3$, mmol = 9.5 (1), 10.6 (2), and 8.0 (4); toluene.

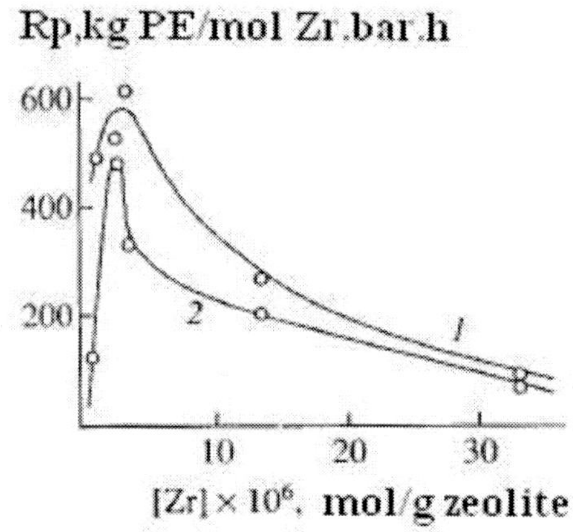

Figure 5. Effect of Zr-cene concentration on the maximum (1) and stationary (2) rate of ethylene polymerization with NaZSM-5(H_2O)/$AlMe_3$/**1** catalyst. Polymerization conditions: [C_2H_4] = 0.054mol/l; T_{pol} = 34^0 C; [$AlMe_3$] = 90 mmol/l.

Polymerization of olefin on *zeolite(H_2O)/$AlMe_3$/Zr-cene* catalysts proceeds with a positive temperature coefficient in the range of 30-75^{0}C. The values of activation energy for ethylene and propylene polymerization with the zeolite supported zirconocene catalysts (Table 5) are typical for processes of polymerization and their monomers with traditional heterogeneous Ziegler–Natta catalysts.

Table 5. Activation energy for ethylene and propylene polymerization with the zeolite supported zirconocene catalysts

Catalyst	Monomer	E_{obs}, kJ/mol
NaZSM-5(H_2O)/$AlMe_3$/1	Ethylene	20
NaZSM-5(H_2O)/$AlMe_3$/2		32
NaZSM-5(H_2O)/$AlMe_3$/2	Propylene	48
NaZSM-5(H_2O)/$AlMe_3$/4		31.3

In the case of homogeneous metallocene catalysts, the length of polyolefin chains is controlled by the main reactions of chain transfer via β-H elimination to Zr and the coordinated monomer. These processes lead to formation of unsaturated end groups in PP. Chain transfer to the organoaluminum cocatalyst occurs only at very high Al : Zr molar ratios (> 8000) [14, 15]. In our studies of propylene polymerization on the immobilized catalysts, the 2 and 4 molar ratio Al : Zr was less 4000. The concentration of end –C = C– groups in PP samples synthesized at various monomer concentrations (0.26-10.0 mol/l) was determined by the method of ozonation of double bonds [16]. Using these data the number-average molecular mass ($M_{C=C}$) of PP and the degree of polymerization of propylene ($P_{C=C}$) can be calculated. It was established, that as in the case of homogeneous catalysts [14], the dependences 1/$P_{C=C}$ of PP vs. 1/[C_3H_6] are linear in a wide range of monomer concentrations (Figure 6). That is, in the polymerization of propylene on supported *ansa*-zirconocene catalysts, reaction of the chain transfer via β-H elimination to Zr and the coordinated monomer are the main ones.

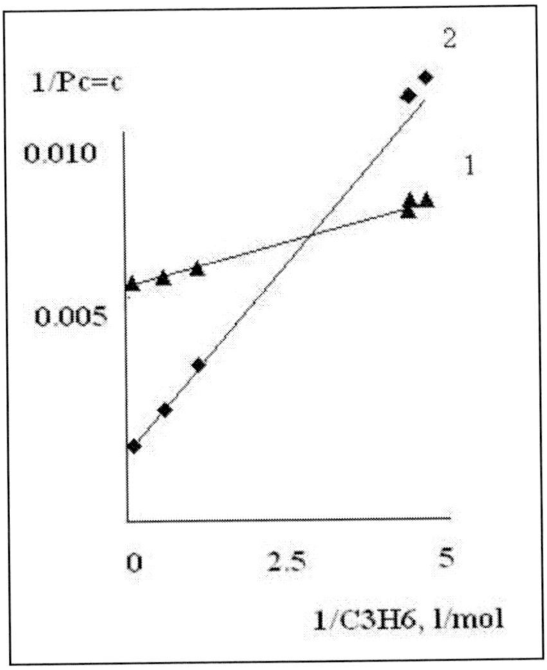

Figure 6. The number-average degree of polymerization of PP vs. monomer concentration. Catalyst: NaZSM-5(H_2O)/$AlMe_3$/**2** (1) and NaZSM-5(H_2O)/$AlMe_3$/**4** (2). Polymerization conditions: $T_{pol} = 40^0C$; amount of Zr-cene = 1.3–2.6 μmol; Al:Zr = 4000-3000, mol/mol.

Comparison of the Properties of Polyethylene and Polypropylene Obtained Using Supported Zeolite(H_2O)/AlR_3/Zr-Cene and Homogeneous Zr-Cene/MAO Catalysts

The use of *zeolite(H_2O)/$AlMe_3$ /Zr-cene* catalysts promotes the formation of polyolefins with higher molecular mass and wider molecular mass distribution ($M_w/M_n>>2$) as compared to similar homogeneous systems (Table 6) [5, 12 and 13].

Immobilization of *ansa*-zirconocene catalysts on the zeolite support can affect their stereospecific properties, and the character of effect depends on the composition and structure of metallocenes. The isotacticity index of PP [12] and content of isotactic pentad mmmm in PP increases on polymerization with **2** and **3** catalysts of C_2 symmetry immobilized on NaZSM-5 and NaY zeolites compared to homogenous system (Table 7). With the use of an immobilized "dual side" catalyst **4** of C_1 symmetry the content of isotactic pentads mmmm in PP decreases, and the number of stereo errors (mmmr, mmrr, and mrrm) increases [12].

Thus, the supported zirconocene catalysts with alkylaluminoxanes anchored to zeolites are active in polymerization of olefin without the addition of free MAO and another aluminumorganic cocatalyst. Kinetic regularities of ethylene and propylene polymerization arc typical for olefin polymerization with heterogenized catalysts. It is possible to conclude that the polymerization in the presence of the *zeolite(H_2O)/AlR_3/Zr*-cene catalysts occurs mainly on the support and the transition of active centers into the solution is less likely.

Table 6. Molecular-mass characteristics of PE and PP prepared with homogeneous and zeolite supported zirconocene catalysts

Catalyst	[Zr], mmol/l	[Al], mmol/l	Poly-olefin	Mw	Mw / Mn
1 – MAO	0.17	100	PE	87870	4.8
*NaZSM-5(H_2O)/AlMe$_3$/ **1***	0.21	90	PE	150500	18.7
2 – MAO	0.40	200	PP	8800	2.0
*NaZSM-5(H_2O)/AlMe$_3$/ **2***	0.19	160	PP	58200	18.9
*NaZSM-5(H_2O)/AlMe$_3$/ **3***	0.06	70	PP	58750	15.3

Conditions of olefin polymerization: [C_2H_4] = 0.06 mol/l, 34^{0}C; [C_3H_6] = 0.26 mol/l, 40^{0}C.

Table 7. Stereospecific properties of homogeneous and zeolite supported *ansa-Zr*-cene catalysts

Catalyst	[Zr], μmol	Al/Zr	$T_{pol,}$ ^{0}C	[C_3H_6], mol/l	mmmm, %
2 – MAO	3.3	4000	40	11.3	73.7
*NaZSM-5(H_2O)/AlMe$_3$/ **2***	2.6	3040	40	11.3	75.6
*NaY(H_2O)/AlMe$_3$/ **2***	2.1	4700	40	11.3	84.8
3 – MAO	1.5	2600	40	0.26	48.0
*NaZSM-5(H_2O)/AlMe$_3$/ **3***	3.3	1300	40	0.33	85.0
*NaY(H_2O)/AlMe$_3$/ **3***	1.9	3600	40	0.26	78.3

The Composite Material Developed on the Base of Natural Zeolites and Ultrahigh-Molecular Mass Polyethylene (UHMPE)

High nature reserves and low cost zeolite-containing tuffs including quartz, montmorillonite, mica and other minerals apart from the main component (clinoptilolite, chabazite, mordenite), may be used as the fillers of polymers to obtain the polymer composite materials. Finely dispersed (powdered) fractions of natural zeolites are applied in rubber mixtures for intensification of the vulcanization process in the production of artificial leather, as a filler of polyethylene (LDPE and HDPE), polypropylene, polystyrene and other polymers. These composites are usually prepared by the traditional method of mechanical blending of filler and polymer in melt.

In 1975, an essentially new procedure for the polymer filling was suggested in Semenov Institute of Chemical Physics of Russian Academy of Sciences [17]. The process of polymerization filling comprises three major stages: (1) preparation of filler, (2) catalytic activation of filler surface and (3) monomer polymerization on the surface of the filler, resulting in the formation of a polymer shell of controlled thickness around the particles of filler [18]. The method of polymerization filling has no alternatives in the development of composites with high- and ultrahigh-molecular-mass polymer matrixes and ultrahigh-filled composites.

Finely dispersed (powdered) fractions of natural zeolites from various deposits in Russia (Kholinskoye, Pegasskoye, and Shivyrtuyskoye) with the particle size from 0.5 to 60 μm were used for synthesis of composite materials on the base on ultrahigh-molecular-mass polyethylene (UHMPE). Vanadium and zirconocene catalysts were applied for activation of zeolites [1, 18, and 19]. The composite material developed on the basis of natural zeolites and UHMPE combines enhanced wear resistance properties with a low friction coefficient and has good deformation-strength characteristics (Tables 8 and 9) [20]. The material may find an application in the manufacturing of units working in conditions of dry friction at high loading: in construction of movable supports of cross-country gas pipe-lines; facing plates for protection of transport and excavation mining equipment from adhesion of rocks.

Table 8. Comparative tribological characteristics of the natural zeolite filled UHMPE, unfilled UHMPE and fluoroplastic (F-4) (surface roughness of counterpart, 0.35μm; loading, 47N)

Material	Wear coefficient with respect to steel, K· 10^6, mm^3 / N m	Friction coefficient
UHMPE-zeolite(30 wt.%, fraction with 50-60 μm particles)	0,6/1,1	0,17
UHMPE-zeolite (40 wt.%, fraction with 50-60 μm particles)	1,1/1,3	0,17
UHMPE-zeolite (30 wt%, nonfractioned with the particles smaller than 60 μm)	0,6/0,3	0,17
UHMPE (M = 1,2 X 10^6)	1,23/1,2	0,17
Fluoroplastic (F-4)	0,1/1,0	0,19

The unique properties of natural zeolites are implemented in the ion-exchange processes related to protection of the environment. Zeolites with particle sizes from 10-15 mm, to finely dispersed fractions with particles of micrometer size, are produced in development of zeolite deposits. At present, sorbents are used only with the large particles (1-3 mm) in the ion-exchange processes. The powdered fractions of natural zeolites lose their filtering ability very quickly when aqueous solutions are passed through them due to high zeolite material compaction.

While studying the ion-exchange processes related to water purification it was found [21] that crushing of natural zeolite material (chabazite) at mechanical activation is an effective method of increasing its selectivity. In work [22], regimes of activation of natural zeolites were found that allow one to obtain the finely dispersed mechanically activated zeolites with enhanced selectivity.

The modification of finely dispersed natural zeolites was carried out for their application as an organomineral sorbent for purification of environmental water and sewage from radionuclides of strontium and cesium, from heavy metals and ammonium under static and dynamic conditions [19, 23]. A water-permeable polyolefin (PE) coating on the surface of zeolite particles was formed by ethylene polymerization (Scheme 2). The material represents ultrahigh-filled polyolefin compositions with the content of the finely dispersed natural zeolite of 95-97 wt.%. The water-permeable polyolefin coating on zeolite particles is inert, characterized by high chemical stability, and transforms fine (with the size of 0.5-60 μm) particles of zeolite into larger water-permeable agglomerates of particles (> 150 μm).

Table 9. Stress-strain properties of UHMPE and the natural zeolites filled UHMPE

Content of zeolite in composite, wt%	Temperature of zeolite dehydration, ^0C	Yield point in elongation, MPa	Tensile strength, MPa	Elongation at break, %	Tensile elastic modulus, MPa	Specific impact toughness by Izod, at 22 °C, kJ/m^2
Unfilled UHMPE	–	25	40	360	600	116
22	150	24	38	270	700	115
30	300	22	31	320	740	111
40	300	23	22	300	770	60

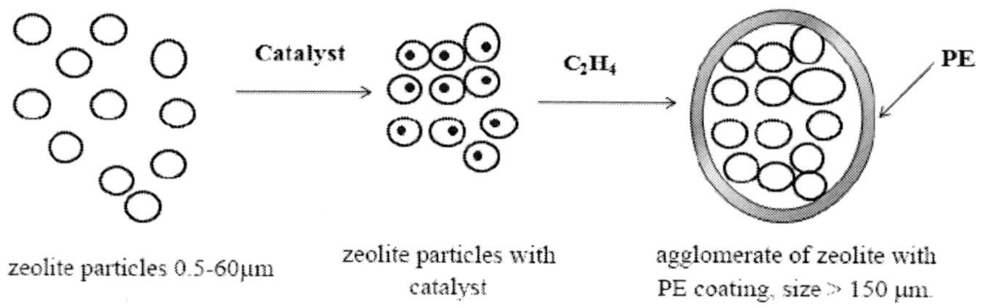

Scheme 2. The modification of zeolite by polyethylene.

Cation-exchange properties of natural zeolite are retained after its modification by polyolefin. As a result of encapsulation, finely dispersed zeolite acquires the filtering capacity which is necessary for its use in sorption processes both under static and dynamic conditions (Table 10). No analogues of such composite sorbents are known.

Table 10. Total cation-exchange capacity (TCEC) and dynamic cation-exchange capacity (DCEC) *) by NH_4^+ of initial CLT) and PE modificated CLT (PE-CLT)**

Sorbent	TCEC, meq /g	DCEC, meq /g*)
CLT	1.37	It is impossible to determine
CLT + 3 mass%PE	1.32	1.12
CLT + 5 mass%PE	1.24	1.10

CLT**) -The clinoptilolite-bearing tuff from Shivyrtuiskoye deposit (Russia) contained 60 ±6wt% of clinoptilolite, the fraction with the 50-60 μm particles. *) DCEC was determined for solution flowrate V = 0.016 cm^3/s.

REFERENCES

[1] Enikolopov, N.S.; D'yachkovskii, F.S.; Meshkova, I.N.; Ushakova, T.M.; Dubnikova, I.L. USSR Inventor's Certificate No. 1 066 193, 1982, *Byull. Izobret.*, 1985, 11, 199.

[2] Gul'tseva, N.M.; Ushakova, T.M.; Krasheninnikov, V.G.; Meshkova, I.N.; D'yachkovskii, F.S. *Polimery.,* 1989, 6-7, 247-257.

[3] Kaminsky, W. US Pat. 4 431 788, 1984; *Chem. Abstr.*,1981, 95, 151475.

[4] Weiss, K. In Book *Metallorganic Catalysts for Synthesis and Polymerization*; Ed. Kaminsky, W.: Springer, Berlin, 1999, 97-101.

[5] Meshkova, I.N.; Ushakova, T.M.; Ladygina, T.A.; Kudinova, O.I.; Novokshonova, L.A. *Polym.Bull.*, 2000, 5-6, 461-468.

[6] Novokshonova, L.A.; Meshkova, I.N.; Kovaleva, N.Yu.; Ladygina, T.A.; Kudinova, O.I.; Leipunskii, I.O. In Book *Future Technology for Polyolefin and Olefin Polymerization Catalysis;* Ed. Terano M. and Shiono T.: Tokyo, 2002, 162-167.

[7] Breck, D. *Zeolite Molecular Sieves*, Wiley, New York, 1974.

[8] Akaki, T.; Aoyagi, T.; Ueuama, N. *J. Polym. Sci., Part A, Polym. Chem.*, 1973, 11, 699-703.

[9] Korneev, N.N.; Khrapova, I. M.; Polonskii, A.V. *Izv. Akad.Nauk SSSR, Ser. Khim.*, 1993, 8,1453-1457.

[10] Lelyukhina, Yu.L.; Sakharovskaya, G.B.; Kurilenko, G.N.; Skulkova, D.P. In Book *Metallorganic Components of Catalysts.*Ed.GNIIKhTEOS/NIITEKhim, Moscow, 1986, 68-79.

[11] Novokshonova, L.A.; Kovaleva, N.Yu.; Ushakova, T.M. *Kinet. Katal.*, 2005, 46, 904-910.

[12] Meshkova, I.N.; Kudinova, O.I.; Kovaleva, N. Yu.; Grinev V.G.; Ladygina, T.A.; Kiseleva E.V.; Novokshonova, L.A. *Polym. Sci. B*, 2009, 51, 401-408.

[13] Meshkova, I.N.; Ladygina, T.A.; Ushakova, T.M.; Novokshonova, L.A. *Polym. Sci. A*, 2002, 44, 1310-1318.

[14] Jungling, S.; Mulhaupt, R.; Stehling, U. *J. Polym. Sci., Part A: Polym. Chem.*, 1995, 33, 1305-1310.

[15] Resconi, L.; Fait, A.; Piemontesi, F.; Colonnesi, M.; Rychlicki, H.; and Zeigler, R. *Macromolecules*, 1995, 28, 6667-6676.

[16] Razumovskii, S.D.; Lisitsyn, D.M.. *Polym.Sci. A*, 2008, 50, 1187-1197.

[17] Kostandov, L.A.; Enikolopov, N.S.; D'yachkovskii, F.S., and Novokshonova L.A. *US Pat.* 4241112, 1979.

[18] Novokshonova L.A., and Meshkova I.N. *Polymer Science*, 1994,36, 517-525.

[19] Meshkova, I.N., Nikashina V.A., Grinev V.G., Ushakova, T.M.; Serova I.V.; Novokshonova L.A., RU Pat. 2284857, B 01J20/16, 2005, *Byul. Izobr.*, 2006, no 28.

[20] Ushakova, T.M.; Meshkova, I.N.; Guruli N.T.; KovalevaN.Yu.; Gul'tseva, N.M.; Grinev V.G., and Novokshonova L.A. *Polymer Science*, 1998, A40, 7, 658-662.

[21] Yusupov, T.S.; Nikashina, V.A.; and Shumskaya, L.G. In *Materials of 6 Congress Dressers CIS*, 2007, Moscow, 201-202.

[22] Nikashina V.A.; Streletsky A.N.; Kolbanev, I.V.; Meshkova, I.N.; Grinev V.G.; Serova I.B.; Yusupov, T.S.; and Shumskaya, L.G. *Clay Minerals*, 2011, 46, 329-337.

[23] Meshkova, I.N.; Nikashina V.A.; Ushakova, T.M.; Grinev V.G.; KovalevaN.Yu.; and Novokshonova L.A. *Vysokomolekulyarnye Soedineniya*, 2005, B 47, 9, 1755-1760.

In: Zeolites: Synthesis, Chemistry and Applications
Editors: Moisey K. Andreyev and Olya L. Zubkov
ISBN: 978-1-61942-861-4

Chapter 13

NORMALIZED KNUDSEN-BASED PERMEANCE FOR EVALUATION OF MICROPORE SIZES OF ZEOLITE AND SILICA MEMBRANES

Hye Ryeon Lee, Gang Li, Masakoto Kanezashi, Tomohisa Yoshioka and Toshinori Tsuru*

Department of Chemical Engineering,
Hiroshima University, Japan

ABSTRACT

The determination of pore sizes is very important for the preparation, characterization, and application of porous membranes. Normalized Knudsen-based permeance (NKP) was proposed for pore-size evaluation of microporous membranes, which is based on the gas translation model.

In this article, NKP was briefly summarized and verified using the permeance of zeolite membranes such as MFI (0.55×0.56 nm) and DDR (0.36×0.44 nm). Moreover, we proposed the use of structured alkoxides for the control of amorphous silica networks with loose structure. Silica membranes were prepared via a sol-gel process using two kinds of organoalkoxysilanes with pendant group (methyltriethoxysilane (MTES) and phenyltriethoxysilane (PhTES)) and three kinds of bridged alkoxides (hexaethoxysiloxane (HEDS), tetraethoxydimethyl disiloxane (TEDMDS) and bis (triethoxysilyl) ethane (BTESE)).

Judging from the NKP fitting curves, the order of average pore size for silica membranes by organoalkoxysilanes and bridged alkoxides could be estimated as follows: TEOS < MTES < PhTES and TEOS < HEDS < TEDMDS≈BTESE, respectively. It can be concluded that the membrane pore size or structure of silica membrane can be controlled for larger pore sizes with various types of structured alkoxides.

---

* Email: tsuru@hiroshima-u.ac.jp.

INTRODUCTION

Microporous inorganic membranes are promising for separation of molecular mixtures including gas and liquid phase with energy saving, since they can be used at high temperatures and in severe conditions. Inorganic membranes can be categorized into zeolites and ceramic membranes such as silica membrane. TEOS-derived silica membranes consist of amorphous silica networks with micropores of approximately 0.3 nm that allow small molecules such as helium (kinetic diameter: 0.26 nm) and hydrogen (kinetic diameter: 0.289 nm), to permeate [1-2]. The pore size is appropriate for the separation of hydrogen from small molecules such as N_2 (kinetic diameter: 0.36 nm) and CO_2 (0.33 nm), but appears to be too small for separation of mixtures with large molecules such as n-C_4H_{10} (0.43 nm) and i-C_4H_{10} (0.50 nm). Therefore, the control and characterization of the pore size for SiO_2 membranes is quite an important issue.

Pore size control of a sol-gel-derived SiO_2 membrane can be achieved by using structured alkoxides which are typically categorized into pendant-type alkoxides [3-9] and bridged alkoxides [10-15]. Pendant-type alkoxides have organic functional groups such as methyl and phenyl, directly attached to a Si atom, while bridged alkoxides contains functional groups between two silicon atoms, such as bis(triethoxysilyl) ethane (BTESE, ≡Si-C-C-Si≡ unit) or bis(triethoxysilyl) methane (BTESM, ≡Si-C-Si≡ unit). The determination of pore sizes is very important, for the preparation, characterization, and application of porous membranes. In general, for the estimation of pore size distribution less than 1 nm, the permeance of several gases with different molecular sizes were measured at high temperatures to reduce the effect of surface flow. And the pore sizes were determined only qualitatively, based on permeance ratios such as He/N_2 and He/SF_6, since no theoretical model is available for the determination of subnano-meter pore sizes.

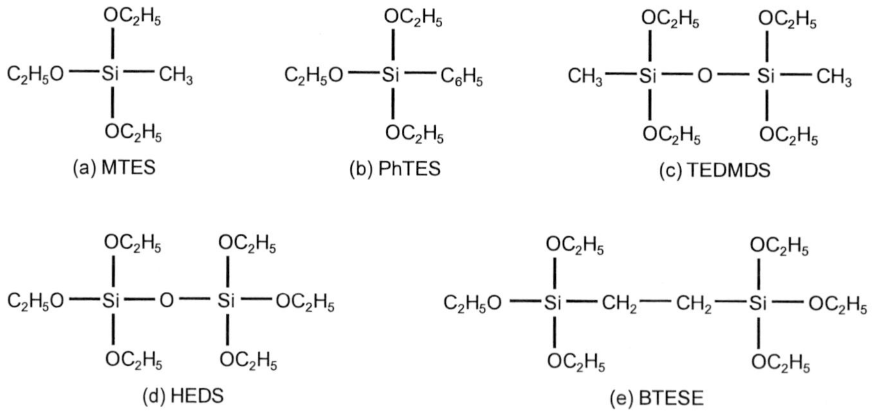

Figure 1. The structure of silica precursors in this work. (a) methyltriethoxysilane (MTES), (b) phenyltriethoxysilane (PhTES), (c) tetraethoxydimethyl disiloxane (TEDMDS), (d) hexaethoxy disiloxane (HEDS), (e) bis(triethoxysilyl) ethane (BTESE).

In this study, the normalized Knudsen-based permeance (NKP) was proposed for the determination of membrane pore sizes less than 1 nm [14, 16, 17], and MFI and DDR zeolite membranes were used to verify the theory, since they had intrinsic pore sizes. Moreover, the NKP was also used to characterize the pore sizes of silica membranes which were prepared via a sol-gel process using two kinds of organoalkoxysilanes with pendant group (methyltriethoxysilane (MTES) and phenyltriethoxysilane (PhTES)) and three kinds of bridged alkoxides (hexaethoxysiloxane (HEDS), tetraethoxydimethyl disiloxane (TEDMDS) and bis(triethoxysilyl) ethane (BTESE)) as showon in Figure 1. And the pore sizes of silica membranes were also evaluated by the normalized Knudsen-based permeance (NKP).

PROPOSAL OF NORMALIZED KNUDSEN-BASED PERMEANCE (NKP) FOR PORE-SIZE DETERMINATION

The normalized Knudsen-based permeance (NKP) method, which is based on the gas translation (GT) model originally proposed by J. Xiao and J. Wei [16] and by A.B. Shelekhin *et. al.* [17], was proposed for the determination of membrane pore sizes of less than 1 nm [14]. NKP is the permeance ratio of the i-th component (P_i), to that predicted from the j-th component based on the Knudsen diffusion mechanism ($P_j\sqrt{M_j/M_i}$). When He, the smallest molecule, is taken as a reference (j-th component), is the permeance of the i-th component predicted from helium permeance under the Knudsen diffusion mechanism. Therefore, NKP corresponds to the ratio of experimentally-obtained permeance to predicted ones based on helium and is shown in the following equation.

$$NKP = \frac{P_i}{P_{He}\sqrt{M_{He}/M_i}} \quad (1)$$

If the permeation mechanism follows Knudsen diffusion, NKP is equal to 1. In the original GT as shown in Eq. (2), the molecular size is not explicitly incorporated based on the assumption that the size of the gas is small enough.

$$P_i = d_p\rho_{g,i}\cdot\sqrt{\frac{8}{\pi M_i RT}}\cdot\frac{\varepsilon_i}{\tau_i L_i}\cdot\exp\left(-\frac{E_{p,i}}{RT}\right) \quad (2)$$

When the diffusion distance of the i-th component in a cylindrical pore with the diameter of d_p is assumed to be (d_p-$d_{k,i}$) for the i-th component (molecular size: $d_{k,i}$) due to the steric hindrance, the permeance, P_i, through microporous inorganic membranes (membrane porosity, ε; pore diameter, d_p; tortuosity, τ; thickness, L) can be formulated in Eq. (3), using the molecular weight of the permeating component, M_i, gas constant, R, and the temperature, T. It should be noted that probability of the i-th component, $\rho_{g,i}$, corresponds to the geometrical probability, that is, the product of effective area for permeation, $(d_p - d_{k,i})^2/d_p^{\ 2}$, and the probability of 1-dimensional diffusion through the membrane, 1/3 [16, 17].

$$P_i = \frac{(d_p - d_{k,i})}{3}\left(\frac{(d_p - d_{k,i})^2}{d_p^{\ 2}}\right)\sqrt{\frac{8}{\pi M_i RT}}\cdot\frac{\varepsilon_i}{\tau_i L_i}\cdot\exp\left(-\frac{E_{p,i}}{RT}\right) \quad (3)$$
$$= \frac{k_o}{\sqrt{M_i RT}}\cdot\exp\left(-\frac{E_{p,i}}{RT}\right)$$

Therefore, by inserting Eq. (3) into Eq. (1), NKP can be expressed as follows, with the assumption that ε_i, τ_i, L_i are the same irrespective of the types of gases and assuming the activation energies, $E_{P,i}$, are the same for any type of gas.

$$NKP = \frac{k_{o,i}}{k_{o,He}} \cdot \exp\left(-\frac{E_{p,i} - E_{p,He}}{RT} \right) = \frac{(d_p - d_{k,i})^3}{(d_p - d_{k,He})^3} \cdot \exp\left(-\frac{E_{p,i} - E_{p,He}}{RT} \right) \quad (4)$$

Eq. (4) indicates that NKP can be expressed with the configurational factors and the activation energy of the permeances of He and the i-th component, which require a set of experimental data on the temperature dependence of permeances.

In this section, for simplicity, the assumption has been made that the activation energies, $E_{p,i}$, are the same for any types of gas, to easily evaluate the pore sizes of microporous membranes, leading to the following expression for NKP.

$$NKP \approx \frac{(d_p - d_{k,i})^3}{(d_p - d_{k,He})^3} \quad (5)$$

VERIFICATION OF NORMALIZED KNUDSEN-BASED PERMEANCE (NKP) BY ZEOLITE MEMBRANES

Figure 2 (a) and (b) show normalized Knudsen-based permeance (NKP) as a function of molecular size, for MFI [18, 19] and DDR [20, 21] zeolite membranes, respectively. MFI and DDR zeolite membranes were reported to have intrinsic pores of 0.55×0.56 nm and 0.36×0.44 nm, respectively [18-21].

By assuming d_p=0.55 for MFI membranes, NKP experimentally obtained at 200°C were predicted well by a curve calculated by Eq. (5). NKP for DDR membranes were calculated with d_p=0.36 and 0.44, and are shown with solid and dotted curves, respectively.

Van den Bergh *et al.* [21] measured the permeances of He, H_2 and N_2 in the range of 30-600°C, while the CO and CO_2 permeances through DDR membrane were measured up to 400°C.

As shown in Eq. (4), NKP consists of interaction between permeating molecules and the membrane (expressed as the activation energy, temperature-dependent).

CO_2 showed larger NKP than the calculated values even at 400°C, probably due to the contribution of surface diffusion and larger interaction with the membrane.

Since the exponential term in Eq. (4) approaches to unity and the contribution of surface diffusion decreased at high temperatures, permeation experiments at high temperatures are preferable to obtain pore size based on Eq. (5).

As a whole, the calculated curve for d_p=0.44 nm showed good agreement with experimental data measured at high temperatures. The proposed NKP offers a convenient and easy-to-use measurement to evaluate the pore sizes of microporous membranes.

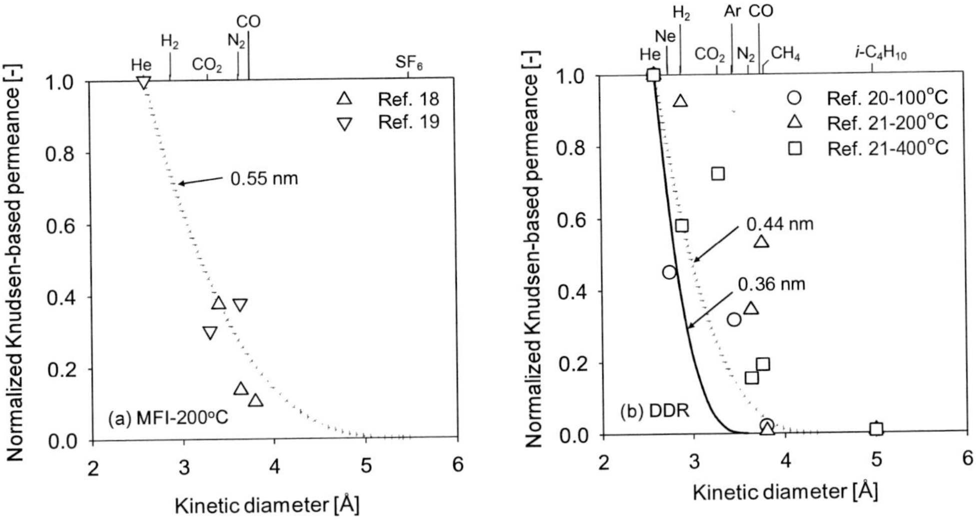

Figure 2. Normalized Knudsen-based permeance as a function of kinetic diameter; points are experimental and curves are calculated based on Eq. (4) for (a) MFI and (b) DDR zeolite membrane.

PREPARATION AND GAS PERMEATION OF PORE-SIZE-CONTROLLED SILICA MEMBRANES

Silica sols were prepared by the hydrolysis and polymerization reaction of each silica precursors in ethanol with water and an acid catalyst [7, 15]. The composition of the solution was ethoxy: H_2O = 1: 10 in a molar ratio and total weight of the solution was 50.0g. The composition for each silica sols was summarized in Table 1.

A porous α-alumina support (NOK Corp., Japan, O.D. = 3.0 mm, pore size = 100–150 nm) was used as a support in this work. The intermediate layer was coated with SiO_2-ZrO_2 (Si/Zr=1) sol at 180 °C and fired at 550 °C in an air atmosphere.

The separation layer was fabricated with 0.5 wt% of silica sol at 180°C and followed by firing in the range of 350-550°C in air or N_2 atmosphere [7, 15]. Single gas of industrial grade (He, H_2, CO_2, N_2, CH_4, C_3H_8 and SF_6) was fed on the outside (upstream) of a cylindrical membrane at 200-500 kPa, keeping the downstream at atmospheric pressure.

Figure 3 shows the gas permeance at 200°C for (a) *mono*-Si membranes and (b) *di*-Si membranes as a function of the gas kinetic diameter. In general, silica membranes having large average pore sizes show a high H_2 gas permeance through large membrane pores. TEOS-derived silica membrane showed clearly narrow pore size distribution, and gases with molecular sizes larger than N_2 could hardly permeate the membrane compared with MTES and PhTES in Figure 3 (a).

The silica membranes from bridged alkoxides (BTESE, TEDMDS, HEDS) showed a high hydrogen permeance of $1–3\times10^{-6}$ $mol\cdot m^{-2}\cdot s^{-1}\cdot Pa^{-1}$ with a low H_2/N_2 perm-selectivity (~20) and sufficiently high selectivity for H_2/SF_6 (>1,000). However, a TEOS-derived silica membrane showed a high H_2/N_2 perm-selectivity of approximately 1000 with a low hydrogen permeance in Figure 3 (b).

Table 1. The composition of each silica sols in this study

Si precursor	Sol molar composition Si/H_2O/ catalyst
MTES	1/30/0.12
PhTES	1/30/0.16
TEDMDS	2/40/0.2
HEDS	2/60/0.2
BTESE	2/60/0.1

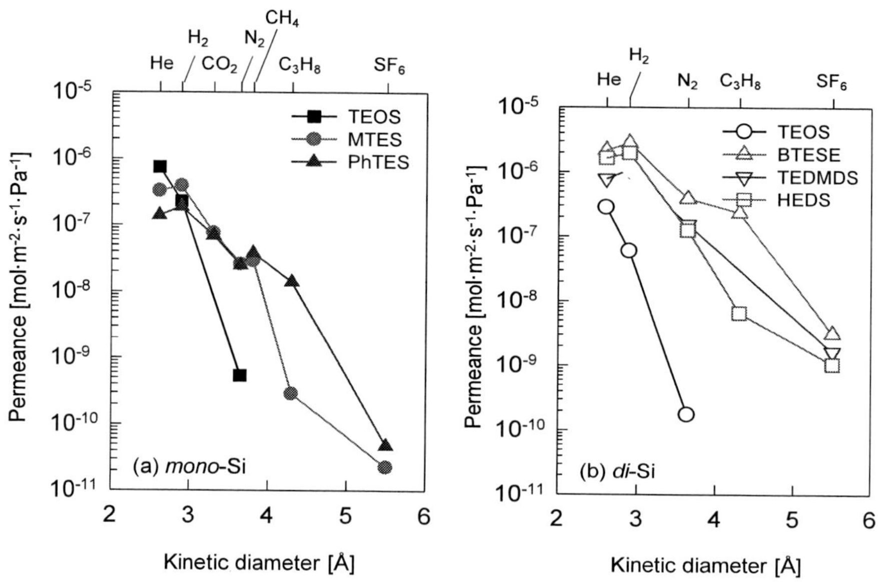

Figure 3. The gas permeance at 200°C for (a) *mono*-Si membranes and (b) *di*-Si membranes as a function of the gas kinetic diameter.

From these results, it is qualitatively suggested that structured alkoxides (organoalkoxysilanes and bridged alkoxides)-derived silica membranes have larger pore sizes than TEOS-derived silica membranes.

EVALUATION OF PORE-SIZE-CONTROLLED SILICA MEMBRANES BY NORMALIZED KNUDSEN-BASED PERMEANCE (NKP)

The NKP plots and fitted curves, with pore sizes based on Eq. (5) for (a) mono-Si membranes and (b) *di*-Si membranes derived from the different Si precursors, are shown in Figure 4. In *mono*-Si membranes, TEOS-derived silica membrane had the smallest pore size, 0.34 nm, and MTES and PhTES-derived silica membrane showed relatively larger pore sizes, 0.52 and 0.76 nm, respectively. Judging from NKP for *di*-Si membranes, the order of pore size can be estimated as follows: TEOS-derived silica (0.34 nm) < HEDS-derived silica (0.50 nm) < BTESE-derived silica (0.69 nm) and TEDMDS-derived silica membrane (0.70 nm). The bridged alkoxide-derived silica membranes (*di*-Si alkoxides) showed larger pore sizes than mono-Si alkoxide-derived silica membranes. This demonstrates that different Si precursors led to different pore sizes, and that the pore sizes of silica membranes increased with the pendant group size. The pore sizes determined by NKP are consistence very well with their gas permeation properties.Figure 5 shows the H_2 permeance of silica membranes derived from various Si precursors (TEOS [2], MTES, PhTES, BTESE, TEDMDS and HEDS) as a function of their average pore sizes, as determined by the NKP method. For alkoxides without pendant groups, the order of H_2 permeance and average pore size for TEOS, HEDS, and BTESE-derived silica membranes was as follows: TEOS < HEDS < BTESE-derived silica membranes in Figure 7. For structured alkoxides with methyl pendant groups, H_2 gas permeance for MTES and TEDMDS-derived silica membranes also increased with increasing average pore size. Silica membranes with structured alkoxides of *di*-silicon showed higher gas permeances and larger pore sizes than those of *mono*-silicon for both alkoxides with and without pendant groups. This suggested that the pore size of silica membrane could be controlled by *di*-Si precursors as a "spacer" technique.

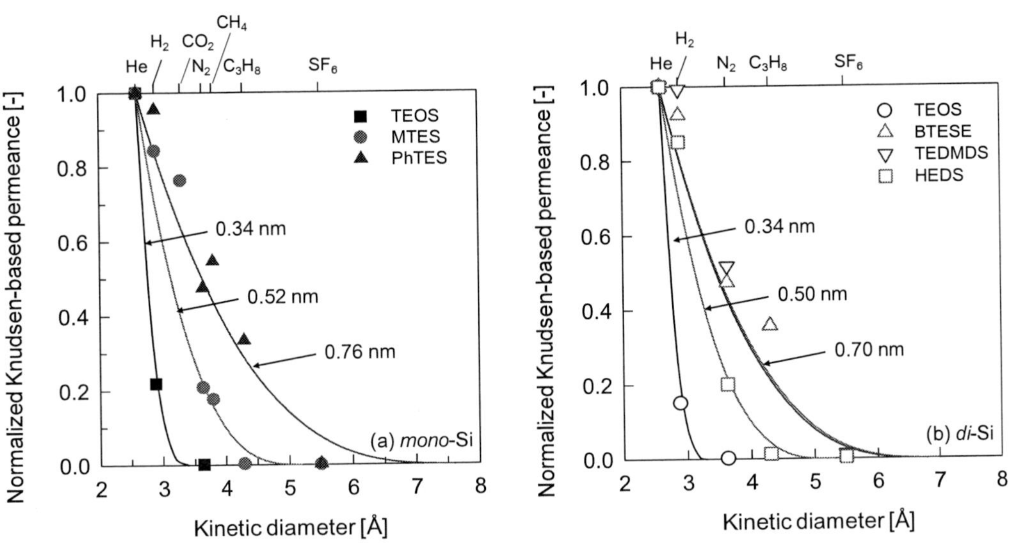

Figure 4. The effect of Si precursors on the average pore size of (a) *mono*-Si membranes and (b) *di*-Si membranes.

MTES, PhTES and TEDMDS-derived silica membranes, which were prepared from structured alkoxides with pendant groups, had relatively large pore sizes, but low H_2 permeances. In the case of silica membranes with pendant-type alkoxides, several possible reasons could account for their large pore sizes and low gas permeance. First, the minimum unit for the construction of silica networks for silica membrane with pendant-type alkoxides is larger than that of silica membranes without pendant-type alkoxides, because the methyl and phenyl groups in the silica precursors are non-hydrolyzable, which results in large pore sizes due to steric hinderance of the pendant groups. Second, the degree of cross-linking in pendant-type alkoxides could be smaller than nonpendant-type alkoxides, due to the number of ethoxy groups that initiate cross-linking for silica networks, resulting in a looser structure. Moreover, pendant groups may aggregate inside the silica networks during the sol-gel process, forming a larger space for gas permeation. However, the effective pore space of the silica matrix available for gas permeation was decreased since the pendant groups do not allow the permeation. The reduced hydrogen permeance for silica membranes with pendant-type alkoxides could be ascribed to the reduced pore space due to the incorporation of pendant groups into the silica network.

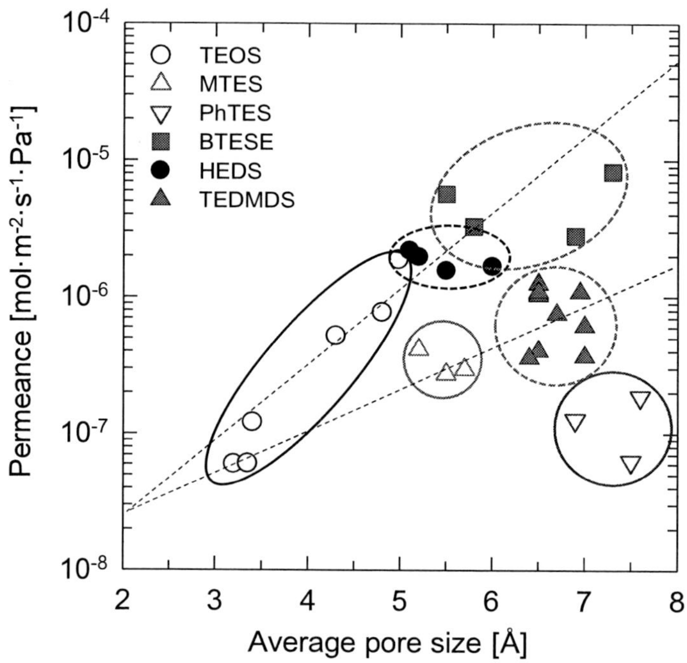

Figure 5. Comparison of H_2 permeance at 200°C for various silica membranes and their average pore size by NKP.

CONCLUSION

A normalized Knudsen-based permeance (NKP), which was derived from gas translation (GT) model with proper assumptions, was proposed as a convenient and easy-to-use measurement technique to evaluate the pore sizes of microporous membranes. NKP was first verified using the permeance of MFI and DDR zeolite membranes with intrinsic pore sizes. Furthermore, NKP was used to evaluate the pore sizes of silica membranes, which were prepared via sol-gel process with two types of structured alkoxides (organoalkoxysilanes and bridged alkoxides) for the control of silica networks with loose structure.

Based on the NKP fitting curves, the order of average pore size for silica membranes by organoalkoxysilanes and bridged alkoxides could be estimated as follows: TEOS < MTES < PhTES and TEOS < HEDS < TEDMDS≈BTESE, respectively, which were consistent very well with their gas permeation properties. The good agreement between the results of gas permeation characteristics and pore sizes determined by NKP suggests that the NKP can be applicable to evaluate the micropore sizes of zeolite and silica membrane efficiently and quantitatively.

References

[1] R.M. de Vos, Henk Verweij, High-selectivity, high-flux silica membranes for gas separation, *Science*, 279 (1998) 1710.

[2] T. Tsuru, Nano/subnano-tuning of porous ceramic membranes for molecular separation, *J. Sol-gel Sci. Technol.*, 46 (2008) 349.

[3] N.K. Raman, C.J. Brinker, Organic "template" approach to molecular sieving silica membranes, *J. Mem. Sci.*, 105 (1995) 273.

[4] M.C. Duke, J.C. Diniz da Costa, G.Q. (Max) Lu, M. Petch, P. Gray, Carbonised template molecular sieve silica membranes in fuel processing systems: permeation, hydrostability and regeneration, *J. Mem. Sci.*, 241 (2004) 325.

[5] J.-H. Moon, Y.-J. Park, M.B. Kim, S.-H. Hyun, C.H. Lee, Permeation and separation of a carbon dioxide/nitrogen mixture in a methyltriethoxysilane templating silica/α-alumina composite membrane, *J. Mem. Sci.,* 250 (2005) 195.

[6] K. Kusakabe, S. Sakamoto, T. Saie, S. Morooka, Pore structure of silica membranes formed by a sol-gel technique using tetraethoxysilane and alkyltriethoxysilanes, *Sep. Purif. Technol.*, 16 (1999) 139.

[7] G. Li, M. Kanezashi, T. Tsuru, Preparation of organic-inorganic hybrid silica membranes using organoalkoxysilanes: The effect of pendant groups, *J. Mem. Sci.*, 379 (2011) 287.

[8] H.L. Castricum, R. Kreiter, H.M. van Veen, D.H.A. Blank, J.F. Vente, J.E. ten Elshof, High-performance hybrid pervaporation membranes with superior hydrothermal and acid stability, *J. Mem. Sci.*, 324 (2008) 111.

[9] H.L. Castricum, A. Sah, R. Kreiter, D.H.A. Blank, J.F. Vente, J.E. ten Elshof, Hydrothermally stable molecular separation membranes from organically linked silica, *J. Mater. Chem.*, 18 (2008) 2150.

[10] R. Kreiter, M.D.A. Rietkerk, H.L. Castricum, H.M. van Veen, J.E. ten Elshof, J.F. Vente, Stable hybrid silica nanosieve membranes for the dehydration of lower alcohols, *Chem. Sus. Chem.*, 2 (2009) 158.

[11] M. Kanezashi, K. Yada, T. Yoshioka, T. Tsuru, Design of silica networks for development of highly permeable hydrogen separation membranes with hydrothermal stability, *J. Am. Chem. Soc.,* 131 (2009) 414.

[12] M. Kanezashi, K. Yada, T. Yoshioka, T. Tsuru, Organic-inorganic hybrid silica membranes with controlled silica network size: Preparation and gas permeation characteristics, *J. Mem. Sci.*, 348 (2010) 310.

[13] H.R. Lee, M. Kanezashi, T. Yoshioka, T. Tsuru, Preparation of hydrogen separation membranes using disiloxane compounds", *Desal. Water Treatm.*, 17 (2010) 120.

[14] H.R. Lee, M. Kanezashi, Y. Shimomura, T. Yoshioka, T. Tsuru, Evaluation and fabrication of pore-size-tuned silica membranes with tetraethoxydimethyl disiloxane for gas separation, *AIChE J.*, 57 (2011) 2755.

[15] H.R. Lee, T. Shibata, M. Kanezashi, T. Mizumo, J. Ohshita, T. Tsuru, Pore-size-controlled silica membranes with disiloxane alkoxides for gas separation, *J. Mem. Sci.*, 383 (2011) 152.

[16] J. Xiao, J. Wei, Diffusion mechanism of hydrocarbons in zeolites –I. Theory, *Chem. Eng. Sci.*, 47 (1992) 1123.

[17] A.B. Shelekhin, A.G. Dixon, Y.H. Ma, Theory of gas diffusion and permeation in inorganic molecular-sieve membranes, *AIChE J.*, 41 (1995) 58.

[18] M.C. Lovallo, A. Couzinis, M. Tsapatsis, Synthesis and characterization of oriented MFI membranes prepared by secondary growth, *AIChE J.*, 44 (1998) 1903.

[19] J. Lindmark, J. Hedlund, Modification of MFI membranes with amine groups for enhanced CO_2 selectivity, *J. Mater Chem.*, 20 (2010) 2219.

[20] J. van den Bergh, W. Zhu, F. Kapteijn, K. Yajima, K. Nakayama, T. Tomita, S. Yoshida, Separation of CO_2 and CH_4 by DDR membrane. *Res. Chem. Intermed.*, 34 (2008) 467.

[21] J. van den Bergh, A. Tihaya, F. Kapteijn, High temperature permeation and separation characteristics of an all-silica DDR zeolite membrane. *Microporous Mesoporous Mater.*, 132 (2010) 137.

In: Zeolites: Synthesis, Chemistry and Applications
Editors: Moisey K. Andreyev and Olya L. Zubkov
ISBN: 978-1-61942-861-4

Chapter 14

PHYSICAL AND CHEMICAL CHARACTERISTICS OF ZEOLITE FROM KHOTYNETZ DEPOSIT (OREL REGION)

S. M. Motyleva**, M. N. Kuznetsov and L. I. Leontieva
SSI All Russian Research Institute of Horticultural breeding, Orel, Russia

ABSTRACT

The problem of soil pollution and obtaining of safe production in conditions of techno- and agrogenic pollution of the environment is urgent at present. Application of natural nanoporous minerals – zeolites is a prospective direction for solving this problem. Zeolites have high adsorption abilities and ion-selective features; they can irreversibly adsorb cations of heavy metals. The investigation of the peculiarities of zeolite thin structure and its sorbing features is of practical importance – it allows calculating doses necessary for lowing toxic admixtures in soil and other objects. In our opinion, this direction of the nanotechnology in agriculture and horticulture is turned to the development of environment protecting agrotechnologies, conditions and methods which allow conducting of selective adsorption, stabilizing of soil processes, obtaining of ecologically safe production and neutralizing toxicants, allergens and pathogens in the environment.

Our research work is devoted to the investigation of zeolite of Khotynetz deposit in Orel region (Russia). Clinoptilolite, 34% and montmorillonite, 12% are parts of this zeolite. Clinoptilolite is a highly siliceous zeolite. Its crystalline grating consists of aluminosiliconoxygenous framework with nanochanals (pores). Crystals of montmorillonite have a laminated structure.

Montmorillonite and clinoptilolite cause high adsorption features of the rock. The crumbling up significantly increases specific and exchanging volume.

The investigation of the surface structure of different particles of the mineral (0,05-0,02 mm fraction) by scanning electronic microscope made it possible to reveal the peculiarities of the surface microstructure and define pore size. By energy and dispersion spectrometry it has been stated that a mass part of silicon in Khotynetz zeolite is 30,41%

---

* info@vniispk.ru.

on the average. The mineralogical composition stipulates unique features of Khotynetz zeolite as a nanocomposition material with high contents of amorphous silicon.

The process of selective adsorption was studied in the laboratory model experiments by nanoporous natural polymer of lead ions with 6 and 12 mg/l concentration and of nickel ions with 4 and 8 mg/l concentration under constant temperature and different pH and exposition time. The residual quantity of metal ions in the model solutions was determined by HPLC.

It has experimentally been stated that the adsorption of lead and nickel ions in the studied concentration diapasons depends on pH solution and depends little on ion concentration in the solution.

The obtained results were used in the establishing of field experiments with berry crops.

Keywords: Zeolite, appropriateness of the absorption of heavy metals, ecological situation, scanning electron microscopy, local energetic dispersion spectrometry

INTRODUCTION

The preservation of soil resources and diminution of consequences of technogenic pollutions with heavy metals at the soil-plant agro-ecosystem are a fundamental problem of the ecology. The matters relating to a level of pollution of the environment with heavy metals in the regions with large-scale industrial enterprises have already been the focus of attention of investigators for some ten years. There is a large factual material confirming that side by side with the natural redistribution of chemical elements in the soil profile there are processes of anthropogenic degradation of soils under the influence of the industrial pollution (Paramonova, 1991; Goreshnikova, 1995; Kuznetsov, 2009). The ecological situation in the countries of the world is mainly conditioned by a pollution of soils and waters both with industrial and agricultural waste. It leads to the disturbance of the ecological balance. The agriculture of Orel region is developed in the fields of grain farming, live-stock farming and fruit-growing. At the present ecological situation it is necessary to reduce the pollution of agricultural soils with heavy metals.

A promising direction of soil detoxication is an application of natural nanoporous minerals – zeolite. The world science names a natural zeolite as a mineral of the XXI century. The zeolite has high absorption ability, ion-exchange features and can irreversibly absorb cations of heavy metals.

In Orel region there is a Khotynetz deposit where the located reserves of zeolite are tens of mln.m^3. Clinoptilolite, 34% and montmorillonite, 12% form a part of the Khotynetz zeolite. Clinoptilolite is of special value as it is a highly siliceous mineral, has nano-canals and contents up to 40% of amorphous silicon. Montmorillonite is a schistous clayey mineral and has high absorption features. Such a mineralogical composition conditions a unique feature of the Khotynetz zeolite as a natural system with nano-features.

A successful use of the Khotynetz zeolite in different fields of agriculture has been demonstrated in papers of Ivanov G.I. (1970), Jarovan N.I. (2008), Gromova V.S. (2002), Dvoriankin E.A. (2003), Leontieva L.I. (2008), Kuznetsov M.N. (2009, 2010).

However, the microstructure of the surface and the absorptoin features of the Khotynetz zeolite are insufficiently studied. A versatile investigation of its useful features is necessary for the purpose of zeolite involving in the agriculture (innovation and nanotechnology application). Lead and nickel as well as other heavy metals are recognized as priority toxicants-polluters, which dominate in the list of carcinogenic and harmful substances of a cumulative character (Alikbaeva et al., 2005). Therefore, the goal of the work was to obtain new fundamental knowledge of the composition, structure and absorbing features of the Khotynetz zeolite relative to lead and nickel. The investigations are original and done with up-to-date analytical equipment.

EXPERIMENTAL PART

Objects of the investigation were air-dry samples of zeolite from the Khotynetz deposit with granules 0,5-0,2 mm in size, of roundish form, grey color and insoluble in water. The morphology of the surface of particles was studied by scanning electron microscopy (SEM) in conditions of secondary electrons with work tension 15-20 kW. The chemical composition of the zeolite – basic petrogenic elements (Si, Al, Ti, Fe, Mg, Ca, K, P, S) were determined by local energetic dispersion spectrometry (EDS). The locality of the analysis was 3 mkm, the sections not less than 12 mkm were scanned.

Relative errors of the chemical analysis were the following: with the element content from 1 up to 5% - not more than 10%; with the content from 5 up to 10% - not more than 5%; with the content over 10% - not more than 2%. 500 particles of zeolite were investigated.

The determination of absorption parameters of cations Pb^{2+} and Ni^{2+} was carried out in the model solutions containing cations in concentrations 6 and 12 mg/1 (for Pb^{2+}) and 4 and 8 mg/1 (for Ni^{2+}) under a constant temperature T=293° K and different pH (5,6 and 7). These concentrations of Pb^{2+} and Ni^{2+} enter the interval of concentrations of mobile forms of metals under different rates of the soil pollution. The speed of absorption of metal cations was studied in the static conditions under the relationship of the volume of the liquid phase to the mass of solid phase, $V/m=100dm^3 \cdot kg^{-1}$. After certain time intervals (0,5; 1; 1,5; 2; 4; 6; 24 hours) the investigated solution was taken in fives ml and centrifuged with 5 000 revolutions per minute during five minutes. The contents of the metal were determined by HLPC.

The absorption process was being studied under the constant mixing and with the speed 400 revolutions per minute.

DISCUSSING OF THE RESULTS

Natural zeolites bear on carcass silicates of alumina of alkali and alkaline-earth metals. Physical and chemical characteristics of the Khotynetz zeolite determining its absorbing and kinetic characteristics (according to data of http www.promc.ru/zeolite/index.php) are presented in Table 1.

The zeolite chemical composition and structure have been studied. The chemical composition is given in Table 2. The Khotynetz zeolite is a highly siliceous and acid-proof mineral. The overall appearance of the zeolite surface and sizes of separate particles are illustrated in Figure 1. The character of zeolite porosity determines its absorption features. The total porosity characterizes the sum of all pores irrespective of the morphology.

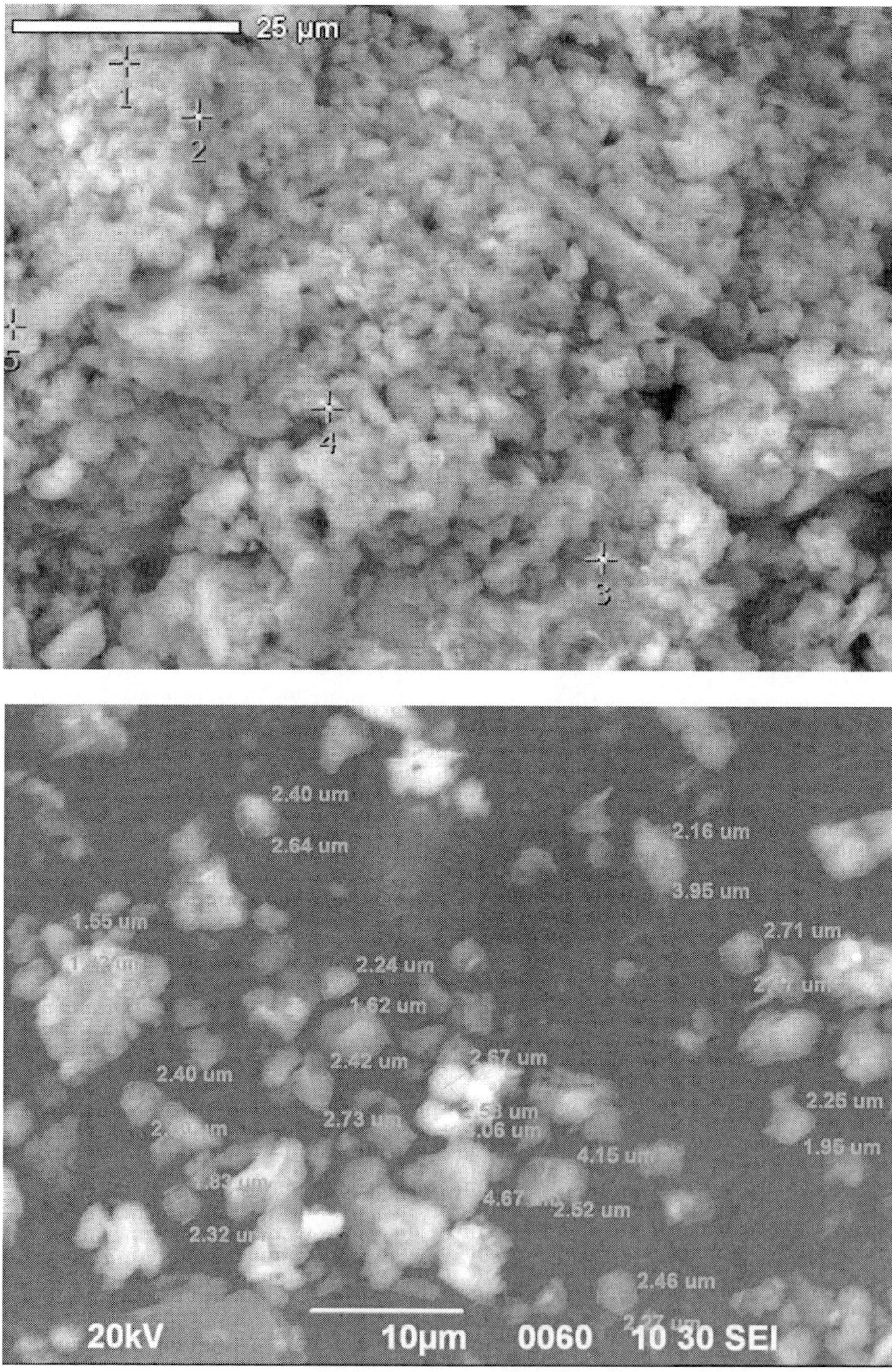

Figure 1. Overall appearance of zeolite surface and particle size.

Table 1. Characteristics of the Khotynetz zeolite

Physical characteristics	
Density, g/sm^3	2,49
Volume mass, g/sm^3	1,2
Total porosity, %	55
Specific capacity, sm^2/g	4500-6000
Mechanical strength under 20°C	19
Waterproof, %	89
Exchange capacity with Na, cmol+/kg	10
Phasic composition of zeolite	
Clinoptilolite - $(Na,K)_4Ca[Al_6Si_{30}O_{72}]\cdot 20\ H_2O$, %	34
Mordenit - Ca 2(Na,K)$[Al_8Si_{40}O_{96}]\cdot 28\ H_2O$, %	4
Montmorillonite - $Na_{0.3}(OH)_4Al_{1.6}Mg_{0.3}(Si_4O_{10})$,%	44
Quartz + Crystoballite (SiO_2), %	1
Mica - K1-x(H2O)[Al2Si3O10 (OH)2]nH2O %	5
Calcite - $CaCO_3$, %	

Porosity is conditioned by a presence of blank spaces, canals, cavities of molecular size in the crystal frame and depends on the crystal structure of the frame. The absorption features depend not only on the total volume of pores but also on their size and morphology. Values of the specific capacity (Table 1) are indicative of the high specific surface conditioned by a porous structure of zeolite. We first have obtained the results of morphological analysis of the Khotynetz zeolite surface.

Table 2. Chemical composition of zeolite, per cent from mass

Element	mass %						Average
O	41,29	44,6	49,38	44,18	47,51	49,38	38,95
Mg	1,66	2,2	3,66	2,09	11,57	3,61	2,03
Al	6,62	4,87	6,86	43,7	27,99	6,86	10,17
Si	22,59	29,15	34,41	12,4	11,51	34,41	17,69
P	0,37	0,42	0,51	0,62	0,89	0,5	0,87
S	0,44	0,79	0,93	1,05	1,67	0,93	0,91
Ca	0,62	1,17	4,33	2,28	1,94	4,32	2,04
K	3,14	2,54	3,26	2,22	1,67	3,61	4,48
Ti	0,92	1,32	1,08	1,18	1,24	1,11	1,44
Fe	5,53	4,12	4,56	3,48	3,47	2,48	3,29
Oxides of elements	mass %						Average
MgO	2,76	2,21	5,97	2,51	3,11	4,01	3,42
Al_2O_3	12,48	10,24	12,95	30,11	19,91	15,24	16,82
SiO_2	48,34	52,12	63,11	58,31	52,86	48,18	53,82
P_2O_5	3,14	4,11	2,45	3,11	3,77	3,05	3,27
SO_3	1,11	1,42	1,31	1,52	1,07	1,12	1,26
Ca	1,23	1,41	1,27	1,54	1,54	1,21	1,37
K_2O	3,25	4,01	5,04	3,14	1,36	3,31	3,35
TiO_2	1,53	2,04	1,48	1,58	1,54	1,42	1,60
FeO	7,11	6,24	5,97	5,46	3,74	5,12	5,61

Figure 2. Pictures of a zeolite system: a,c – clinoptilolite surface; b,d,e,g – montmorillonite surface.

Figure 2 demonstrates the surface of zeolite particles with 5 000, 10 000 and 100 000 magnification. A detailed morphological analysis of zeolite particles shows the presence of 2 different relieves of the surface. Clinoptilolite inclusions are characterized with a comparatively uniform surface (a) having nanocanals - pores at a diapason 80-150 nm (c), which form meso- (2-50 nm) and micro pores (<2nm). The surface of montmorillonite is not uniform at macro- (b,e) and micro level (g). Here and there a schistous surface is observed. Clusters of roughness and bulges are well visible. If their diameter is over 50 nm., they are macro pores. Just the presence of macro-, meso- and micro pores and a relationship between them form the activity of the zeolite system. A compound relief of the zeolite surface may affect the geometry of absorbing ions, their energetic characteristics and therefore the process of absorption (Kunilova, 2007). Besides, in nanostructures a role of the surfaces of the division increases and size effects are characteristic for them (Bulygina et al., 2006). In the research of Zhadovsky I. T. (2010) it is shown that cations of heavy metals in solutions form hydrated complexes, the radii of which are 2-3 times larger than a radius of a not hydrated cation, e.g. for Pb^{2+} r_{cr} = 125 *nm*, but r_g = 262 *nm*; for Ni^{2+} r_{cr} = 74 *nm*, but r_g = 340 *nm*. These distances are in proportion with sizes of zeolite pores.

Our investigation of the morphology of the surface of the Khotynetz zeolite and analysis of the scientific data confirm that the Khotynetz zeolite has a surface with a lot of meso- and micro pores.

While studying absorption features of the Khotynetz zeolite kinetic, the dependences of the absorption process of Pb^{2+} cations were investigated on the co-ordinates: the concentration of the cations (2+) and the time of the absorption process (Figure 3). It has been stated that absorption processes are more active under acid values of the medium (pH = 5).

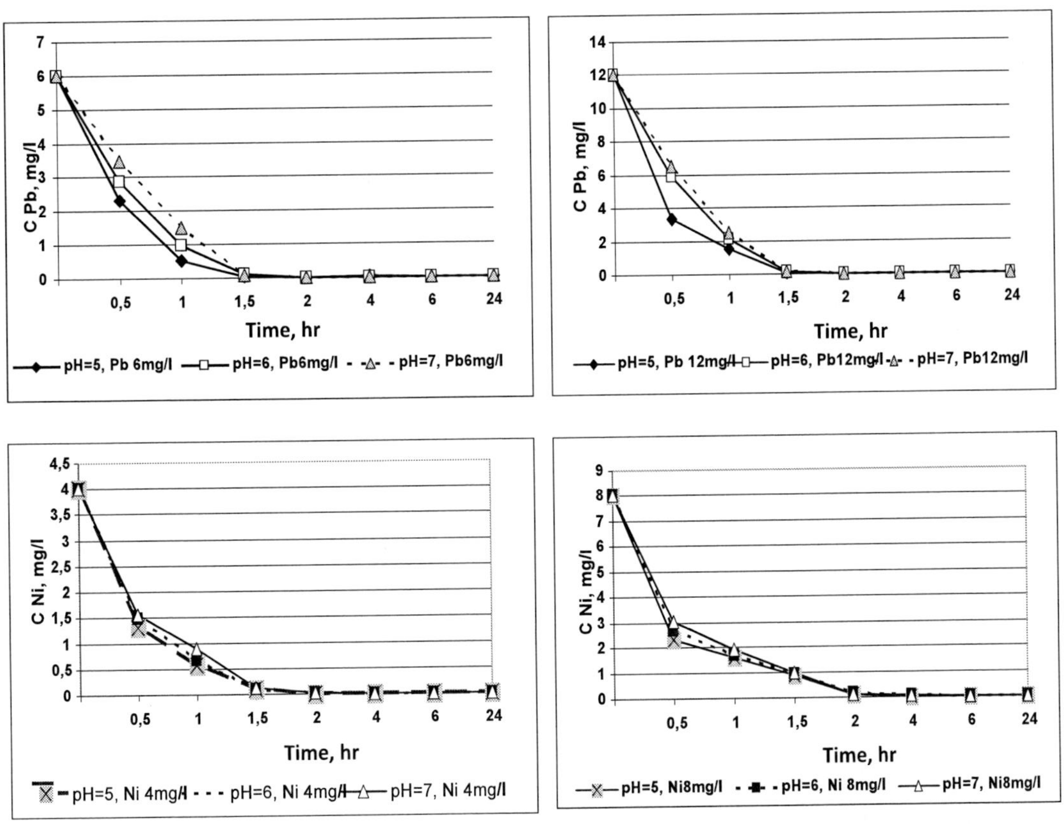

Figure 3. Experimental dependences of Pb^{2+} and Ni^{2+} concentrations in model solutions with different pH values on the time of the absorption process. Speed of mixing – 400 revolutions per minute.

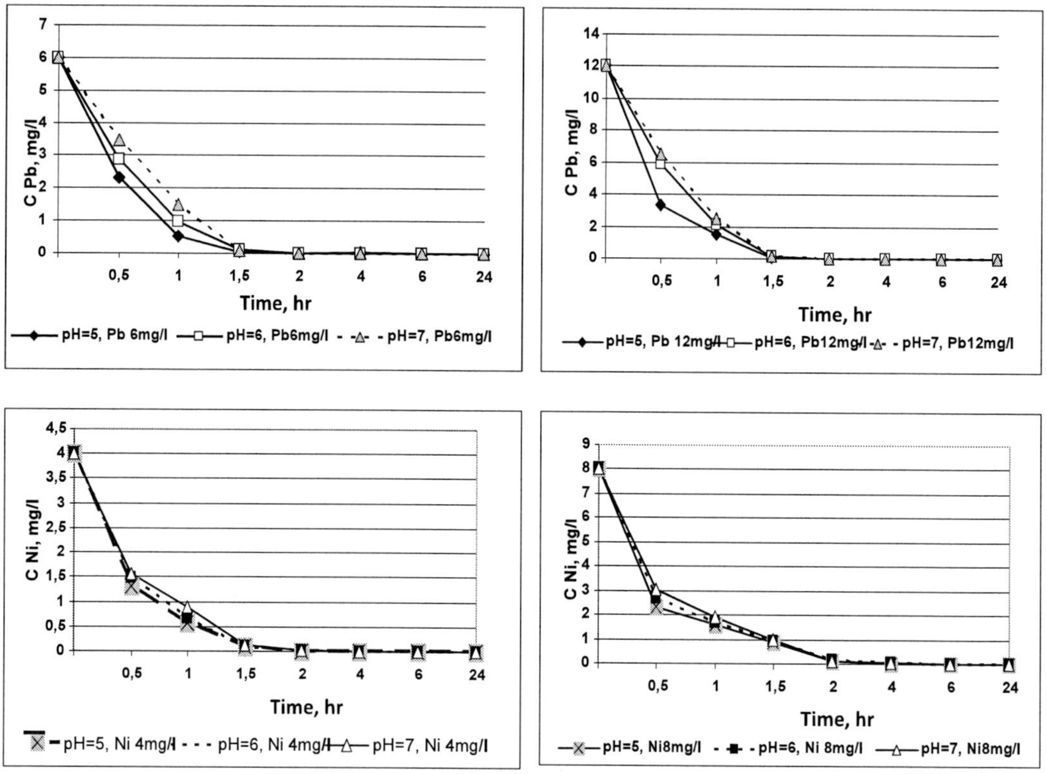

Figure 4. Influence of the concentration and pH of the medium on the absorption of Pb^{2+} (1, 2) and Ni^{2+} (3, 4) by zeolite.

It appears from the experimental data that during the first half an hour over 60% of Pb^{2+} and Ni^{2+} ions in the solution are absorbed, but by 2 hours only 0,3% ions are left in the solution. Higher concentration of $Pb2^{+}$ and Ni^{2+} cations insignificantly reduces an absorption process. When pH=5, the absorption of Pb^{2+} is by 16-20% more rapid than when pH=7. The absorption of Ni^{2+} cations is little dependent on pH of the solution. The extraction rate (absorption rate) has been calculated. The extraction rate is a quantity of the absorption process; it is expressed in per cents and calculated by a formula:

$$R = \frac{g_k}{g_{np}} \cdot 100,$$

where

g_k is a mass quantity of a cation in the control solution;

g_{nn} is a mass quantity of a cation in the sample.

Table 3. A range of metal cation mobility (according to Zhadovsky I. T., 2010)

	Pb^{2+}	$< Hg^{2+}$	$< Mn^{2+}$	$< Ni^{2+}$
E_a, *kDg/mol*	14.6 ± 0.6	9.2 ±1.2	7.43	7.15
k, c^{-1}	$9.9 \cdot 10^{-5}$	$3.9 \cdot 10^{-4}$	$0.93 \cdot 10^{-3}$	$1.7 \cdot 10^{-3}$

The curves of the rate of extraction of cations Pb^{2+} and Ni^{2+} have a common character, but the "steepness" of curves is specific. Just in a minute up to 80% cations of metals are absorbed irrespective of their initial concentration in a model solution and pH of the medium. The rate of Ni^{2+} extraction does not depend on the investigated intervals of pH and concentration of a cation (Figure 4 – 3,4). The rate of Pb^{2+} absorption depends on the concentration and pH of the medium. When increasing Pb^{2+} concentration a twice and shifting pH of medium towards alkali, the absorption processes become slower. The speed of absorption of nickel is higher than of lead, especially in the first 30 minutes. This is accounted by significant differences in mobility of cations $Pb2^{+}$ and Ni^{2+} (Table 3) and in radii of hydrated complexes mentioned above.

A high value of a constant of external diffusion speed, low energy of activation and a large radius of the hydrated complex favour the acceleration of the absorption process of cation Ni^{2}. A range of absorption of ions of lead and nickel by natural zeolite of the Khotynetz deposit has been determined: $Pb^{2+} < Ni^{2+}$.

Figure 4 illustrates graphic dependences of an extraction rate of cations Pb^{2+} and Ni^{2+} (R,%) on pH of the solution and concentration of the element (C, mg/l).

Conclusion

New data on the morphology of the Khotynetz zeolite surface have been obtained.

The appropriateness of the absorption of heavy metals (lead and nickel) has been revealed. This allows developing technologies of absorbing extraction of metals from different natural mediums – soil, water.

The dependences of the concentrations of lead cations (2+) and nickel cations (2+) in solutions on the time of absorption course on the surface of the Khotynetz zeolite have been found.

Acknowledgments

The authors are grateful to RFFI for the support of the present work (project № 11-04-97559 р-центр-а).

References

Alikbaeva, L.A. et al. Harmful substances in the environment. *Elements of Groups I-IV of the Periodical System and their inorganic combinations:* Reference book edited by Filov V.A et al.; Rus. Acad. Of Natural Sciences. – Saint-Petersburg: Professional, 2007. – 461 p.

Bulygina, E.V., Makarchuk, V.V., Panfilov Yu.V., Shakhnov V.A. *Nanosize structures: classification, formation and investigations.* – Educational textbook for universities. – M.: SCIENCE-PRESS. – 2006. – 80 p.

Dvoryankin, E.A. Combined application of zeolite and herbicides on sugar-beet/ E.A Dvoryankin// *Agriculture.* – 2003. - № 1. – P. 9-10.

Goreshnikova, E.V. The effect of soddy podzolic soil and liming on plant intake of cadmium, zinc and lead / *Candidate dissertation*/ E.V. Goreshnikova. – M., 1995. – 145 p.

Gromova, V.S. The effect of natural and anthropogenic factors on cesium-134 accumulation in fruit-berry production/ V.S. Gromova/ / *Storage and Processing of Agricultural Raw Material.* – 2002. - № 10. – P. 41-43.

Ivanov, G.I. Zeolite preparations for livestock and poultry/ *Mixed fodder production.* – 1997. - № 2 . – P.21-22.

Kunilova, I.V. The development of method of extraction of nonferrous metal and silver ions from copper containing technogenic material on the ground of use of chemically modified natural zeolites. *Author's abstract of candidate dissertation*/I.V. Kunilova. – M., 2007. – 19 p.

Kuznetsov, M.N. Plant communities in conditions of land pollution by heavy metals: *Monography*/ M.N. Kuznetsov. – Orel : VNIISPK, 2009. – 232 p.

Kuznetsov, M.N. Problems of biosphere pollution by heavy metals: *Monography* / M.N. Kuznetsov. – Orel: VNIISPK, 2011. 384 p.

Leontyeva, L.I. Zeolite application efficiency in raspberry and gooseberry cultivation. *Author's abstract of candidate dissertation*/ L.I. Leontyeva. – Orel. – 2008. – 23 p.

Natural and synthetic zeolite. *Application of zeolite.* http://www.promc.ru/zeolite/index.php (date of visiting 12.08.2011).

Paramonova, E. A. Biogenic and toxic elements in the agrophytocenosis under intensive chemicalization. *Author's abstract of candidate dissertation*/ E.A. Paramonova. –M., - 1991. – 23 p.

Yarovan N.I. The effect of Khotynetz natural zeolite on the processes of adaptation of cows/ N.I. Yarovan // *Reports of RAAS.* – 2008. – 2008. - № 2. – P. 87.

Zhadovsky, I.T. Thermodynamics and kinetics of ion exchange of nonferrous metals on the surface of iron-manganese ore/ *Author's abstract of candidate dissertation,* Saint-Petersburg, 2010. – P. 19.

In: Zeolites: Synthesis, Chemistry and Applications
Editors: Moisey K. Andreyev and Olya L. Zubkov
ISBN: 978-1-61942-861-4

Chapter 15

DEALUMINATED ZEOLITES IN BIOLOGICAL SYSTEMS

Zoltan Blum and Håkan Eriksson*
Department of Biomedical Laboratory Science,
Faculty of Health and Society, Malmö University, Malmö, Sweden

INTRODUCTION

Recent direct and indirect evidence have pointed to both biological reactivity and toxicity of micro- and nano-particles [1, 2, 3]. Particle size appears to be a decisive factor and particles can enter cells in two ways; either through endocytosis, were particles enter the cell in an endosome; or by otherwise passing the cell membrane directly into the cytoplasm. Endocytosis covers sampling of small volumes of fluids and soluble molecules from the milieu surrounding the cell, as well as internalization of large particulate matter such as whole cells or cell fragments from the immediate vicinity of the cells. With respect to toxicity, it is important to identify the size and the chemical composition at which biological systems no longer regard the extraneous matter as particles and hence the material enters the cells directly, through other mechanisms than endocytosis.

Zeolites are microporous aluminium silicates, from the tecto (framework) silicate group of minerals. The structures are built from corner sharing SiO_4 and AlO_4 tetrahedra and in most natural and also synthetic zeolites, the Si to Al ratio is close to unity; the high abundance of negatively charged AlO_4 tetrahedra will establish a polyanion with expressed hydrophilic properties. If the Si to Al ratio is increased, the overall charge of the framework is decreased and the framework property gradually shifts from hydrophilic to hydrophobic. Removal of aluminium, i.e. dealumination, can be accomplished in different ways, eventually resulting in a siliceous replica of the original zeolite, with the framework more or less intact. Also, some non-mineral zeolites can directly be made essentially aluminium free.

Endocytosis of extra cellular fluid and soluble molecules is referred to as pinocytosis while endocytosis of non-solutes, such as microorganisms and large cell fragments, is called phagocytosis. Phagocytosis denotes internalization of relatively large objects with sizes larger than a few hundred nm, up to several μm. Most cells are able to perform pinocytosis whereas

* Corresponding author.

phagocytosis is restricted to specialized cells, i.e. neuthrophiles, monocytes/macrophages and dendritic cells. Based on the above, particle toxicity will be discussed in terms of particle size and *i*, toxicity and adverse effects on phagocytosing cells *ii*, toxicity and adverse effects on non-phagocytosing cells.

PHAGOCYTOSING CELLS

Dealuminated zeolites are commercially available as preparations containing particles in the nm to µm range. These particles can be size fractionated by sedimentation and by allowing a suspension of dealuminated zeolite Y (USY) to sediment for one hour, a preparation of particles with an average hydrodynamic diameter of 600 nm is obtained [4]. Particles with a smaller average size can be obtained by prolonging the sedimentation time and figure 1 shows the toxic effect of various particle sizes with respect to phagocytosing and non-phagocytosing cells. THP-1, a cell line of human monocytic origin (American Tissue Culture Collection, ATCC, TIB-202), is shown as an example of phagocytosing cells and a mouse hybridoma, γ3-11.1 (ATCC, HB-8700), producing monoclonal antibodies against γ-IFN, as a non-phagocytosing cell (equal results have been obtained using isolated peripheral monocytes and B-lymphocytes).

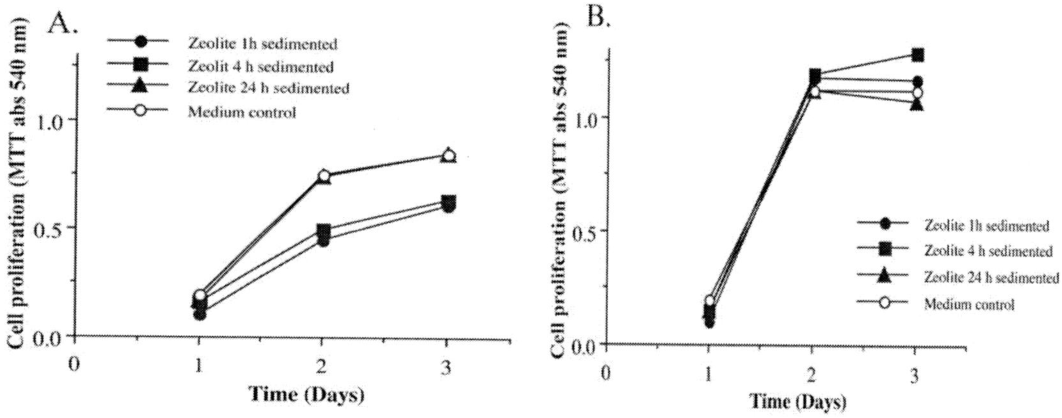

Figure 1.Toxicity of USY particles on phagocytosing and non-phagocytosing cells. THP-1 cells (A) or hybridoma cells (B) were co-cultured with dealuminated zeolite Y and cell proliferation was measured by the MTT assay [5]. Medium control (○), supernatant containing zeolite particles after 1h (●), 4h (■) and 24 h (▲) of sedimentation, respectively. Light scattering at 500 and 800 nm was used to determine the zeolite concentrations in the supernatants after sedimentation for 1 and 4 h [6] and cells were co-cultured with a final concentration of 40 µg/ml. Co-culture with the 24 h supernatant was done using the same dilution factor as with the 4 h supernatant in order to obtain a final concentration of 40 µg/ml.

Figure 1 clearly shows that USY particles at the investigated concentration are non-toxic to non-phagocytosing cells. However, in the case of phagocytosing cells, particles in the supernatant after 1 and 4 h sedimentation are toxic. No toxicity was observed by particles remaining in suspension after 24 h of sedimentation indicating these particles to be too small to be sensed as particles and internalized by phagocytosis. In the presence of an unlimited supply of particles, phagocytosing cells engage in unrestricted phagocytosis, and not only

USY particles are toxic to those cells. In fact, all kinds of particle in unlimited amounts are toxic to phagocytosing cells [7, 8] including particles of $Al(OH)_3$ used as an adjuvant in vaccine preparations, which are otherwise regarded as being safe and non-toxic (Figure 2).

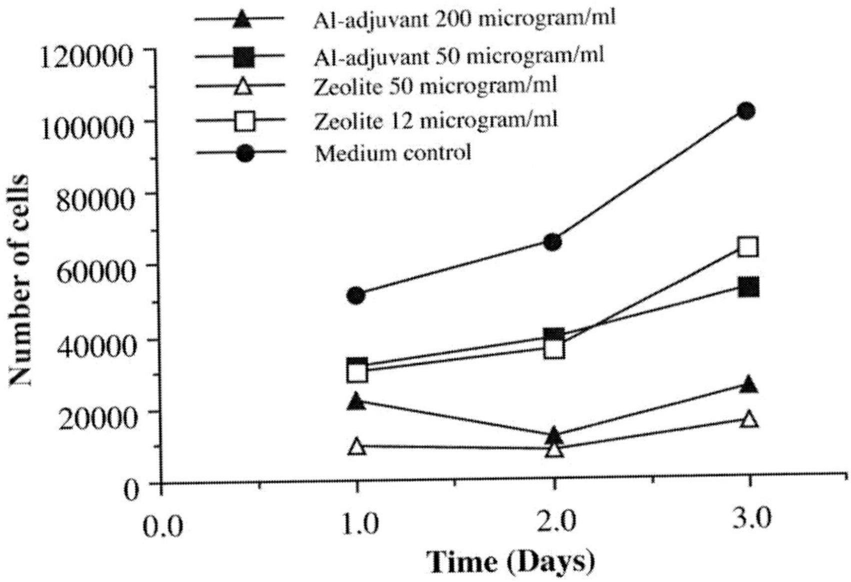

Figure 2. Toxicity of Al-adjuvant and USY particles on phagocytosing cells. THP-1 cells and $Al(OH)_3$ (Al adjuvant) or USY (1h supernatant) particles were co-cultured and viable cells were counted at various time. Medium control (●), final concentration of Al-adjuvant 200 µg/ml (⑦), 50 µg/ml (■), USY particles 50 µg/ml (Δ) and 12 µg/ml (□).

Particles are often used to study the phagocytosing mechanism and to be able to endocytose or engulf a particle, cells and particles have to be in close vicinity. Surface contact between cells and particles is necessary to induce phagocytosis and this can be achieved merely by co-sedimentation of cells and particles. However, in biological systems, close encounters between cells and the material to be endocytosed is mediated by bio-molecules acting as receptors and ligands inducing the surface contact. Receptor mediated endocytosis of particles can be studied by attaching bio-molecules acting as ligands that will be recognized by receptors expressed on the surface of the endocytosing cells. Receptor mediated association resulting in an increased endocytosis is seen in many biological systems and in the field of immunology this is known as opsonization.

Compared to other particulate matter, dealuminated zeolite particles do not show higher toxicity per se to phagocytosing cells. Instead dealuminated zeolites have, compared to particles of different chemical composition, a very high surface adsorption capacity of bio-molecules such as proteins and lipids. Covalent attachment is not needed; proteins and lipids are adsorbed with high affinity, merely by incubation with zeolite particles [6, 9]. Due to the straightforward adsorption procedure, investigations of receptor-mediated endocytosis can be easily done using various ligand densities and ligand combinations . Figure 3 shows the co-adsorption of human immunoglobulin G (IgG) or ovalbumin (OA) onto USY particles that already carry (adsorbed) fluorescence labelled OA (Alexa-OA) on the surface.

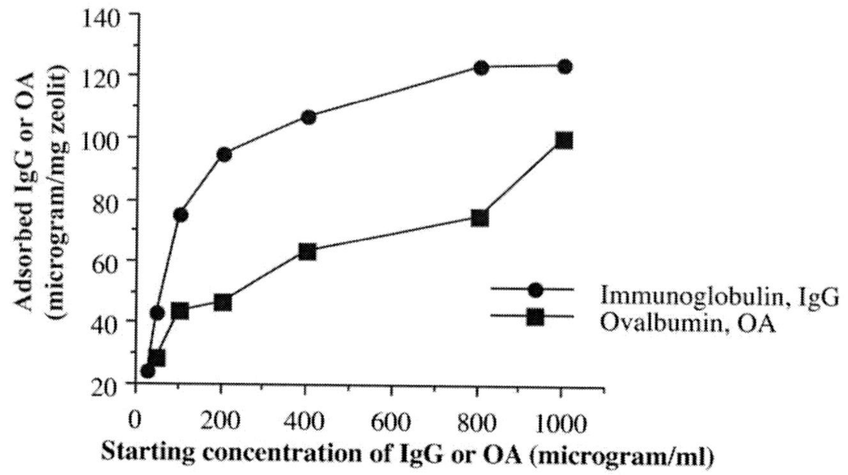

Figure 3. Co-adsorption of IgG or OA on Alexa-OA coated USY particles. Dealuminated zeolite Y particles from the supernatant after sedimentation for 1h were fluorescent labelled by pre-adsorbing Alexa Fluor® 488 labelled ovalbumin (10 μg Alexa-OA/mg zeolite particles) onto the particles before incubation with various amounts of human IgG (●) or ovalbumin (□).

Non-specific adsorption of proteins onto surfaces may result in a change of the 3-dimensional structure and hence denaturation of the protein. However, proteins are by and large adsorbed onto zeolite particles in a biologically active form, as shown by the residual antigen binding capacity of antibodies adsorbed on dealuminated zeolite particles (Figure 4). It can be assumed that a protein contains several regions that will mediate the adsorption onto the zeolite surface and IgG for instance will be randomly adsorbed on dealuminated zeolite Y, exposing the antigen-binding site or Fc region. Several varieties of receptors are involved in receptor mediated phagocytosis [10] and adsorption of various proteins and lipids on USY particles have been used to investigate phagocytosis facilitated by Fc-receptors, lectin like receptors, scavenger receptors, and toll-like receptors (TLRs) on human peripheral monocytes and dendritic cells [9, 11]. The facile adsorption of ligands on USY particles has proven a useful tool establishing a clear correlation between the amount of ligand adsorbed on the particles and cellular phagocytosing capacity.

Furthermore, zeolite particles contain channels and cavities that is an integral part of the crystal structure, making them unique in being all-surface materials with no dead bulk. Once phagocytosed, the zeolite particle becomes localized in an endosome and thus exposed to various biological microenvironments, dependant on the cell's endosomal mechanisms and pathways. High amounts of both high and low molecular weight probes can be adsorbed onto the zeolite surface and/or cavities and the probes are thus delivered into the endosomes of viable cells. In this respect, USY particles have been used to deliver an inhibitor of inducible nitric oxide synthase, iNOS, an enzyme catalyzing the production of NO [12], into viable cells [4].

The immune system protects the body from pathogenic microorganisms such as viruses and bacteria and the immune system is also capable of discriminating between infectious agents and endogenous cells, resulting in the specific elimination of pathogens upon infection. During an immune reaction, several distinct cell types, each having a specific function, are involved and collaborate to efficiently eliminate the pathogen. An immune reaction is initiated by special cell types called antigen-presenting cells (APCs) and these cells are capable of recognizing the infectious agent and alerting other cells of the immune system. Thus, they initiate the pathogen-specific immune activation and are pivotal for the development of a protective immune reaction.

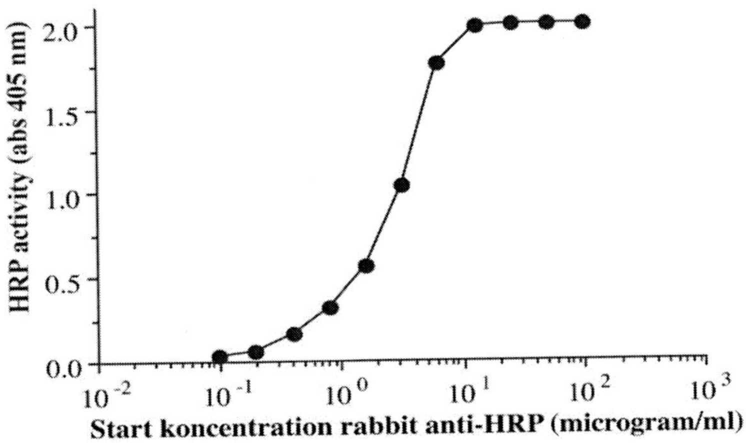

Figure 4. Binding of HRP by USY particles containing adsorbed antibodies against the enzyme. USY particles were incubated with various concentrations of the IgG fraction of a rabbit antiserum against horseradish peroxidase (HRP). After incubation the particles were washed and free adsorption sites were blocked by incubation with human IgG 1 mg/ml. Particles containing adsorbed antibodies against HRP were then incubated with HRP (2.5 µg HRP/ ml in PBS containing 0.5 % BSA (w/v) and washed before the HRP activity was assayed using ABTS and H_2O_2.

APCs, often referred to as "professional phagocytosing cells", sample their environment by phagocytosis, thereby capturing antigens to be presented and used to alert the immune system. Given the central role for APCs and their phagocytosing capacity in the induction of immune responses, it is of utmost interest to investigate how APCs phagocytose and present foreign material to immune-reactive cells and accordingly initiate pathogen-specific immune activations.

Normally a very low number of circulating antigen specific T-lymphocytes are found in peripheral blood. However the number of antigen specific T-lymphocytes can be increased by pre-immunization and due to vaccination, in Sweden the entire population can be regarded as immunized against the tetanus toxin. Tetanus toxoid (TT) is the inactivated form of tetanus toxin used in vaccines and TT can be used as a re-call antigen and exposure to immune cells from most blood donors will result in T-lymphocyte activation and proliferation. TT can be adsorbed onto dealuminated zeolites and in our experiments USY showed the highest adsorption capacity of TT (Figure 5A).

Antigen adsorbed on USY particles is an efficient immune stimulating vehicle as shown by challenging human peripheral mononuclear cells with TT loaded USY particles (Figure 5B). A preparation of peripheral mononuclear cell contains both APCs and T-lymphocytes and by adsorbing the antigen, TT, onto zeolite particles, the APCs are efficiently targeted and T-cell proliferation due to a secondary immune response is achieved. It should also be stressed that the amount of TT adsorbed onto the USY particles corresponds to a free concentration of 24 ng/ml, i.e. too low to induce any re-call response from a preparation of mononuclear cells (data not shown).

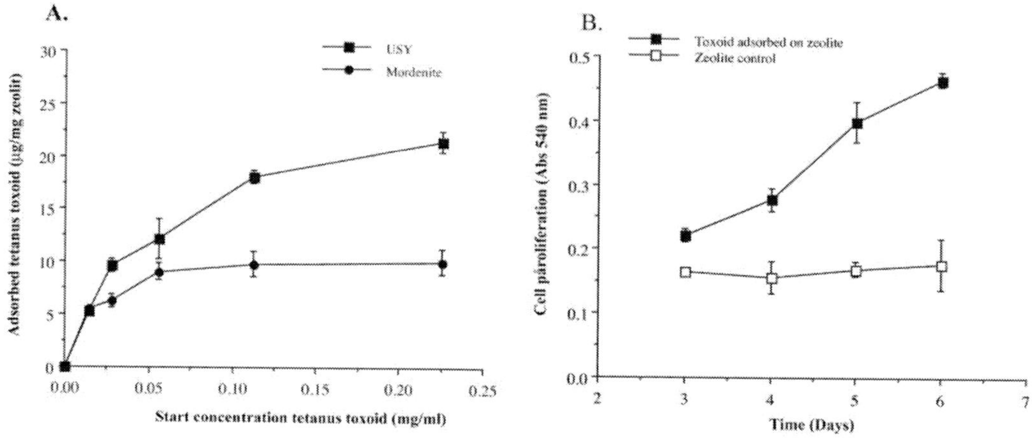

Figure 5. Adsorption of TT onto dealuminated zeolites and immune stimulating properties.

(A) Zeolite suspensions after 1 h sedimentation containing dealuminated zeolite Y (□) or dealuminated mordenite (❻) were incubated with various concentrations of TT. Adsorbed TT was indirectly calculated by measuring the amount of TT remaining in the supernatant after incubation.
(B) Human peripheral mononuclear cells were challenged with USY containing adsorbed TT (□), corresponding to 24 ng TT/ml or, as a control, USY with adsorbed serum proteins (☐). Cell proliferation was measured by the MTT assay [5].

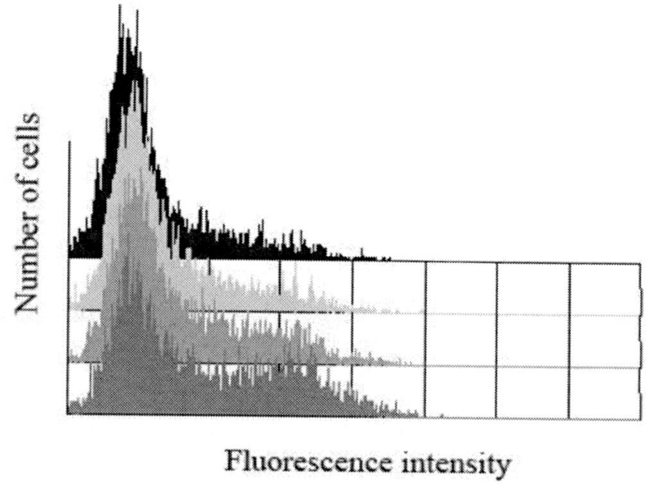

Figure 6. Oxidative burst by human peripheral monocytes after co-culture with USY particles. Isolated monocytes were co-cultered with various amounts of OxyBurst® loaded USY particles and after one hour at 37°C the endosomal oxidative burst was analyzed by flow cytometri. Histogram after incubation with control zeolite particles (zeolite particles without any adsorbed OxyBurst, black histogram) and incubation with OxyBurst zeolite 25 µg/ml (light grey histogram), 50 µg/ml (medium grey histogram), 100 µg/ml (dark grey histogram).

A specific endosomal delivery of probes detecting enzyme activity, ion concentration, pH or redox potential is promising applications of dealuminated zeolites. The probes are often of low molecular weight and diffuse into the cells after incubation and accumulate in the endosomes. Alternatively a zeolite/probe complex can be used to deliver the probes directly into the endosomes and thus increase the specificity. In the case of no or insufficient adsorption of the probe by zeolites, conjugating the probe with a protein can be used as an alternative. The probe H_2DCF is used to detect reactive oxygen species, ROS, and this probe can also be obtained as a conjugate with BSA, commercially available as OxyBurst®-BSA [13, 14]. H_2DCF can be oxidized to the green fluorescent molecule dichloroflourescein (DCF) and by adsorbing the reagent OxyBurst®-BSA onto USY, the oxidative burst after phagocytosis can be investigated and analysed (Figure 6).

Size Dependent Toxicity of Dealuminated Zeolites

The recent, exceptional development in the field nanotechnology has raised serious concerns regarding the impact of various nano-sized objects on e.g. health and environmental issues. Nano particle based probes and techniques have found increasing use in various aspects of molecular biology and medicine and studies regarding the possible toxicity of the nano objects employed are urgently needed.

As stated above, below a certain size regime, the particle notion is suspended and common cellular uptake mechanisms are bypassed. Countless studies have shown that, based on size only and not chemical composition, nano objects easily traverse biological boundaries. Given the non-conform physico-chemical properties of nano objects (owing to the exceptional surface to bulk atom ratio), access to the interior of whichever biological compartment thus exposed, may unexpectedly trigger events ranging from the trivial to the fatal. Hence, toxicology studies regarding the most common nano objects e.g. metals, alloys, metal oxides, semiconductors, silicates, carbon, have been undertaken. Results, however, are at best inconclusive, but it appears that in general small (<5-10 nm) objects present more of a health hazard than larger objects.

In the literature, no studies dedicated to the evaluation of cellular uptake/toxicity of zeolites in this size regime can be found. In fact, even in studies regarding chemically similar materials, i.e. asbestos, mesoporous silicates, clays, quartz, and glasses, particle sizes generally fall in the regime where phagocytosis will dominate [1, 2, 3]. Hence, one can only speculate, based on the findings for large siliceous particles, that small zeolite particles are not cytotoxic; the obvious exception to the non-toxicity of siliceous materials is asbestos and also other fibrous silicates.

References

[1] Zhao, J. and V. Castranova. 2011. Toxicology of nanomaterials used in nanomedicine. *J. Tox . Env. Health*. 14: 593-632.

[2] Galagudza, M. M., Korolev, D. V., Sonin, D. L., Postnov, V. N., Papayan, G. V., Uskov, I. S., Belozertseva, A. V. and Shlyakhto, E. V. 2010. Targeted drug delivery

into reversibly injured myocardium with silica nanoparticles: surface functionalization, natural biodistribution, and acute toxicity. *Int. J. Nanomedicine*. 5: 231–237.

[3] Slowing, I. I., Vivero-Escoto, J. L., Wu, C.-W. and Lin, V. S.-Y. 2008. Mesoporous silica nanoparticles as controlled release drug delivery and gene transfection carriers. *Adv. Drug Deliv. Rev*. 60: 1278-1288.

[4] Andersson, L. I. and H. Eriksson. 2007. De-aluminated zeolite Y as a tool to study endocytosis, a delivery system revealing differences between human peripheral dendritic cells. *Scand. J. Immunol*. 66: 52-61.

[5] T. Mossman. 1983. Rapid colorimetric assay for cellular growth and survival: application to proliferation and cytotoxicity assays. *J. Immunol Methods* 65: 55-63.

[6] Dahm, A. and H. Eriksson. 2004. Ultra-stable zeolites--a tool for in-cell chemistry. *J. Biotechnol*. 111: 279-290

[7] Fenoglio, I, A. Croce, F. Di Renzo, R. Tiozzo and B. Fubini. 2000. Pure-Silica Zeolites (Porosils) as Model Solids for the Evaluation of the Physicochemical Features Determining Silica Toxicity to Macrophages. *Chem. Res. Toxicol*. 13: 489-500

[8] Ghiazza, M. M. Polimeni, I. Fenoglio, E. Gazzano, D. Ghigo and B. Fubini. 2010. Does Vitreous Silica Contradict the Toxicity of the Crystalline Silica Paradigm? *Chem. Res. Toxicol*. 23: 620-629.

[9] Hellman,P., L. Andersson and H. Eriksson. 2009. Ligand surface density is important for efficient capture of immunoglobulin and phosphatidylcholine coated particles by human peripheral dendritic cells. *Cellular Immunology* 258: 123-130.

[10] Platt, C.D., J. K. Ma, C. Chalouni, M. Ebersold, H. Bou-Reslan, R. A. D. Carano, I. Mellman, and L. Delamarre. 2010. Mature dendritic cells use endocytic receptors to capture and present antigens. *PNAS*. 107: 4287-4292.

[11] Andersson, L. I., P. Hellman, and H. Eriksson. 2008. Receptor-mediated endocytosis of particles by peripheral dendritic cells. *Hum. Immunol*. 69: 625-633.

[12] Lowenstein, C. J. and E. Padalko. 2004. iNOS (NOS2) at a glance. *J. Cell Science*. 117: 2865-2867.

[13] Ryan, T.C., G. J. Weil, P. E. Newburger, R. Haugland and E. R. Simons. 1990. Measurement of superoxide release in the phagovacuoles of immune complex-stimulated human neutrophils. *J Immunol Methods* 130: 223-233.

[14] Product information; OxyBurst®H_2HFF Green BSA, O-13291. Molecular Probes, Life Technologies, CA, USA.

INDEX

A

B

C

D

E

F

G

H

I

J

K

L

M

N

O

P

Q

R

S

T

U

V

W